CAMBRIDGE TRACTS IN MATHEMATICS

General Editors

142 Harmonic Maps between Riemannian Polyhedra

J. Eells
Cambridge/Warwick

B. Fuglede
Copenhagen

With a preface by M. Gromov

Harmonic Maps between Riemannian Polyhedra

CAMBRIDGE
UNIVERSITY PRESS

PUBLISHED BY THE PRESS SYNDICATE OF THE UNIVERSITY OF CAMBRIDGE
The Pitt Building, Trumpington Street, Cambridge, United Kingdom

CAMBRIDGE UNIVERSITY PRESS
The Edinburgh Building, Cambridge CB2 2RU, UK http://www.cup.cam.ac.uk
40 West 20th Street, New York, NY 10011-4211, USA http://www.cup.org
10 Stamford Road, Oakleigh, Melbourne 3166, Australia
Ruiz de Alarcón 13, 28014 Madrid, Spain

First published 2001

Printed in the United Kingdom at the University Press, Cambridge

A catalogue record for this book is available from the British Library

Library of Congress Cataloguing-in-Publication data
Eells, James, 1926–
Harmonic maps between Riemannian polyhedra / J. Eells & B. Fuglede; with a preface
by M. Gromov
p. cm. – (Cambridge tracts in mathematics)
Includes bibliographical references.
ISBN 0 521 77311 3
1. Harmonic maps. 2. Riemannian manifolds. I. Fuglede, Bent. II. Title. III. Series.
QA614.73.E353 2001
514′.74–dc21 00-054670

ISBN 0 521 77311 3 hardback

Contents

Gromov's Preface

Harmony and Harmonicity

If you fall in love with harmonic functions your mathematician's soul will never come to rest unless you comprehend the origin of their irresistible appeal and beauty. And if you are bent on spaces, manifolds and maps you start researching for the geometric habitat of harmonicity.

In 1964 Eells and Sampson found the promised land not limited to functions but encompassing harmonic maps between (almost) arbitrary Riemannian manifolds. Yet one's greed for generality has not been quenched and the urge to extend harmonicity to rugged terrains of singular spaces could not be contained for long.

Here the story of this begins. There are two players in the harmonic mappings game: the source space X and the target Y. Suppose we are granted a *harmonic structure* on X, that is a distinguished space (or rather a sheaf) of $\mathbb{R}$-valued functions on X regarded as "harmonic". Then one can, under suitable assumptions, define another space (sheaf) consisting of corresponding "subharmonic functions" on X. Similarly, one needs distinguished functions on Y: these should be thought of as "convex functions". A map $f : X \to Y$ is declared "harmonic" if the pull-back of every "convex function" on Y is "subharmonic" on X.

Now a hard choice is to be made: how much of "Riemannian" is one willing to sacrifice for the sake of generality (singularity) and harmonicity? The authors decide in favour of "measurable Riemannian" for both X and Y, where X is a (rather general) topological polyhedron and Y is allowed an arbitrary local topology modified by the negative sign restriction on the curvature. In other words, the dimensions of X and Y are limited, for most part, to be finite, fractality (e.g. subellipticity) is not admitted and foliated structures are not allowed. By paying this price one arrives at a full fledged harmonic theory on X, extending Nash–De Giorgi–Moser, which then perfectly welds with the negative curvature on Y. It is some three hundred pages of smooth ride.

Misha Gromov
May 15, 2000

Authors' Preface

During the past thirty-five years harmonic maps between smooth Riemannian manifolds have played a significant role in geometry and elliptic analysis. They provide an especially rich mixture of classical potential theory and the Riemannian geometry of maps. Relevant guidelines are described in [EL 1978, 1988], mostly without proof, but with full references.

About eight years ago it became apparent that the notion of harmonicity should be expanded to include maps between certain singular spaces. In particular, (a) for applications to rigidity Gromov (see [GS 1992]) has shown that Riemannian polyhedra, and more generally the geodesic spaces of Alexandrov and Busemann (Definition 2.7), are natural targets; (b) certain Riemannian polyhedra (e.g., normal complex analytic spaces) are natural domains for harmonic maps.

This monograph is a research essay on harmonic maps between admissible Riemannian polyhedra (definitions in Chapter 4), these being prototypes of the relevant singular spaces. While harmonic spaces (in the sense of Brelot) are natural domains for harmonic functions, we have not been able to study harmonic maps in that generality, not even when adding a suitable Dirichlet space structure to obtain a notion of energy. We have discovered, however, that admissible Riemannian polyhedra are both geodesic, harmonic and Dirichlet spaces, more precisely hypoelliptic Dirichlet spaces in the sense of Feyel and de La Pradelle [FP 1978]. These polyhedra illustrate clearly our main ideas, and provide a wealth of examples as well. Thus harmonic maps (especially when presented in their variational context, via the Dirichlet integral) between Riemannian polyhedra are our main object of study.

A particular novelty in our presentation is the use of the fine topology of H. Cartan (the weakest topology in which all subharmonic functions are continuous), and its intimate relation to quasitopological concepts (defined in terms of capacity). This leads among other things to the quasicontinuity of finite energy maps in the sense of Korevaar and Schoen [KS 1993] into geodesic spaces, also in our setting of maps with polyhedral domain.

In spite of their importance, we do not treat applications in detail, for that simply would take us too far afield. However, here are three such – the first having smooth Riemannian domains and singular targets, the second having singular domains and smooth Riemannian targets:

- p-adic super-rigitity for lattices of rank one in the isometry groups of the quaternionic hyperbolic space and the Cayley plane, [GS 1992] and [Co 1992].
- Representation of integral homology classes of a compact Riemannian manifold by harmonic maps of compact oriented normal circuits [E 1997]. The proof is based on the method in [EF 1991] – except for the case of 2-dimensional homology classes, which can be so represented, using [Cha 1988].
- Reduction methods, as in [ER 1993, Chapter IV]. The idea (Examples 12.2 and 13.5) is to obtain an equivariant harmonic map $\psi : M \to N$ between smooth Riemannian manifolds, starting from a harmonic map $\varphi : M/K \to N/L$ between orbit spaces, where K (resp., L) is a compact group of isometries of M (resp., N); and ψ covers φ.

We have discussed various aspects of our text with many colleagues and with much profit; in particular, S. Hildebrandt, J. Jost, N. Korevaar, B. Lackey, L. Lemaire, C. Plaut, M. Ramachandran, R. Schoen, Richard Stong, K.–T. Sturm, D. Toledo and M. Wolf. Hereby we record our special thanks to all!

It is our pleasure to record our appreciation of the effort of M. Gromov, who graciously responded to our invitation to write a preface. And to T. Serbinowski for providing Chapter 15, a version of an unpublished part of his Thesis.

We further thank the referees, whose constructive comments have substantially improved the final version.

The second named author is grateful for all the facilities at the Department of Mathematics, University of Copenhagen, made available to him in his retirement. In this connection we are indebted to Anders Thorup and Lene Kørner, who helped us with the AmSTeX file.

Cambridge/Warwick
Copenhagen
September 2000

James Eells
Bent Fuglede

1
Introduction

This chapter consists of two parts: (1) A brief summary of the theory of smooth harmonic maps between Riemannian manifolds; that should provide background comparison to the main theme of the monograph. (2) A description of Riemannian polyhedra as harmonic and geodesic spaces; and energy-minimizing maps between them.

The following terminology for maps $f : X \to Y$ between metric spaces is adopted throughout this monograph (cf. Chapter 4):

- f is said to be *Lip continuous* if f is *locally* Lipschitz;
- f is called a *Lip homeomorphism* if f is bijective and if f and f^{-1} are both Lip continuous. Equivalently, f is a *locally* bi-Lipschitz bijection.

Similarly, f is said to be *Hölder continuous* if f satifies a *local* Hölder condition.

If X is a manifold or polyhedron with boundary, that boundary is denoted bX. The topological boundary of a subset A of a topological space is denoted ∂A.

The smooth framework

Let M be a smooth manifold without boundary, endowed with a smooth positive definite Riemannian metric g. The pair (M, g) is called a *Riemannian manifold.* We shall only consider *connected* Riemannian manifolds. Each tangent vector space $T_x(M)$ is Euclidean, with inner product $\langle \cdot, \cdot \rangle_x = g(x)(\cdot, \cdot)$. The volume measure μ_g on (M, g) has local representation of the form $\sqrt{\det g(x)}\, dx$. Associated to g is a metric d_M on M compatible with its underlying topology – defined before Definition 2.7 in terms of the Riemannian length of paths in (M, g).

For simplicity of exposition, assume here that M is compact. Define the Hilbert space $W^{1,2}(M)$ (a Sobolev space) as the completion of the space of smooth functions $u : M \to \mathbb{R}$ with respect to the inner product

$$\int_M \{u(x)v(x) + \langle \nabla u(x), \nabla v(x) \rangle_x\} \, d\mu_g(x),$$

where ∇u denotes the gradient field of u, and

$$\langle \nabla u(x), \nabla v(x) \rangle_x = g^{ij}(x) \frac{\partial u(x)}{\partial x^i} \frac{\partial v(x)}{\partial x^j}.$$

The *Dirichlet integral*, or *energy functional*, of u is

$$E(u) = \frac{1}{2} \int_M |\nabla u(x)|_x^2 \, d\mu_g(x).$$

Its *Euler–Lagrange operator* is the Laplace–Beltrami operator Δ, characterized by

$$\frac{dE(u_t)}{dt}\bigg|_{t=0} = - \int \Delta u_t(x) \frac{du_t(x)}{dt}\bigg|_{t=0} d\mu_g(x).$$

Here (u_t) is a 1-parameter deformation of $u = u_0$. Locally,

$$\Delta u = (\det g)^{-1/2} \partial_i (g^{ij} (\det g)^{1/2} \partial_j u),$$

where $\partial_i = \partial / \partial x^i$, $i = 1, \cdots, n$.

A function u on an open set $U \subset M$ is *harmonic* if $\Delta u = 0$ in the distributional sense. *Such a function is smooth* (after correction on a null set). For our main theme, two fundamental properties are emphasized:

(a) *Dirichlet problem.* For any small closed ball B in (M, g) and smooth function $f : \partial B \to \mathbb{R}$ on its boundary, there is a unique smooth function $u : B \to \mathbb{R}$ such that $u_{|\partial B} = f$ and $u_{|\operatorname{int} B}$ is harmonic.

(b) *Harnack's monotone convergence property.* For any increasing sequence (u_n) of harmonic functions on a connected open subset U of M, the pointwise limit function $u = \sup_n u_n$ is either harmonic on U, or $u \equiv \infty$.

Most of the qualitative aspects of smooth linear potential theory (as in [GT 1998, Part I], [Hel 1969]) can be derived from those two properties. That is the essence of Brelot's harmonic spaces, see Chapter 2.

Now take a second smooth Riemannian manifold (N, h), and consider a continuous map $\varphi : M \to N$ of class $W^{1,2}(M, N)$, i.e., the components φ^k in terms of local coordinates y^k, $k = 1, \cdots, n$, are of class $W^{1,2}$. The *energy density* of φ is the integrable function $e(\varphi) : M \to \mathbb{R}$ given almost everywhere by

$$e(\varphi)(x) = \tfrac{1}{2}\,\mathrm{trace}_{g(x)}(\varphi^* h) = \tfrac{1}{2} g^{ij}(x) \frac{\partial \varphi^\alpha(x)}{\partial x^i} \frac{\partial \varphi^\beta(x)}{\partial x^j} h_{\alpha\beta}(\varphi(x)).$$

In other words, $2e(\varphi)(x)$ is the square of the Hilbert–Schmidt norm of the differential $d\varphi(x)$, viewed as a linear map $T_x(M) \to T_{\varphi(x)}(N)$ between the indicated Euclidean tangent spaces. The *energy* of φ is defined as the integral

$$E(\varphi) = \int_M e(\varphi) d\mu_g.$$

If φ is smooth, the Euler–Lagrange operator of that energy functional E evaluated on the map φ is the *tension field* $\tau(\varphi)$, which is a vector field along the map φ:

$$\frac{dE(\varphi_t)}{dt}\Big|_{t=0} = -\int_M \Big\langle \tau(\varphi)(x), \frac{d\varphi_t(x)}{dt}\Big|_{t=0} \Big\rangle_x d\mu_g(x).$$

Again, (φ_t) is a 1-parameter deformation of $\varphi = \varphi_0$.

A continuous map $\varphi \in W^{1,2}(M, N)$ is said to be *weakly harmonic* if it satisfies $\tau(\varphi) \equiv 0$, or in local coordinates

$$\Delta \varphi^k = -(\Gamma^k_{\alpha\beta} \circ \varphi) \langle \nabla \varphi^\alpha, \nabla \varphi^\beta \rangle, \tag{1.1}$$

in the weak sense. It follows that φ (after correction on a null set) is C^∞-smooth [ES 1964, §8]. Here is a simple bootstrap argument for this: The right hand member of (1.1) is integrable (over the pre-image of the coordinate patch in question), and it follows from (1.1) that φ is of class $C^{1,\lambda}$ for some $\lambda > 0$. The right hand member of (1.1) is therefore of class C^λ, and so, by (1.1), φ is $C^{2,\lambda}$, etc.

There are weakly harmonic maps which are not continuous, provided that $\dim M > 2$ [EL 1978, §3.5].

A continuous map $\varphi \in W^{1,2}(M, N)$ is weakly harmonic iff φ is *harmonic*, i.e., φ is *bi-locally E-minimizing* in the sense described below in the section Harmonic maps (even for maps from an admissible Riemannian polyhedron, cf. Definition 12.1).

Continuous maps which minimize E in a given homotopy class of maps $M \to N$ are clearly bi-locally E-minimizing and hence harmonic.

A continuous map $\varphi : M \to N$ between smooth Riemannian manifolds is said to be *totally geodesic* if it maps geodesics of M linearly to geodesics of

N. Equivalently, if φ preserves connections [EL 1983, Proposition 2.21]. *A totally geodesic map is harmonic.*

Note that if $N = \mathbb{R}$ then the harmonic maps $M \to \mathbb{R}$ are just the harmonic functions. More generally, a special case of harmonic map $\varphi : M \to N$ arises when φ is *horizontally weakly conformal*; i.e., if for any point $x \in M$ at which the differential $d\varphi(x) \neq 0$, its restriction to the orthogonal complement of $\ker(d\varphi(x))$ in $T_x(M)$ is conformal and surjective. *A map $\varphi : M \to N$ is harmonic and horizontally weakly conformal iff it is a harmonic morphism*; i.e., for every function v which is harmonic on an open subset V of N, the composition $v \circ \varphi$ is a harmonic function on $\varphi^{-1}(V) \subset M$.

T. Ishihara has characterized harmonic and totally geodesic maps in a similar spirit: Say that a smooth function $v : V \to \mathbb{R}$ is *convex* if its Hessian $\nabla^2 v$ is positive semidefinite. And is *subharmonic* if Δv $(= \operatorname{trace} \nabla^2 v) \geqslant 0$. Then: *A continuous map $\varphi : M \to N$ is harmonic, resp. totally geodesic, iff it pulls germs of convex functions on N back to germs of subharmonic, resp. convex, functions on M* [Ish 1979].

The remaining sections of this Introduction describe the broad lines of development in our text – first for Riemannian polyhedra as harmonic and geodesic spaces; and then for maps, especially for various E-minimizers.

Harmonic and Dirichlet spaces

There are several natural frameworks for the qualitative aspects of potential theory on locally compact spaces. For instance (Chapter 2), [Br 1957; 1958a,b; 1959]:

1. *Brelot harmonic spaces*, identified and developed [Br 1969] as spaces characterized (as already indicated) through
   - (a) local unique solvability of the Dirichlet problem for continuous boundary functions, and
   - (b) Harnack's monotone convergence property.

A more general notion, for somewhat different purposes, has been studied by Bauer [Bau 1966]. See [CC 1972], [Bau 1984] for comparisons.

2. *Dirichlet spaces of Beurling and Deny* [BD 1959], [De 1970], [FOT 1994]. Here the starting point is a Dirichlet form E, abstracting key properties of the classical Dirichlet integral D (in particular, Beurling's observation that D decreases under normal contractions).

By a theorem of Feyel and de La Pradelle [FP 1978], certain so-called hypoelliptic Dirichlet spaces determine Brelot harmonic spaces.

It has been established in important work by Biroli and Mosco [BM 1991; 1995] that for certain so-called admissible Dirichlet spaces, the extremals of the Dirichlet form E are Hölder continuous, in the presence of a suitable Poincaré inequality. Admissibility requires that

(a) a certain Carathéodory distance d on the underlying space X be a pseudometric whose topology is the given one, and
(b) the Radon measure of the Dirichlet space satisfy a ball doubling condition.

Sturm [St 1995b] has shown (even without (b)) that (X, d) (if complete) *is a geodesic space* (meaning briefly that any two points in the same component can be joined by a geodesic, i.e., a rectifiable path whose length is the distance between the points).

The traditional examples of Brelot and Dirichlet spaces are analytical in character: Spaces of solutions of second order linear elliptic operators, possibly with discontinuous coefficients ([Her 1964; 1965], [HH 1969; 1972], [Bo 1967], [La 1980]).

Riemannian polyhedra

By way of contrast, this monograph is primarily concerned with harmonic spaces derived from and motivated by geometric considerations. In more detail: A connected locally finite n-dimensional simplicial polyhedron X is called *admissible* (cf. [Chen 1995]) if

(i) X is dimensionally n-homogeneous, i.e., every simplex is a face of an n-simplex, and
(ii) X is locally $(n-1)$-chainable, i.e., relative to some triangulation, for any simplex s, any two n-simplexes containing s are joinable by a chain of contiguous $(n-1)$- and n-simplexes containing s.

Examples 8.2 and 8.3 indicate that both these conditions are necessary for our purposes.

A polyhedron X becomes a *Riemannian polyhedron* when endowed with a Riemannian structure g, defined by giving on each maximal simplex s of X a Riemannian metric g_s equivalent to a Euclidean metric on s. The components of g_s need not be smooth, but merely *bounded measurable.*

Adapting constructions by De Cecco and Palmieri [CP 1988; 1990], an *intrinsic distance* d_X (Carathéodory distance) is defined on X, which thereby becomes a *length space*, and hence a geodesic space, if complete (Proposition 4.1).

On an admissible Riemannian polyhedron X the Sobolev space $W^{1,2}(X)$, defined as the completion of a suitable space of Lipschitz continuous functions in the Sobolev norm, *is a Dirichlet space* (Proposition 5.1).

Harmonic functions on X

On an admissible Riemannian polyhedron X define the *energy functional* $E : W^{1,2}_{\text{loc}}(X) \to \mathbb{R}$ by

$$E(u) = \frac{1}{2} \int_X |\nabla u|^2 d\mu_g.$$

Weakly harmonic (and subharmonic) functions are defined using the energy functional, and characterized variationally (Proposition 5.2, Theorem 5.2).

Theorem 5.3 is a maximum principle for weakly subharmonic functions of class $W^{1,2}(X)$, of the sort due to Hervé [Her 1964] in the case where X is a domain in $\mathbb{R}^n$ carrying a Riemannian metric with bounded measurable coefficients; our proof is different.

Harnack's inequality for weakly harmonic functions on an admissible Riemannian polyhedron X is established (Theorem 6.1), involving a careful and leisurely adaptation of the argument of Moser [Mos 1961]. In the spirit of De Giorgi [Gi 1957] and Moser [Mos 1960], it is shown in Theorem 6.2 that *every weakly harmonic function on X is Hölder continuous.* (Simple examples illustrate that such functions may not be Lipschitz continuous.) Furthermore (Theorem 6.3), *every locally uniformly bounded family of harmonic functions on X is locally uniformly Hölder equicontinuous.*

Continuous, weakly harmonic functions are called *harmonic.* Altogether we are led to Theorem 7.1 which states that *an admissible Riemannian polyhedron has a natural Brelot harmonic space structure; and its harmonic funtions are Hölder continuous* (i.e., have Hölder continuous versions). It was first proved by Hervé [Her 1964] in the case where X is a domain in $\mathbb{R}^n$ with a bounded measurable Riemannian structure.

Now suppose that X satisfies the following Poincaré inequality (always fulfilled locally):

$$\left(\int_K |u|\, d\mu_g \right)^2 \leqslant c(K) E(u)$$

for any $u \in \text{Lip}_c(X)$ (the compact Lip functions $u : X \to \mathbb{R}$). Then $\sqrt{E(u)}$ is a *norm* on $\text{Lip}_c(X)$, and *the completion* $L^{1,2}_0(X)$ *of* $\text{Lip}_c(X)$ *in this norm is a regular Dirichlet space of diffusion type* (Proposition 7.3). In view of the continuity of weakly harmonic functions the theorem of Feyel and de La Pradelle [FP 1978] applies to the Dirichlet space $L^{1,2}_0(X)$, and produces the Brelot harmonic space of Theorem 7.1. That and constructions of Hervé [Her 1962] combine to ensure (always in the presence of the above Poincaré inequality) that *every admissible Riemannian polyhedron has a unique sym-*

metric Green function G, Hölder continuous off the diagonal (Theorem 7.3, Propositon 7.4).

We show that $G(x, y)$ is bounded above and below by constant multiples of $d_X(x, y)^{2-n}$ when x and y vary in a compact set and $n > 2$ (Theorem 7.4). Our proof of these estimates exploits the full local polyhedral structure of an admissible Riemannian polyhedron, thus permitting us to adapt a method used by Littman, Stampacchia and Weinberger [LSW 1963] in their proof of Lemma 7.2 and Theorem 7.4 for the particular case of a ball X in $\mathbb{R}^n$ endowed with a Riemannian metric with bounded measurable components. A key ingredient in their proof is the use of a suitable homothetic transformation in order to reduce the case of a variable point y to that of a fixed y. When X is an admissible Riemannian polyhedron, a finite number of such homotheties are applied in our proof of Theorem 7.4. The same technique is further used in the proof of the Poincaré inequality for functions (Theorem 5.1), and in a Harnack inequality for positive functions harmonic off a point (Proposition 6.2).

In their paper [BM 1995] Biroli and Mosco studied potential theory on a very general type of metric space endowed with a Dirichlet form of diffusion type, and even allowing for degeneracy. As an alternative approach to the Harnack inequality and the Hölder continuity of weakly harmonic functions on an admissible Riemannian polyhedron, one could use our preparatory results to verify the remaining axioms in the Biroli–Mosco theory in the present case of an admissible Riemannian polyhedron (X, g). This leads to the existence of the Green function on small balls in (X, g) and to our estimates of it, invoking a uniform estimate of the volume of balls with variable centre in a compact subset of X (Lemma 4.4).

Geometric examples

Various admissible Riemannian polyhedra are presented in Chapter 8. Included are the following examples which are also *circuits* (i.e., satisfying (i) and a global version of (ii) above, and also requiring that every $(n-1)$-simplex is a face of exactly one or two n-simplexes, see Chapter 4):

- Smooth Riemannian manifolds, with or without boundary. (The associated harmonic functions are the local solutions u of the Laplace–Beltrami equation $\Delta u = 0$ in the interior, and having vanishing normal derivative at the boundary, see Remark 5.3.) Also triangulable Riemannian Lipschitz manifolds.
- Riemannian joins of smooth Riemannian manifolds.
- Conical singular Riemannian spaces (Cheeger [Chee 1980; 1983]).

- Normal complex analytic spaces (Giesecke [Gie 1964] and Łojasiewicz [Ło 1964]).

In a somewhat different context (Examples 8.12 and 8.13): Let K be a compact group of isometries of a complete smooth Riemannian manifold M, and $\pi : M \to M/K$ the orbit projection. The Brelot harmonic sheaf $\mathcal{H}_M$ of M determines the direct image sheaf $\pi_*\mathcal{H}_M$ on M/K, a Brelot harmonic sheaf there.

Analogously, a Riemannian orbifold has such a harmonic space structure. For instance, the leaf space $M/\mathcal{F}$ of a Riemannian foliation $\mathcal{F}$ with closed leaves of a Riemannian manifold M (Reinhart [Rei 1961]).

Maps between polyhedra

Suppose that (X, g) is an admissible Riemannian polyhedron of dimension m (compact, for simplicity) and that g is simplexwise smooth. Let (Y, d_Y) be any separable metric space (to begin with).

A *Riemannian domain* is understood to be a connected open subset of a Riemannian manifold (M, g) whose metric completion is compact in M.

For suitable maps φ of a Riemannian domain into Y an energy density $e(\varphi)$ was defined by Korevaar and Schoen [KS 1993], via a deep subpartitioning lemma, as a certain limit of approximate energy densities $e_\varepsilon(\varphi)$; the energy $E(\varphi)$ then equals $\int e(\varphi)\, d\mu_g$. See Chapter 14 for a summary of that construction.

Building on this concept of energy, and its properties, we consider analogously the following approximate energy density of a measurable map $\varphi : (X, g) \to (Y, d_Y)$:

$$e_\varepsilon(\varphi)(x) = \int_{B_X(x,\varepsilon)} \frac{d_Y^2(\varphi(x), \varphi(x'))}{\varepsilon^{m+2}}\, d\mu_g(x').$$

The *energy* $E(\varphi)$ of φ is defined as

$$E(\varphi) = \sup_{f \in C_c(X,[0,1])} \left(\limsup_{\varepsilon \to 0} \int_X f\, e_\varepsilon(\varphi)\, d\mu_g \right) \quad (\leqslant \infty).$$

It is shown that a map $\varphi : X \to Y$ has finite energy iff there is an integrable function $e(\varphi)$ on X (called the *energy density* of φ) such that $e_\varepsilon(\varphi) \to e(\varphi)$ as $\varepsilon \to 0$, in the sense of weak convergence of measures. This is also equivalent to φ *having a quasicontinuous version with restrictions of finite energy in the sense of Korevaar–Schoen to the top-dimensional simplexes s of X*. The sum of these energies $E(\varphi_{|s})$ then equals $E(\varphi)$ (Theorem 9.1). (A map $\varphi : X \to Y$ is said to be quasicontinuous if φ has continuous restrictions to closed sets with complements of arbitrarily small capacity.)

In case Y is a Riemannian C^1-manifold (N, h) without boundary, we also have the elementary concept of energy of a quasicontinuous map $\varphi : X \to N$, the energy density being given by $e(\varphi) = \mathrm{trace}_g\, \varphi^* h$, in coordinate patches V on N:

$$e(\varphi) = (h_{\alpha\beta} \circ \varphi)\langle \nabla\varphi^\alpha, \nabla\varphi^\beta \rangle \qquad \text{in } \varphi^{-1}(V).$$

This makes immediate sense, by covariance, if φ is continuous or if N is itself a patch. In any event, the pre-images $\varphi^{-1}(V)$ are *quasiopen*, and the components $\varphi^1, \cdots, \varphi^n$ are of class $W^{1,2}(U)$ in the sense of Kilpeläinen and Malý, [KilM 1992, §2] (cf. the next to last section of Chapter 7 below). The energy of φ is now defined by $E(\varphi) = \int_X e(\varphi)\, d\mu_g$.

This concept of energy of maps φ of (X, g) into a Riemannian manifold N is identified (Lemma 9.3) with the one suggested by Nash's isometric embedding theorem. The previous requirement that the Riemannian metric g of X be simplexwise smooth is not needed for this energy concept. With that requirement, however, *the two concepts of energy of maps into manifolds are identical*, provided e.g., that φ is continuous or that N is compact. In either of these cases, $e_\varepsilon(\varphi) \to e(\varphi)$ as $\varepsilon \to 0$ holds both in L^1-norm and pointwise almost everywhere in X, when φ has finite energy (Theorem 9.2).

There is a similar result for maps into a Riemannian polyhedron Y with continuous Riemannian metric h (Theorem 9.3). When (Y, h) admits a Riemannian embedding in some $\mathbb{R}^q$ we recover the elementary expression for energy used in [GS 1992] and [Chen 1995].

The energy functional E on $W^{1,2}(X, Y)$ is lower semicontinuous, and $W^{1,2}(X, Y)$ has a Rellich-style precompactness property if Y is also compact.

Assuming that Y admits a bi-Lipschitz embedding in a Euclidean space (this holds for any compact Riemannian polyhedron), Poincaré's inequality for maps $X \to Y$ of finite energy is derived from that for functions (Proposition 9.1).

This allows us to establish the Hölder continuity of locally E-minimizing maps into suitable targets, by adapting a line of arguments of Jost [J 1997a], with underlying work by Jäger–Kaul [JK 1979], Caffarelli [Caf 1982], Giaquinta–Giusti [GG 1982], Giaquinta–Hildebrandt [GH 1982], Meier [Mei 1984] and Sturm [St 1995a].

With our first concept of energy for maps from an admissible Riemannian polyhedron X with simplexwise smooth Riemannian metric g we obtain the following local regularity property, assuming that Y is a simply connected complete Riemannian polyhedron of nonpositive curvature (in the sense of A.D. Alexandrov): *If $\varphi : X \to Y$ is a local E-minimizer, then φ is Hölder continuous* (Theorem 10.1).

Likewise, assuming that (Y, d_Y) is a complete Riemannian polyhedron with $\text{curv}(Y, d_Y) \leqslant K$ for some $K > 0$: Let $\varphi : X \to Y$ be a local E-minimizer, and suppose that the range of φ is contained in a compact convex set $V \subset Y$ of diameter $< \pi/(2\sqrt{K})$ such that geodesics in V are uniquely determined by their endpoints. Then φ is Hölder continuous (Theorem 10.2).

In particular, every continuous locally E-minimizing map φ of X into a Riemannian polyhedron Y with upper bounded curvature is Hölder continuous (Corollary 10.3).

Similar results with the second energy concept and hence with a manifold target N are obtained in Propositions 12.1, 12.2 and Corollary 12.2. Some limitation on the size of the range of φ is necessary, even for maps between manifolds, see [HKW 1977, §6] (cf. Example 12.3 below).

With the above regularity theorems at hand it is possible to extend, by known direct methods of variational theory, classical results on existence of harmonic maps between Riemannian manifolds to the present setting of maps between Riemannian polyhedra.

Consider first the case of a target Y of nonpositive curvature. A key fact is that this curvature restriction implies strong convexity properties of E [GS 1992, §4], [KS 1993, Chapter 2] and [J 1994, §2].

For the case of *free homotopy* we obtain, assuming that X and Y are compact Riemannian polyhedra, X being admissible and Y of nonpositive curvature: *Every homotopy class of continuous maps $X \to Y$ has an E-minimizer, and any such is Hölder continuous* (Theorem 11.1).

A more detailed analysis gives the following uniqueness property: *If $\varphi_0, \varphi_1 : X \to Y$ are homotopic E-minimizers which agree at some point of X, then φ_0 coincides with φ_1 on X* (Corollary 11.1).

Theorem 11.1 is due to

(a) Eells and Sampson [ES 1964] in case X and Y are both smooth Riemannian manifolds; the E-minimizers are smooth.

(b) Gromov and Schoen [GS 1992] when X is a smooth Riemannian manifold. The E-minimizers are Lipschitz continuous.

Korevaar and Schoen [KS 1993] have extended [GS 1992], dropping the polyhedral restriction on Y, permitting Y to be any geodesic space (compact and of nonpositive curvature). Then, as mentioned above, E is defined as a limit of approximating energy integrals.

(c) Chiang [Chi 1990], in case X is a Riemannian orbifold and Y a smooth Riemannian manifold.

(d) Chiang and Ratto [CR 1992], in case X is a Riemannian manifold with conical singular points and Y a smooth Riemannian manifold.

Suppose now that in the hypotheses of Theorem 11.1 the polyhedron X has a *boundary* $bX \neq \emptyset$, and denote by $W^{1,2}_{\psi}(X,Y)$ the class of those quasicontinuous maps $\varphi \in W^{1,2}(X,Y)$ for which $\varphi_{|bX} = \psi_{|bX}$ a.e. Then the proof of Theorem 11.1 can be adjusted to give existence and inner regularity for the *Dirichlet problem: A given continuous map $\psi \in W^{1,2}(X,Y)$ can be deformed to a unique E-minimizer in $W^{1,2}_{\psi}(X,Y)$, without altering the boundary map $bX \to Y$, and that minimizer is Hölder continuous on $X \setminus bX$* (Theorem 11.2).

There are several special situations in which the curvature restrictions on Y can be relaxed; e.g., if it is required that the curvature of Y be bounded above, and that φ have small range (Theorem 11.4). Chapter 15, by Serbinowski, describes (Hölder and Lipschitz) regularity of energy minimizers from Riemannian domains into geodesic spaces with curvature bounded above. Also, in analogy with a theorem of Lemaire [Lem 1978], Sacks and Uhlenbeck [SaU 1981] and Jost [J 1994], we obtain, when M denotes a closed Riemann surface without boundary, and Y a compact Riemannian polyhedron with curvature bounded above and second homotopy group $\pi_2(Y) = 0$: *Every homotopy class of continuous maps $M \to Y$ contains a Lipschitz continuous E-minimizer* (Theorem 11.5).

The main results stated in this section are valid more generally if Y (instead of being a Riemannian polyhedron) is a geodesic space which locally is bi-Lip embeddable in a Euclidean space. In fact, the methods of [KS 1993; 1997] provide extensions not even requiring that the geodesic space Y be locally compact; however, that is not our primary concern here, and *it will be understood throughout that our geodesic spaces are locally compact* (in the absence of other indications). See, however, Chapter 11, Questions and remarks, Item 7.

Harmonic maps

A continuous map φ of an admissible Riemannian polyhedron X into a metric space Y will (tentatively) be termed *harmonic* if φ is *bi-locally E-minimizing* in the sense that every point of X has a relatively compact neighbourhood U to which there corresponds an open set $V \supset \varphi(U)$ in Y such that $E(\varphi_{|U}) \leqslant E(\psi_{|U})$ for every continuous map $\psi : X \to Y$, locally of finite energy and with $\psi(U) \subset V$ and $\psi = \varphi$ in $X \setminus U$. Every continuous locally E-minimizing map $\varphi : X \to Y$ is harmonic. The converse holds e.g., if Y is a simply connected complete Riemannian polyhedron of nonpositive curvature (Lemma 12.1).

When Y is a smooth Riemannian manifold (always without boundary), harmonicity of φ is found (Theorem 12.1) to be equivalent to φ being weakly harmonic (i.e., E-critical), and also to φ pulling germs of convex functions on Y back to germs of subharmonic functions on X, thus extending Ishihara's result mentioned earlier.

There is also a version of this characterization in which X is a 1-dimensional Riemannian polyhedron (Proposition 12.4). That is based on an extension of Jensen's inequality to continuous convex functions on the target Y, which is supposed simply connected complete and of nonpositive curvature. Furthermore, Ishihara's pull-back characterization of *totally geodesic maps* is extended to maps between geodesic spaces (Theorem 12.2).

Harmonic morphisms

A continuous map φ between admissible Riemannian polyhedra is said to be a *harmonic morphism* if φ pulls germs of harmonic functions on Y back to germs of harmonic functions on X. Using the Green kernel of Y, natural conditions have been determined to ensure that φ is an open map; it suffices that $\dim Y > 1$ (Proposition 13.1).

In case N is a smooth Riemannian manifold without boundary, horizontal conformality of maps $X \to N$ (X an admissible Riemannian polyhedron) is defined as in the smooth case, and there is a generalization of the characterization of harmonic morphisms of Fuglede [Fug 1978] and Ishihara [Ish 1979]: *A continuous map $\varphi : X \to N$ is a harmonic morphism iff φ is harmonic and horizontally conformal* (Theorem 13.2).

As an example, suppose that $\pi : M \to N$ is a smooth branched cover of a complete Riemannian manifold $N = (N, h)$, as in Example 8.11. With distance function on M induced from $\pi^* h$, both M and N are admissible Riemannian polyhedra. And, relative to these structures, π is a harmonic morphism.

Singular frameworks

By way of a summary: The energy $E(\varphi)$ of suitable maps $\varphi : X \to Y$ of an admissible Riemannian polyhedron into a metric space is defined in the spirit of Korevaar and Schoen [KS 1993], drawing heavily on their study of energy of maps from a Riemannian domain.

For functions (Chapters 2–3 and 5–8) the Riemannian metric g of X is assumed to be measurable and locally with elliptic bounds. For maps (Chapters 9–13) g is further restricted to be simplexwise smooth, although that is unnecessary if the target is a smooth Riemannian manifold N. In that case, and with N compact, the energy agrees with the Dirichlet type integral associated covariantly with an atlas on N (using the theory of Sobolev (1,2)-functions on quasiopen sets due to Kilpeläinen and Malý [KilM 1992]). Similarly if the target is a compact Riemannian *polyhedron* with continuous Riemannian metric.

There is a good existence theory of energy minimizing maps $X \to Y$ in the spirit of Eells and Sampson [ES 1964], both in the case of "free homotopy" and for given Dirichlet data on a boundary bX; and these E-minimizers are Hölder continuous, provided the target has nonpositive curvature. (Hölder continuity is best possible here; however, if X is a smooth Riemannian manifold, then E-minimizers are locally Lipschitz [KS 1993, Section 2.7].) Nonpositive curvature of Y can be weakened to upper bounded curvature in the case of maps with suitably small range.

Harmonic maps between admissible Riemannian polyhedra X, Y are defined, tentatively, as continuous bi-local E-minimizers (a concept slightly weaker than local E-minimizers). No general theory has been developed for them – except that there are versions of Ishihara's pull-back characterization of harmonic maps in the cases where Y is a Riemannian manifold or where X or Y is 1-dimensional. Aside from its own interest this characterization implies, in particular, that the limit of a locally uniformly convergent E-bounded sequence of harmonic maps is a harmonic map; and also that closed polar subsets of X are removable singularity sets for harmonic maps $X \to Y$.

Part I

Domains, targets, examples

2

Harmonic spaces, Dirichlet spaces, and geodesic spaces

This chapter summarizes briefly (with references) the three topics in the chapter heading. Our aim now is to provide some general background for our presentation.

Harmonic spaces

General references: [Br 1960; 1969], [Her 1962], [CC 1972]. For a survey, see [Bau 1984].

As a frame for axiomatic potential theory Brelot introduced the following notion of harmonic space [Br 1957; 1958a,b; 1959], cf. [CC 1972, p. 61]. On a locally compact, locally connected Hausdorff space X let there be given a (complete) sheaf $\mathcal{H}$ which to every open set $U \subset X$ assigns a linear subspace $\mathcal{H}(U)$ of $C(U)$ (the continuous functions $U \to \mathbb{R}$). The functions in $\mathcal{H}(U)$ are called *harmonic functions* on U. The sheaf property means, briefly, that harmonicity is a local property. $\mathcal{H}$ is called a (Brelot) *harmonic structure* on X, and $(X, \mathcal{H})$ is called a *harmonic space*, if the following two axioms are satisfied:

1. *The topology on X has a base consisting of regular sets.*

   A *regular set* is, by definition, an open, relatively compact set $V \subset X$ with $\partial V \neq \emptyset$ for which the Dirichlet problem is uniquely and positively solvable:

   For any function $f \in C(\partial V)$ there shall exist a unique harmonic function $H_f^V \in \mathcal{H}(V)$ such that the extension of H_f^V with f is continuous on the closure $\overline{V}$, and furthermore $f \geqslant 0$ (on ∂V) implies $H_f^V \geqslant 0$ (on V).

2. *$\mathcal{H}$ has the Harnack monotone convergence property*: For any increasing sequence of harmonic functions u_n on a connected open subset U of X, the pointwise limit function $u = \sup_n u_n$ is either harmonic on U or else $u \equiv \infty$.

We shall require that the topology on X has a *countable base*, and further that *the constant functions on X are harmonic*: $1 \in \mathcal{H}(X)$. If X is compact, $\mathcal{H}(X)$ consists of the functions which are constant on each component of X. Clearly, a harmonic space X has no isolated points.

Every component W of a regular set V is regular, and $H_f^V = H_f^W$ in W. By Axiom 1 it follows that *the regular domains form a base for the topology of X*.

Axiom 2 extends easily to increasing *nets* of harmonic functions on a domain U.

The (local) *minimum principle* asserts that, if a harmonic function u on a domain U has a smallest value, then u is constant. (Apply Axiom 2 to the increasing sequence $u_n = n(u - \min u)$.)

If $\mathcal{H}$ and $\mathcal{H}'$ are harmonic structures on the same space X and if $\mathcal{H} \subset \mathcal{H}'$ then $\mathcal{H} = \mathcal{H}'$ ([CC 1965, p. 23] and Corollary 13.1 (ii) below).

Clearly, the germs of solutions of Laplace's equation on $\mathbb{R}^n$ form a Brelot harmonic sheaf for which every open ball is regular.

We recall a few more basic notions and results for a Brelot harmonic space $(X, \mathcal{H})$.

Definition 2.1. The *harmonic measure* ω_x^V for a regular set V and a point $x \in V$ is the positive Radon measure on ∂V given by

$$\int f d\omega_x^V = H_f^V(x), \qquad f \in C(\partial V).$$

This measure is well defined since the linear functional $f \mapsto H_f^V(x)$ on $C(\partial V)$ is positive, by definition of a regular set. It is a probability measure because $1_V \in \mathcal{H}(V)$ and hence $\int d\omega_x^V = H_1^V(x) = 1$. We may of course view ω_x^V as a measure on X supported by ∂V.

For a regular domain V *the support of ω_x^V is the whole boundary ∂V*. (If $\omega_x^V(e) = 0$ for some open set e in ∂V then there is $f \in C(\partial V)$ with $f = 0$ off e and $f(y) > 0$ at some $y \in e$, hence $0 = \int_{\partial V \setminus e} f\, d\omega_x^V = H_f^V(x) \to f(y)$ as $x \to y$, which is absurd.)

Given a regular domain V, if a function $f : \partial V \to [-\infty, \infty]$ is integrable with respect to $\omega_{x_0}^V$ for some $x_0 \in V$, then f is integrable for every $x \in V$, and the integral $\int f\, d\omega_x^V$ is a harmonic function of $x \in V$. (If at first f is lower semicontinuous and $> -\infty$, this follows from Axiom 2 when writing f as an increasing pointwise limit of finite continuous functions on ∂V. For general f, the upper integral $\overline{H}_f(x) := \int^* f\, d\omega_x^V$ is harmonic in $x \in V$, being the pointwise infimum of the harmonic functions $\overline{H}_g$ as g ranges over the lower directed family of all l.s.c. (i.e., lower semicontinuous) functions $g \geqslant f$, $g > -\infty$ on ∂V. Similarly, $\underline{H}_f(x) := \int_* f\, d\omega_x^V$ is harmonic; and since $\overline{H}_f - \underline{H}_f \geqslant 0$, with equality at x_0, we conclude from the above minimum principle that $\overline{H}_f = \underline{H}_f$. Hence f is indeed integrable with respect to ω_x for every $x \in V$, and its integral is harmonic in V.)

For a given regular domain V *the harmonic measures* ω_x^V, $x \in V$, *are pairwise absolutely continuous.* (In fact, a set $e \subset \partial V$ has measure 0 with respect to ω_x^V iff the function f equal to ∞ in e and to 0 in $\partial V \setminus e$, is integrable with respect to ω_x^V, and this property is independent of x, see above.)

For given $x \in X$, ω_x^V converges weakly to the Dirac measure ε_x as the regular set V shrinks to $\{x\}$. (Given $f \in C_c(X)$ and $\varepsilon > 0$, we have in fact $|f - f(x)| < \varepsilon$ in some neighbourhood W of x, and so $\left|\int f\, d\omega_x^V - f(x)\right| = \left|\int (f - f(x))\, d\omega_x^V\right| < \varepsilon$ for any regular neighbourhood $V \subset W$ of x.)

Using the above results one may establish *Mokobodzki's theorem*: The family of all positive harmonic functions on a domain $U \subset X$ having the value 1 at a given point of U is *equicontinuous.* (For the proof see [Br 1969, p. 22].)

A consequence of this is the following *Harnack inequality*: For any harmonic function $u > 0$ in a domain $U \subset X$ and any compact set $K \subset U$, we have

$$\max_{x \in K} u(x) \leqslant c \min_{x \in K} u(x),$$

where c is independent of u. (Indeed, for every $x \in U$ and $\varepsilon > 0$ there is, by equicontinuity, a neighbourhood V of x such that

$$\max_{y \in V} u(y) \leqslant \frac{1+\varepsilon}{1-\varepsilon} \min_{y \in V} u(y).$$

As in classical potential theory, a standard covering argument completes the proof, cf. e.g., [Hel 1969, §2.16].)

Definition 2.2. A function s on an open set $U \subset X$ with values in $]-\infty, \infty]$ is *superharmonic* if

(i) s is lower semicontinuous,

(ii) for every point $x \in U$ and every regular set V with $x \in V \subset \overline{V} \subset U$,

$$s(x) \geqslant \int s\, d\omega_x^V, \quad \text{and}$$

(iii) s is not identically ∞ on any component of U.

Here (iii) can be replaced equivalently by s being finite on a dense subset of U [Br 1960, p. 70].

A function $s : U \to [-\infty, \infty[$ is called *subharmonic* if $-s$ is superharmonic. A function which is both super- and subharmonic is the same as a harmonic function. The super/subharmonic functions on U clearly form a convex cone.

The pointwise infimum of a finite family of superharmonic functions on U is obviously superharmonic. The pointwise limit of an increasing net of superharmonic functions on a domain $U \subset X$ is either superharmonic or $\equiv \infty$. This leads to the (local) *minimum principle* [Br 1959]: *If a superharmonic function s on a domain $U \subset X$ has a smallest value then s is constant.* (Proof as above for harmonic functions.)

A function $f : U \to \mathbb{R}$ (U open in X) is harmonic if and only if $g \circ f$ is subharmonic in U for every convex function g on an interval containing $f(U)$. (For the "if part", take $g(t) = \pm t$; for the "only if part", write g as the upper envelope of a sequence of affine-linear functions.)

Definition 2.3. A *potential* on a harmonic space $(X, \mathcal{H})$ is a superharmonic function $p \geqslant 0$ on X having no other harmonic minorant $\geqslant 0$ than the function 0.

The potentials on X form a convex cone. Any superharmonic function $s \geqslant 0$ on X majorized by a potential is itself a potential.

A classical theorem of F. Riesz extends as follows to any harmonic space $(X, \mathcal{H})$: *Every superharmonic function s on X which has a subharmonic minorant can be written uniquely as $s = p + h$, where p is a potential on X and h is harmonic on X*; namely, h is the greatest subharmonic minorant of s, see [Br 1969, p. 38]. In particular, every subharmonic minorant of a potential is $\leqslant 0$.

The *boundary minimum principle*: If a superharmonic function s on an open set $U \subset X$ satisfies

$$\liminf_{U \ni x \to y} s(x) \geqslant 0 \qquad \text{for every } y \in \partial U,$$

and if moreover $s \geqslant -p$ on U for some potential p on X, then $s \geqslant 0$ on U. If $\overline{U}$ is compact and $\partial U \neq \emptyset$, the condition $s \geqslant -p$ can be omitted. If $\overline{U}$ (and hence X) is non-compact, the condition $s \geqslant -p$ can be replaced by $\liminf s(x) \geqslant 0$ as $x \in U$ tends to the point at infinity for X [Br 1959].

Definition 2.4. A subset e of X is *polar* if every point of X has an open neighbourhood U on which there is a superharmonic function which equals ∞ at every point of $e \cap U$.

In a Riemannian manifold M^n every set which is locally of finite $(n-2)$-dimensional Hausdorff measure is polar.

The polar sets are the potential theoretic null sets. Let e be a polar subset of a harmonic space X. Then, as shown in [Br 1959, p. 1.34]:

- For any regular set $V \subset X$ and any $x \in V$, we have $\omega_x^V(e) = 0$.
- If X is connected then so is $X \setminus e$. (See also [CC 1972, p. 145].)
- If two superharmonic functions on X coincide off e then they are identical.
- If e is closed, if s is superharmonic on $X \setminus e$ and if

$$\liminf_{x \to y} s(x) > -\infty \qquad \text{for every } y \in e,$$

then the extension of s by this lim inf is superharmonic on X. In particular, such a set e is a removable singularity set for locally bounded harmonic functions.

A *$\mathcal{P}$-harmonic*, or *strongly harmonic*, space is a harmonic space on which there exists a potential $p > 0$. Every open subset of a $\mathcal{P}$-harmonic space is a $\mathcal{P}$-harmonic space. Every harmonic space can be covered by $\mathcal{P}$-harmonic open subspaces [CC 1972, §2.3].

Further development of potential theory on a Brelot harmonic space has led to generalizations of most qualitative results from classical potential theory. For example, for a connected $\mathcal{P}$-harmonic space X, if any two non-zero potentials which are harmonic off one and the same point are proportional, then there exists a lower semicontinuous function $G : X \times X \to \,]0, \infty]$, finite and continuous off the diagonal in $X \times X$, such that

(1) For every positive measure μ on X the function $G\mu$ defined on X by

$$G\mu(x) = \int_X G(x, y)\, d\mu(y), \qquad x \in X,$$

is a potential on X, unless $G\mu \equiv \infty$.

(2) Every potential p on X has the form $p = G\mu$ for a unique positive measure μ on X. The support of μ is the complement of the biggest open set in which p is harmonic.

Such a function G is called a *Green kernel* for X. See [Her 1962] and also [CC 1972, pp. 332f.].

Remark 2.1. A notion of harmonic space X, more general than that of Brelot, was introduced by Bauer [Bau 1966] (see also [Bau 1984]), who replaced Harnack's monotone convergence property by the weaker convergence property of Doob [Doob 1956]: For any increasing sequence of functions $u_n \in \mathcal{H}(U)$ (U open in X), $u := \sup_n u_n$ is in $\mathcal{H}(U)$ provided that $\{u < \infty\}$ is dense in U. Bauer showed that the germs of solutions to the heat equation on $\mathbb{R}^{n+1}$ form a harmonic space in his sense (this was his primary motivation), and that large parts of the potential theory on Brelot spaces carry over or can be adapted to Bauer spaces, even though the minimum principle for superharmonic functions may not hold. See also [CC 1972].

Dirichlet structures on a space

General references: [BM 1995], [Fuk 1980], [FOT 1994].

Let X be a separable locally compact space, and μ a Radon measure with $\mu > 0$ on every non-void open subset of X, that is, $\operatorname{supp}\mu = X$. Let $\mathcal{D}$ be a linear subspace of $L^2(X,\mu)$.

Definition 2.5. A *Dirichlet form E on X* with domain $\mathcal{D}(E) = \mathcal{D}$ is a symmetric bilinear map $E : \mathcal{D} \times \mathcal{D} \to \mathbb{R}$ which is nonnegative definite. Set $E(f,f) = E(f)$. Assume that

(i) $\mathcal{D}$ is dense in $L^2(X,\mu)$, and

(ii) $\mathcal{D}$ is complete in the inner product

$$\langle f,g\rangle_{\mathcal{D}} = \langle f,g\rangle_{L^2(X,\mu)} + E(f,g).$$

Thus $(\mathcal{D}, \langle\cdot,\cdot\rangle_{\mathcal{D}})$ is a Hilbert space. A *normal contraction* of $\mathbb{R}$ is a map $T : \mathbb{R} \to \mathbb{R}$ with $T(0) = 0$ and $|T(s) - T(t)| \leqslant |s - t|$ for all $s,t \in \mathbb{R}$. We require that

(iii) for any normal contraction T and any $f \in \mathcal{D}$, we have $T \circ f \in \mathcal{D}$ and

$$E(T \circ f) \leqslant E(f).$$

Say that a Dirichlet form E *is local*, or *has local type*, if $E(f,g) = 0$ whenever $f,g \in \mathcal{D}$ have disjoint compact supports. It *is strongly local*, i.e., *has diffusion type*, if $E(f,g) = 0$ whenever $f,g \in \mathcal{D}$ have compact supports and f is constant on some neighbourhood of $\operatorname{supp} g$.

A *core* of E is a linear subspace $\mathcal{C}$ of $\mathcal{D} \cap C_c(X)$ which is dense in $C_c(X)$ (the continuous functions $X \to \mathbb{R}$ of compact support) in its (inductive limit) uniform topology, and in $\mathcal{D}$ in the norm associated with $\langle\cdot,\cdot\rangle_{\mathcal{D}}$. Say E *is regular* if E has a core, or equivalently if $\mathcal{D} \cap C_c(X)$ is a core.

There is a capacity function associated with a regular diffusion Dirichlet form E. The *capacity* $\operatorname{cap} A$ of an open set $A \subset X$ is defined by

$$\operatorname{cap} A = \inf\{E(u) : u \in \mathcal{D}, u \geqslant 1 \text{ a.e. in } A\},$$

interpreted as ∞ if there is no such u. The capacity of an arbitrary set $A \subset X$ is defined as the infimum of the capacities of all open sets containing A. *Quasieverywhere* (q.e.) means: everywhere off some set of zero capacity.

A function $u : X \to \mathbb{R}$ is called *quasicontinuous* if u has continuous restriction to the complement of open sets of arbitrarily small capacity. Every function $u \in \mathcal{D}$ has a *quasicontinuous version*, i.e., u coincides off a μ-nullset with a quasicontinuous function, uniquely determined q.e. [De 1970, Chapter 4], [Fuk 1980, §3.1].

A regular Dirichlet form E of diffusion type has the representation

$$E(f,g) = \int_X \varepsilon(f,g) \qquad \text{for } f,g \in \mathcal{D},$$

where ε, termed *the energy measure of* E, is a symmetric nonnegative definite bilinear map of $\mathcal{D} \times \mathcal{D}$ to the space of Radon measures on X. This map ε has local character; i.e., on an open $U \subset X$, $\varepsilon(f,g)$ depends only on $f\big|_U$ and $g\big|_U$. Thus we can say that $f \in \mathcal{D}$ *is a local E-minimizer* in an open set $U \subset X$ if

$$\int_U \varepsilon(f,f) \leqslant \int_U \varepsilon(f+g, f+g)$$

for every $g \in \mathcal{D}$ with compact support in U.

In addition to the expressions for $\varepsilon(f,g)$ and $E(f,g)$ by the usual polarization, note the following identity, which may serve as a definition of $\varepsilon(f,f)$:

$$\int_X \lambda\, \varepsilon(f,f) = E(f,\lambda f) - \frac{1}{2} E(f^2,\lambda) \tag{2.1}$$

for all $f \in \mathcal{D} \cap L^\infty(X,\mu)$ and $\lambda \in \mathcal{C}$.

By general principles, there is a unique nonpositive definite selfadjoint operator A on $L^2(X,\mu)$ such that $\mathcal{D}$ is the domain of $\sqrt{-A}$, and

$$E(f,g) = \langle \sqrt{-A}\, f, \sqrt{-A}\, g\rangle_{L^2(X,\mu)}$$

for all $f,g \in \mathcal{D}$.

Denote by $P_t = \exp tA$, $t > 0$, the diffusion semigroup on $L^2(X,\mu)$ generated by A. It is strongly continuous:

$$\lim_{t \searrow 0} P_t f = f \qquad \text{for all } f \in L^2(X,\mu).$$

We represent P_t in terms of its symmetric kernel p_t:

$$\langle P_t f, f\rangle_{\mathcal{D}} = \iint_{X\times X\setminus\mathrm{diag}} f(x)f(y)p_t(x,dy)\,d\mu(x).$$

If we write $2f(x)f(y) = |f(x)|^2 + |f(y)|^2 - |f(x)-f(y)|^2$ then

$$\begin{aligned} E^t(f) :=& \frac{1}{t}\langle f - P_t f, f\rangle_{L^2} \\ =& \frac{1}{t}\|f\|_{L^2}^2 - \frac{1}{2t}\Big(2\|f\|_{L^2}^2 - \iint_{X\times X} |f(x)-f(y)|^2 p_t(x,dy)\,d\mu(x)\Big) \\ =& \frac{1}{2t}\iint_{X\times X} |f(x)-f(y)|^2 p_t(x,dy)\,d\mu(x). \end{aligned}$$

$E^t(f)$ increases as $t > 0$ decreases, and the Dirichlet form E has the representation

$$E(f) = \lim_{t\searrow 0} E^t(f) \qquad \text{for } f \in \mathcal{D}(E).$$

(See [BD 1959] and [De 1970].)

A core $\mathcal{C}$ of E is said to be *μ-separating* if for all points $x \neq y$ in X there is a function $g \in \mathcal{C}$ with $\varepsilon(g,g) \leqslant \mu$ on X, and $g(x) \neq g(y)$. Here $\varepsilon(g,g) \leqslant \mu$ means that $\varepsilon(g,g)$ is absolutely continuous with respect to μ, with Radon–Nikodym derivative

$$e(g) := e(g,g) := \frac{d\varepsilon(g,g)}{d\mu} \leqslant 1 \qquad \text{a.e. on } X.$$

($2e(g)$ is interpreted as the analogue of the square of the length of the gradient of g.)

With a μ-separating core $\mathcal{C}$ is associated the *Carathéodory distance* $d_{\mathcal{C}} : X \times X \to [0,\infty]$, defined by

$$d_{\mathcal{C}}(x,y) = \sup\{g(x) - g(y) : g \in \mathcal{C},\ \varepsilon(g,g) \leqslant \mu \text{ on } X\}. \tag{2.2}$$

Clearly, $d_{\mathcal{C}}$ is symmetric, satisfies the triangle inequality (with the usual care about ∞), and $d_{\mathcal{C}}(x,y) = 0$ iff $x = y$.

Remark 2.2. Sturm [St 1998] has introduced the notion of *measure contractions* (in the spirit of curvature bounds) on metric measure spaces (X, d_X, μ); and has shown how such determines an effective energy functional

E which is an admissible Dirichlet form on $L^2(X,\mu)$, with core $\mathcal{C} = \mathrm{Lip}_c(X)$ for which $d_{\mathcal{C}} = d_X$. For this energy functional Poincaré's inequality is valid.

Definition 2.6. An *admissible Dirichlet space* is a space $X = (X, \mu, E)$ with a regular diffusion Dirichlet form E having a μ-separating core $\mathcal{C}$ and associated Carathéodory distance $d_{\mathcal{C}}$. It is required that

(i) $d_{\mathcal{C}}$ is a complete metric whose topology coincides with that of X, and

(ii) locally, μ is doubling with respect to the balls

$$U(x,r) = \{y \in X : d_{\mathcal{C}}(x,y) < r\},$$

i.e., for any compact set $K \subset X$ there are constants $c > 0$ and $r_0 > 0$ such that $0 < \mu(U(x,2r)) \leqslant c\mu(U(x,r)) < \infty$ for all $x \in K$ and $0 < r \leqslant r_0$.

Remark 2.3. The ball doubling property (Definition 2.6 (ii)) of a Radon measure $\mu \geqslant 0$ on a metric space (X,d) was isolated and exploited by Coifman and Weiss [CW 1971]. On such a metric space a Sobolev space of type $W^{1,p}(X) = W^{1,p}(X,d;\mu)$ can be defined ($1 < p \leqslant \infty$), see [Haj 1996] and [KM 1996]. It can be characterized in terms of a Poincaré inequality, as follows: $u \in W^{1,p}(X)$ *iff* $u \in L^p(X)$ *and there is a function* $0 \leqslant g \in L^p(X)$ *and a constant* $C > 0$ *such that*

$$\fint_{B(x,r)} |u - u_{B(x,r)}|\, d\mu \leqslant Cr \fint_{B(x,r)} g\, d\mu$$

for every $x \in X$ *and* $r > 0$. See [FLW 1996], [HK 1998], [FHK 1998] and [HeKo 1998]. As usual, $\fint_B$ denotes the integral over B divided by $\mu(B)$, and $u_B = \fint_B u$ is the meanvalue of u over a ball B (here the closed ball $B(x,r)$). Ball doubling and related matters in metric measure spaces have been thoroughly analyzed by S. Semmes in [Gro 1999, Appendix B] and in [Chee 1999].

With the capacity function on X relative to $W^{1,p}(X)$, defined in analogy with the capacity on a Dirichlet space, every function $u \in W^{1,2}(X)$ has Hölder quasicontinuous versions in the natural sense, and they approximate u [HK 1998]. There is a version of Morrey's growth lemma in this context [MS 1979].

Using the same notation as in Example 3.5 below: Define the space $W^{1,p}_{\mathfrak{X}}(M)$ $(1 < p < \infty)$ via the norm whose p-th power is

$$\|f\|^p_{W^{1,p}_{\mathfrak{X}}(M)} = \|f\|^p_{L^p(M)} + \int_M \Big(\sum_i |\mathcal{L}_{X_i} f|^2\Big)^{p/2} d\mu_g.$$

Then $W^{1,p}_{\mathfrak{X}}(M)$ *is equivalent to the Sobolev space* $W^{1,p}(M, d^{\mathfrak{X}}; \mu_g)$; see [FHK 1998] and the references in [HK 1998, §1]; both of these have extensive bibliographies concerning the $(1,p)$-Sobolev spaces.

Sturm [St 1995b] has shown that *an admissible Dirichlet space* (even without the ball doubling requirement) *is a geodesic space* (in the sense described in the next section) *with respect to the metric* $d_{\mathcal{C}}$. *Its balls are relatively compact.*

Remark 2.4. The present discussion can be extended immediately, replacing $L^2(X,\mu)$ by the Hilbert space $L^2(X,\mu;V)$ of L^2-maps of X into a Euclidean space V. Normal contractions $T : V \to V$ are maps such that $T(0) = 0$ and $|Ts - Tt|_V \leqslant |s-t|_V$ for all $s, t \in V$.

Here is a general relationship between harmonic spaces and Dirichlet spaces, due to Feyel and de La Pradelle [FP 1978], see also [La 1980]. (Compare with Theorem 7.2 below, in the case where X is an admissible Riemannian polyhedron):

Let $X = (X, \mu; E)$ be a regular Dirichlet space of local type. Suppose that (a) for every open $U \subset X$ the local E-minimizers in U have continuous versions in U, and (b) for every point $x \in X$ there is a continuous local E-minimizer $u \geqslant 0$ in some neighbourhood of x such that $u(x) > 0$. *Then the spaces* $\mathcal{H}(U)$ *of (continuous) local* E*-minimizers in open subsets* U *of* X *determine a Brelot harmonic structure on* X.

Geodesic spaces

General references: [Bal 1995], [BNik 1993], [GLP 1981], [Gro 1999], [P 1998].

Let (Y, d_Y) be a metric space and $\gamma : I = [a,b] \to Y$ a path. Its length is

$$L(\gamma) = \sup_\pi \Big\{ \sum_{i=1}^r d_Y\big(\gamma(t_{i-1}), \gamma(t_i)\big) \Big\},$$

the supremum over all partitions $\pi : a = t_0 < \cdots < t_r = b$ of I. Say that γ *is rectifiable* if $L(\gamma) < \infty$.

A *length space* is a metric space (Y, d_Y) such that for any pair of points $y_0, y_1 \in Y$,

$$d_Y(y_0, y_1) = \inf \{L(\gamma) : \gamma \text{ is a rectifiable path joining } y_0 \text{ to } y_1\}.$$

Such a metric is sometimes said to be *intrinsic* (or *inner*).

Every length space (Y, d_Y) is locally connected. Indeed, every open ball $U = \{y \in Y : d_Y(a, y) < r\}$ is *connected* (even path connected). There is in fact a rectifiable path γ joining a and y such that $L(\gamma) < r$, and so γ has range in U because, for any $z \in \gamma$,

$$d_Y(a, z) \leqslant d_Y(a, z) + d_Y(z, y) \leqslant L(\gamma) < r.$$

Definition 2.7. A *geodesic space* is a length space (Y, d_Y) for which any $y_0, y_1 \in Y$ can be joined by a rectifiable path γ with $d_Y(y_0, y_1) = L(\gamma)$.

A (minimizing) *geodesic segment* of (Y, d_Y), parametrized by path length (as will be understood in the absence of any other indication), is a rectifiable path $\gamma : I \to Y$ for which $d_Y(\gamma(s), \gamma(t)) = |s - t|$ for all $s, t \in I$.

For a compact interval $[a, b]$ this requirement reduces to $d_Y(\gamma(a), \gamma(b)) = |b - a|$. The path γ in the above definition is therefore a geodesic segment.

A *geodesic* of (Y, d_Y) is a path $\gamma : J \to Y$ whose restriction to every sufficiently small compact subinterval of J is a geodesic segment. We shall allow J to be open, closed, or half-open.

A function $f : V \to \mathbb{R}$ on a set $V \subset Y$ is *convex* if for all geodesic segments $\gamma : [a, b] \to Y$ with image in V,

$$f\big(\gamma(ta + (1 - t)b)\big) \leqslant t\, f(\gamma(a)) + (1 - t)\, f(\gamma(b)), \qquad 0 \leqslant t \leqslant 1.$$

On an open set $V \subset \mathbb{R}$ a function $f : V \to \mathbb{R}$ is convex iff the restriction of f to each component of V is convex in the usual sense.

Say that a subset $V \subset Y$ is *convex* if any two points of V can be joined by a geodesic segment of (Y, d_Y) lying in V.

A geodesic space (Y, d_Y) is said to be *locally squared-convex* if every point $y \in Y$ has an open convex neighbourhood V such that $V \times V$ is convex; i.e., for any two minimizing geodesic segments $\alpha, \beta : [0, 1] \to V$ parametrized proportionally to path length, the function $f : [0, 1] \to \mathbb{R}$ defined by $f(t) = d_Y(\alpha(t), \beta(t))$ is convex. In other words, (Y, d_Y) *is locally squared-convex iff the function* $d_Y : V \times V \to \mathbb{R}$ *is convex for every such* V ([AB 1990]).

If this applies to $V = Y$, (Y, d_Y) is said to be *globally squared-convex*.

The following result is Gromov's version of the *Hadamard–Cartan theorem* [AB 1990], [Gro 1987]: *A simply connected complete locally squared-convex geodesic space* (Y, d_Y) *is globally squared-convex; hence any two of its points are joined by a unique geodesic, and that geodesic varies continuously with its endpoints.*

In a geodesic space (Y, d_Y) fix a point y_0; and denote by $\big(\mathcal{G}_{y_0}(Y), D\big)$ the metric space of geodesic segments γ *of constant speed* (i.e., parametrized proportionally to path length) and with common domain $I = [0,1]$, for which $\gamma(0) = y_0$, and

$$D(\beta, \gamma) = \sup\left\{d_Y(\beta(t), \gamma(t)) : t \in I\right\}.$$

Then $\mathcal{G}_{y_0}(Y)$ is clearly contractible, and the map

$$\exp_{y_0} : \mathcal{G}_{y_0}(Y) \to Y \tag{2.3}$$

defined by $\exp_{y_0}(\gamma) = \gamma(1)$ is continuous and surjective. *If* (Y, d_Y) *is a complete locally squared-convex geodesic space, then* $\big(\mathcal{G}_{y_0}(Y), D\big)$ *is complete and* (2.3) *is its universal cover. In particular,* (Y, d_Y) *has no conjugate points.* That is, for all $y_0 \in Y$, $\exp_{y_0}$ maps some neighbourhood V of any $\gamma \in \mathcal{G}_{y_0}(Y)$ homeomorphically onto a neighbourhood of $\gamma(1)$ [Gro 1987], [AB 1990, Theorem 4]. Furthermore, if H denotes a homotopy class of paths in Y joining y_0 to y_1, then H *contains exactly one geodesic.* It is *minimizing*; and it *depends continuously on* y_0, y_1. Every point of Y has a convex neighbourhood in which geodesics are uniquely determined by their endpoints and vary continuously with these [AB 1990, §3].

Consider a triangle *pqr* in a geodesic space (Y, d_Y) with minimal geodesic sides. Fix a real number K. We wish to compare *pqr* with the triangle $\tilde{p}\,\tilde{q}\,\tilde{r}$, with corresponding sides of the same lengths, in the 2-dimensional space form $\mathbb{R}^2(K)$ with constant sectional curvature K. Thus

$$d_Y(p,q) = |\tilde{p} - \tilde{q}\,|, \quad d_Y(p,r) = |\tilde{p} - \tilde{r}\,|, \quad d_Y(q,r) = |\tilde{q} - \tilde{r}\,|,$$

the "norm" referring to the distance in $\mathbb{R}^2(K)$. There is such a comparison triangle, and it is unique subject to the condition on the perimeter

$$d_Y(p,q) + d_Y(q,r) + d_Y(r,p) < 2\pi/\sqrt{K} \qquad \text{if } K > 0. \tag{2.4}$$

Write $2\pi/\sqrt{K} = \infty$ if $K \leqslant 0$.

Say that *pqr* is *K-dominated* if it satisfies (2.4) and has the following *triangle comparison property:*

$$d_Y(\alpha(t), p) \leqslant |\tilde{\alpha}\,(t) - \tilde{p}\,| \tag{2.5}$$

for all $\alpha(t)$ on the geodesic segment qr, with $\tilde{\alpha}(t)$ the corresponding point on $\tilde{q}\tilde{r}$.

Definition 2.8. A geodesic space (Y, d_Y) *has curvature* $\leqslant K < \infty$ (written $\operatorname{curv}(Y, d_Y) \leqslant K$) if every point $y \in Y$ has a neighbourhood V such that every geodesic triangle with vertices in V is K-dominated. Similarly for $\operatorname{curv}(Y, d_Y) \geqslant K > -\infty$. ([A 1951; 1957], [BNik 1993].)

A smooth *complete Riemannian manifold* has curvature $\leqslant K$, resp. $\geqslant K$, iff it has sectional curvature $\leqslant K$, resp. $\geqslant K$.

In a geodesic space with $\operatorname{curv}(Y, d_Y) \leqslant K$ for some $0 \leqslant K < \infty$ *every triangle satisfying* (2.4) *is K-dominated iff Y has bi-point uniqueness,* in the sense that geodesics in Y of length $< \pi/\sqrt{K}$ are unique and vary continuously with their endpoints (the continuous dependence being automatic if Y is complete and locally compact) [A 1951; 1957].

A geodesic space with $\operatorname{curv}(Y, d_Y) \leqslant 0$ *is locally squared-convex.* The easy proof is based on characterizing local squared convexity as follows: Every $y \in Y$ has a neighbourhood V in which $d_Y(m(p,q), m(p,r)) \leqslant \frac{1}{2} d_Y(q,r)$ for any geodesic triangle pqr in V; here $m(p,q)$ denotes the midpoint of the geodesic segment pq. Then applying [BNik 1993, Proposition 2.2]:

$$d_Y\big(m(p,q), m(p,r)\big) \leqslant |m(\tilde{p},\tilde{q}) - m(\tilde{p},\tilde{r})\,| = \tfrac{1}{2}\,|\tilde{q} - \tilde{r}\,| = \tfrac{1}{2} d_Y(q,r).$$

A smooth Riemannian *manifold* Y with intrinsic distance d_Y has nonpositive sectional curvature if (and only if) (Y, d_Y) is locally squared convex, [BGS 1985, §1]. The "if part" does not extend to geodesic spaces, cf. [AB 1990, p. 313].

A simply connected complete geodesic space (Y, d_Y) of nonpositive curvature (i.e., $\operatorname{curv}(Y, d_Y) \leqslant 0$) *has globally nonpositive curvature,* in the sense that every triangle in Y is 0-dominated. Indeed, Y is globally squared-convex, by the above Hadamard–Cartan theorem; we may therefore apply the quoted result of Alexandrov, or argue directly as follows:

For a given triangle pqr in Y with midpoint m of pq, let p_t, q_t, r_t denote the images of p, q, r under the geodesic homothety with centre at m and factor $t > 0$ small enough so that the triangle $p_t q_t r_t$ is 0-dominated. Then $d_Y(r, m) \leqslant |\widetilde{r} - \widetilde{m}|$. In fact,

$$\begin{aligned} t^2 d_Y^2(r_t, m) &\leqslant \tfrac{1}{2} d_Y^2(p_t, r_t) + \tfrac{1}{2} d_Y^2(q_t, r_t) - \tfrac{1}{4} d_Y^2(p_t, q_t) \\ &\leqslant \tfrac{1}{2} t^2 d_Y^2(p, r) + \tfrac{1}{2} t^2 d_Y^2(q, r) - \tfrac{1}{4} t^2 d_Y^2(p, q) = t^2 |\widetilde{r} - \widetilde{m}|^2 \end{aligned}$$

because $d_Y(p_t, r_t) \leqslant t d_Y(p, r)$ and $d_Y(q_t, r_t) \leqslant t d_Y(q, r)$, by global squared-convexity, while $d_Y(p_t, q_t) = t d_Y(p, q)$.

Conversely, every *locally compact*, complete geodesic space (Y, d_Y) of globally nonpositive curvature is simply connected. Indeed, geodesic segments in Y are uniquely determined by their endpoints and (by local compactness) depend uniquely on them. Any continuous path in Y is therefore (even geodesically) fixed-end homotopic to a unique constant-speed geodesic segment.

In a complete geodesic space Y with globally nonpositive curvature consider a geodesic quadrilateral with constant-speed geodesic segments γ_1, γ_2 as a pair of opposite sides and vertices ordered $\gamma_1(0), \gamma_1(1), \gamma_2(1), \gamma_2(0)$. There is a convexity inequality, basically due to Reshetnyak [Resh 1968] and applied in [KS 1993, Corollary 2.1.3]:

$$\begin{aligned} d_Y^2(\gamma_1(t), \gamma_2(t)) \leqslant &(1-t) d_Y^2(\gamma_1(0), \gamma_2(0)) + t\, d_Y^2(\gamma_1(1), \gamma_2(1)) \\ &- t(1-t)\big(d_Y(\gamma_1(0), \gamma_1(1)) - d_Y(\gamma_2(0), \gamma_2(1))\big)^2. \end{aligned}$$

The following counterpart to the Hadamard–Cartan theorem in the case where $K > 0$ is due to Alexander and Bishop [AB 1996, §4]. Say that an open set V in a geodesic space (Y, d_Y) has *radial uniqueness* with respect to a point $y \in V$ if there is a unique Y-geodesic in V from y to any other point of V, and if that geodesic is minimizing within V and varies continuously with its right endpoint. *In a complete geodesic space with* $\operatorname{curv}(Y, d_Y) \leqslant K$, *if an open ball* $V(y, r)$ *of radius* $r \leqslant \pi/(2\sqrt{K})$ *has radial uniqueness with respect to its centre* y *then the ball is convex, and geodesic segments in it vary continuously with their endpoints; furthermore, the function* $d_Y(\cdot, y)$ *on* $V(y, r)$, *is convex, and every triangle in* $V(y, r)$ *is* K*-dominated.*

In particular, a complete geodesic space (Y, d_Y) with $\operatorname{curv}(Y, d_Y) \leqslant K$, $0 \leqslant K < \infty$, is locally contractible, and therefore has a good covering space theory – with geodesic covering spaces $(\widetilde{Y}, d_{\widetilde{Y}})$ likewise of curvature $\leqslant K$, and with isometric projection maps [AB 1990].

Although many of the results in this monograph extend to targets which are complete geodesic spaces (see text following Theorem 11.4, together with Remark 11.6), *henceforth we shall consider only geodesic spaces which are locally compact.*

The following version of the Hopf–Rinow theorem is due to Cohn–Vossen, see [GLP 1981, §1], [Bal 1995, Chapter I]: *For a locally compact length space* Y *the following conditions are equivalent:*

(i) *every half-open minimizing geodesic from a given point extends to a closed interval,*

(ii) *every half-open geodesic extends to a closed interval,*
(iii) *closed bounded subsets of Y are compact,*
(iv) (Y, d_Y) *is complete.*

Any of these implies that (Y, d_Y) *is a geodesic space.*

A subset A of a geodesic space Y is called *totally geodesic* if locally each geodesic segment of Y with endpoints in A is contained in A.

Gromov's gluing lemma [Gro 1987, p. 124], [DJ 1991, Lemma (2a.4)]: *Take two geodesic spaces* Y_1, Y_2. *Let* $A_i \subset Y_i$ $(i = 1, 2)$ *be closed totally geodesic subspaces, and* $f : A_1 \to A_2$ *an isometry. Then the identification space* $Y_1 \cup_f Y_2$ *is a geodesic space. If both* $\operatorname{curv} Y_i \leqslant 0$, *then* $\operatorname{curv}(Y_1 \cup_f Y_2) \leqslant 0$.

Similarly, we can glue suitable maps $\varphi_i : X_i \to Y_i$.

In the same spirit, let A be a totally geodesic subspace of a geodesic space Y. Form the double of Y along A: $D(Y, A) = (Y \times]-1, 1[)/\sim$, where $(y, \varepsilon) \sim (y', \varepsilon')$ iff either $y \in A$ and $y = y'$, or $(y, \varepsilon) = (y', \varepsilon')$. Then the metric of Y induces one on $D(Y, A)$ relative to which $D(Y, A)$ is a geodesic space; and its canonical involution with fixed point set A is an isometry.

3

Examples of domains and targets

This chapter contains some geometric examples of harmonic, Dirichlet and geodesic spaces, as described above. Most of them fit easily into the framework of admissible Riemannian polyhedra; but others not so immediately.

Example 3.1. Riemannian manifolds

Let (M, g) be a smooth Riemannian manifold, at first without boundary. Then the metric tensor g determines:

(a) The Laplace–Beltrami operator Δ, with its *Brelot harmonic sheaf* on M, whose sections are the harmonic functions (the local solutions of the Laplace–Beltrami equation). This was obtained by R.–M. Hervé [Her 1962] using results of Morrey, Gilbarg and Serrin. Sufficiently small d_M-balls are regular because their boundaries are smooth. Furthermore, Δ is the Euler–Lagrange operator of the energy functional $E(u) = \frac{1}{2}\int_M |\nabla u|^2 d\mu_g$, and (M, μ_g, E) *is an admissible Dirichlet space* (if complete).

If a Riemannian manifold M is $\mathcal{P}$*-harmonic*, cf. Chapter 2, Harmonic spaces (for $n = 2$ and M oriented this means that the Riemann surface M is hyperbolic), then M has a symmetric Green kernel: $G(x, y) = G(y, x)$ (uniquely determined up to a constant factor > 0).

(b) The *canonical distance metric* d_M. Supposing that (M, g) is *complete*, the Hopf–Rinow theorem insures that (M, d_M) *is a geodesic space*. It has $\operatorname{curv}(M, d_M) \leqslant K$ (resp., $\operatorname{curv}(M, d_M) \geqslant K$) iff the sectional curvature of (M, g) is $\leqslant K$ (resp., $\geqslant K$; the proof in that case relies on the Rauch comparison theorem). Furthermore, (M, g) is nonpositively curved iff d_M is locally squared-convex.

Suppose now that (M, g) is a smooth complete Riemannian manifold *with smooth boundary* bM. Then again, (M, d_M) *is a geodesic space*, and $\operatorname{curv}(M, d_M) \leqslant 0$ iff d_M is locally squared-convex. Also, $\operatorname{curv}(M, d_M) \leqslant K$ iff (i) the sectional curvature of $(M \setminus bM, g)$ and (ii) the outward sectional curvature of the boundary $(bM, g_{|bM})$ are both $\leqslant K$ [ABB 1993; 1987].

The Laplace–Beltrami operator on M again determines a Brelot harmonic sheaf on M, requiring the local harmonic functions to have vanishing normal derivative on bM. If M is $\mathcal{P}$-harmonic then it has a symmetric Green function; see Chapter 7 applied to (M, g) as a Riemannian polyhedron (locally), cf. also Example 8.4.

Remark 3.1. Local minima for certain p-th power variational integrals $(1 < p < \infty)$ over a smooth manifold M form a (nonlinear) sheaf on M, with properties analogous to those of a Brelot harmonic space (Lehtola [Leh 1986]). A thorough treatment of this has been given in [HKM 1993, §16].

Example 3.2. Almost Riemannian spaces

Let (X, d_X) be a geodesic space of finite integral Hausdorff dimension m, with curvature bounded below, and with no geodesic terminals (i.e., there is no point of X beyond which some geodesic segment cannot be extended). Then X is a *C^1-smooth m-manifold* with a *continuous Riemannian metric* g which is *Hölder continuous* of order $1/2$ on a subset of full Hausdorff m-measure on X. Furthermore, the length function of g determines a *metric whose topology on X is the original one.* In particular, (X, g) is a Riemannian Lipschitz manifold in the sense of Example 3.6 below – but rather special, because the underlying space is smooth.

Such an (X, d_X) is called *almost Riemannian* by Plaut. See [P 1991] for the first properties of such spaces, and [OS 1994] for the other main ingredients of the above assertion. (We are much indebted to Conrad Plaut for explaining to us how that comes about.)

If (X, d_X) as defined above has curvature bounded both above and below, then X has a Riemannian metric of class $C^{1,\alpha}$, from which d_X is derived. This is a famous theorem of Berestovskii [Be 1976] and Nikolaev [N 1980]. See also [BNik 1993] and [P 1998]. See [Gro 1999, §8.20] for a discussion of $C^{1,1}$-Riemannian manifolds.

Example 3.3. Finsler structure on a manifold

Suppose M is a smooth n-manifold. A *Finsler structure* on M is a continuous function $f : T(M) \to [0, \infty[$ such that on each fibre $f(v) > 0$ for all non-zero $v \in T_x(M)$; and $f(\lambda v) = |\lambda| f(v)$ for all $v \in T_x(M)$ and $\lambda \in \mathbb{R}$.

Given a C^1-path $\gamma : I \to M$, its length $L_f(\gamma) = \int_I f(\gamma'(t))\, dt$. For $x_0, x_1 \in M$ define

$$d_f(x_0, x_1) = \inf\{L_f(\gamma) : \gamma \text{ is a } C^1\text{-path joining } x_0 \text{ to } x_1\}.$$

Then *d_f is a metric on M whose topology is that of M.* (See for instance Grauert–Reckziegel [GReck 1965].) Thus *(M, d_f) is a length space.*

Say that f is a *Riemann–Finsler structure* if for every $x \in M$ the function $v \mapsto f^2(v)$ on $T_x(M)$ is strictly convex; in other words the unit balls in $T_x(M)$ are strictly convex. Often that amounts to demanding that f^2 is C^2 in the vertical directions, and for each $v \in T_x(M)$ the second vertical differential

$$g_f(v) = \tfrac{1}{2} d^V d^V f^2(v)$$

is positive definite [BC 1993].

Let $P(M) = (T(M) \setminus \{0\})/\mathbb{R}^*$ denote the bundle of line elements of M (i.e., equivalence classes of non-zero vectors, whereby $v \sim v'$ in $T_x(M)$ iff v' is a real multiple of v); and $\pi^*T(M) \to P(M)$ the bundle induced from the tangent bundle $T(M) \to M$ via the projection map $\pi : P(M) \to M$. Then g_f can be interpreted as a symmetric 2-covariant tensor field on $P(M)$, determining a positive definite inner product on the fibres of $\pi^*T(M)$.

Canonically associated with f are several connections (e.g., see [Ch 1948], [BC 1993], [BS 1994]) on the vector bundle $\pi^*T(M)$:

$$\nabla : C(TP(M)) \times C(\pi^*T(M)) \to C(\pi^*T(M)).$$

These connections all determine the same splitting μ of the exact sequence

$$0 \longrightarrow T^V(P(M)) \overset{\mu}{\rightleftarrows} T(P(M)) \overset{d\pi}{\longrightarrow} \pi^*T(M) \longrightarrow 0,$$

where $T^V(P(M)) = \ker d\pi$, the vertical bundle of $T(P(M))$. Thus *we obtain a Riemannian metric on* $P(M)$, defined by

$$\hat{g}_f(v,w) = g_f(d\pi(v), d\pi(w)) + g_f(\mu(v), \mu(w)).$$

Observe that $\hat{g}_f$ can also be viewed as a Riemannian metric on the manifold $T(M) \setminus \{0\}$.

As noted in [BL 1996] – with thanks to both authors – it is natural here to use a weighted and normalized scalar density on M: Let $P_x(M)$ denote the fibre of $P(M)$ over the point $x \in M$, and dv its $\hat{g}_f$-volume element. Write

$$G_f^{ij} = (\det G_f)^{-1/2} \int_{P_x(M)} g_f^{ij} (\det G_f)^{1/2} dv \Big/ \operatorname{vol} P_x(M).$$

Then define the *f-Dirichlet integral* of a function $u : M \to \mathbb{R}$ by

$$E_f(u) = \frac{1}{2} \int_M (\partial_i u)(\partial_j u) G_f^{ij} (\det G_f)^{1/2} dx.$$

Its Euler–Lagrange operator is the *f-Laplacian*

$$\Delta_f u = (\det G_f)^{-1/2}\partial_i\big(G_f^{ij}(\det G_f)^{1/2}\partial_j u\big),$$

a linear elliptic operator whose local solutions form a Brelot harmonic sheaf on M [Her 1962].

Example 3.4. Metric associated to a holomorphic quadratic differential

Let N be a compact smooth Riemann surface of genus $q \geqslant 2$. Relative to any compatible Hermitian metric h on N the energy functional E is defined on functions, and is a conformal invariant of h. Its local extrema are the harmonic functions (on open subsets) of N; therefore these determine a Brelot sheaf. Furthermore, relative to the area form μ of h, (N, μ, h) is an admissible Dirichlet space.

Now take a holomorphic quadratic differential $\beta \not\equiv 0$ on N. Then locally $\beta(w) = b(w)\, dw^2$ and

$$h^\beta = |b(w)|\, dw\, d\bar{w}$$

is a singular Hermitian metric on N: smooth, positive definite and flat outside the $4q-4$ zeros of β. Using h^β to define path length, and thereby a metric d^β, (N, d^β) *is a geodesic space* [Bers 1960, §11]. It has the following properties:

(a) The geodesics are those paths on which $\arg(b(w)\, dw^2)$ is constant.

(b) A geodesic through a zero of β of order k can have a corner, with angle $\geqslant 2\pi/(k+2)$.

(c) The sum of the angles in a geodesic triangle is $\leqslant \pi$. That insures that $\operatorname{curv}(N, d^\beta) \leqslant 0$.

Example 3.5. Lie algebras of vector fields on a manifold

Let $\mathfrak{X} = (X_0, X_1, \cdots, X_p)$ denote a finite set of smooth vector fields on a smooth manifold M, and $\operatorname{Lie}(\mathfrak{X})$ the Lie algebra generated by $\mathfrak{X}$. Thus every vector field in $\operatorname{Lie}(\mathfrak{X})$ can be written as a finite sum of terms of the form

$$a[X_{i_1}, [X_{i_2}, \cdots, [X_{i_{k-1}}, X_{i_k}]\cdots]]$$

with $a : M \to \mathbb{R}$ a smooth function. Such an $\mathfrak{X}$ determines both analytic and geometric structures on M, some of which we describe here.

If $\mathfrak{X} = (X_0, X_1, \cdots, X_p)$ *is a system of smooth vector fields for which* Lie($\mathfrak{X}$) *spans* $T_x(M)$ *at each point* $x \in M$*, then for any smooth function* $c : M \to \mathbb{R}$*, the differential operator*

$$A = \sum_{i=1}^{p} \mathcal{L}^2_{X_i} + \mathcal{L}_{X_0} + c$$

is hypoelliptic. (That is, u is a smooth function whenever Au is smooth.) (Hörmander [Hö 1967].)

Here $\mathcal{L}_Z$ denotes the Lie derivative on functions, associated with a vector field Z.

The functions on open subsets of M *which are in the kernel of* A *form a harmonic sheaf in the sense of Bauer*, cf. Remark 2.1. *Its sections are smooth. If already* Lie$(X_1, \cdots, X_p)$ *spans* $T_x(M)$ *at each point, then these germs form a Brelot harmonic sheaf.* (Bony [Bo 1970], R.–M. Hervé and M. Hervé [HH 1972].)

We now suppose that (M, g) is a *Riemannian* manifold of dimension n. Let $\mathcal{L}^*_Z$ denote the adjoint (with respect to (M, g)) operator of $\mathcal{L}_Z$ on functions; in charts, writing $\partial_k = \partial/\partial x^k$,

$$\mathcal{L}_{X_j} u = \sum_{k=1}^{n} b^k_j \partial_k u; \quad \mathcal{L}^*_{X_j} v = \frac{1}{\sqrt{\det g}} \sum_{k=1}^{n} \partial_k (b^k_j \sqrt{\det g}\, v).$$

If $\mathfrak{X} = (X_1, \cdots, X_p)$ *is a system of smooth vector fields for which* Lie($\mathfrak{X}$) *spans* $T_x(M)$ *at each point* $x \in M$*, the* $\mathfrak{X}$*-energy* ($\mathfrak{X}$*-Dirichlet integral*) *of* $u : M \to \mathbb{R}$ *is*

$$E^{\mathfrak{X}}(u) = \frac{1}{2} \int \sum_{i=1}^{p} |\mathcal{L}_{X_i} u|^2 \, d\mu_g.$$

Its Euler–Lagrange operator is the $\mathfrak{X}$-*Laplacian*

$$\Delta^{\mathfrak{X}} = \sum_{j=1}^{p} \mathcal{L}^*_{X_j} \mathcal{L}_{X_j} = \frac{1}{\sqrt{\det g}} \sum \partial_i (a^{ij} \sqrt{\det g}\, \partial_j),$$

where $a^{ij} = \sum_{k=1}^{p} b^i_k b^j_k = a^{ji}$, and (a^{ij}) is positive semidefinite. Clearly $\Delta^{\mathfrak{X}}$ can be expanded in the form A above. Therefore, $\Delta^{\mathfrak{X}}$ *is also hypoelliptic*, and so determines a harmonic space in the sense of Brelot.

We now turn our attention to various metric structures associated with such systems $\mathfrak{X} = (X_1, \cdots, X_p)$ of vector fields; the subbundle $H^{\mathfrak{X}} \subset T(M)$ spanned by these consists of vectors called *horizontal.* Suppose again that

$\mathrm{Lie}(\mathfrak{X})$ spans $T_x(M)$ at each point $x \in M$; and for simplicity that M is connected. By a theorem of Chow [Chow 1939] (see also Herm [1977]), any two points of M can be joined by a horizontal path (i.e., by a piecewise smooth path, each segment of which has horizontal tangent vectors).

Fix a Riemannian metric g on M, and define the distance function

$$d^{\mathfrak{X}}(x_0, x_1) = \inf\{g\text{-length of } \gamma : \gamma \text{ a horizontal path joining } x_0 \text{ to } x_1\}.$$

Then $d^{\mathfrak{X}}$ is a metric on M (finite, by Chow's theorem), often called the *Carnot–Carathédory metric of the structure* $(M, g, \mathfrak{X})$. In fact, $(M, d^{\mathfrak{X}})$ *is a length space* (so in particular, its topology is that of M); and is a *geodesic space* if (M, g) is complete [Ham 1990], [Str 1986].

Remark 3.2. Gromov [Gro 1996] has made a far-reaching study of such Carnot–Carathéodory manifolds (CC manifolds). Special attention is given to ball doubling (ball-box equivalence), Riemannian approximation of CC metrics, Hölder equivalence, maps to and from CC manifolds.

Now suppose that there is an integer r such that the set of j-fold iterated brackets of vector fields of $\mathfrak{X}$ ($j \leqslant r$) spans $T_x(M)$ at each point of M; this is *Hörmander's condition on the iterated brackets.*

The system $\mathfrak{X}$ determines several different locally equivalent metrics on M; (two such ϱ_1, ϱ_2 are *locally equivalent* if every point of $M \times M$ has a neighbourhood in which every compact subset K has an associated constant $C > 0$ such that

$$C^{-1}\varrho_1(x,y) \leqslant \varrho_2(x,y) \leqslant C\varrho_1(x,y) \qquad \text{for all } (x,y) \in K.$$

Locally equivalent metrics give the same topology on M). Thorough discussion and comparison is given in [NSW 1985]. Here is the metric we shall use (ϱ_4 in the notation of [NSW 1985]):

Given $\delta > 0$, let $\mathcal{H}(\delta)$ denote the space of piecewise smooth horizontal paths $\gamma : I = [0,1] \to M$ satisfying (a.e.) the differential system

$$\gamma'(t) = \sum_{j=1}^{p} a_j(t) X_j(\gamma(t))$$

for suitable smooth functions $a_j : I \to [-\delta, \delta]$. For $x_0, x_1 \in M$, define the distance

$$d^{\mathfrak{X}}(x_0, x_1) = \inf\{\delta > 0 : \exists \gamma \in \mathcal{H}(\delta) \text{ with } \gamma(0) = x_0, \gamma(1) = x_1\}.$$

Then $d^{\mathfrak{X}}$ *is a metric on* M *giving its original topology* [NSW 1985].

The distance d^g associated to any smooth Riemannian metric g on M is locally equivalent to $d^{\mathfrak{X}}$.

Furthermore, $d^{\mathfrak{X}}$ can be characterized as follows [S–C 1984], [JeS–C 1987]: Firstly, for any $\varepsilon > 0$ take the Riemannian metric g^{ε} on M with local representation

$$(g^{\varepsilon}_{ij}) = \left(a^{ij} + \varepsilon g^{ij}\right)^{-1}.$$

Its Laplace–Beltrami operator Δ^{ε} is viewed as a perturbation of $\Delta^{\mathfrak{X}}$. Bony [Bo 1970] has used such Δ^{ε} to obtain local unique solvability of the Dirichlet problem for $\Delta^{\mathfrak{X}}$ on regular domains. Let d^{ε} be the distance metric associated with g^{ε}. *Its distance function* $d^{\varepsilon}(x_0, x_1)$ *is monotone in* $\varepsilon : 0 < \varepsilon' \leqslant \varepsilon$ *implies* $d^{\varepsilon'}(x_0, x_1) \geqslant d^{\varepsilon}(x_0, x_1)$; *and we have*

$$d^{\mathfrak{X}}(x_0, x_1) = \lim_{\varepsilon \to 0} d^{\varepsilon}(x_0, x_1). \tag{3.1}$$

Thus $(M, d^{\mathfrak{X}})$ *is a limit of the length spaces* (M, d^{ε}), *in a regular way.*

Remark 3.3. We do not know when $(M, d^{\mathfrak{X}})$ itself is a length space, i.e., when

$$d^{\mathfrak{X}}(x_0, x_1) = D^{\mathfrak{X}}(x_0, x_1) := \inf_{\gamma}\Big\{\sup_{\pi} \sum d^{\mathfrak{X}}\big(\gamma(t_{i-1}), \gamma(t_i)\big)\Big\}.$$

We seem to require an approximation stronger than that in (3.1), namely, that for all paths γ and all $\eta > 0$ there is an $\varepsilon > 0$ such that

$$\sum d^{\mathfrak{X}}\big(\gamma(t_{i-1}), \gamma(t_i)\big) - \sum d^{\varepsilon}\big(\gamma(t_{i-1}), \gamma(t_i)\big) < \eta$$

for all sufficiently fine partitions (t_i) of γ. If that is the case, then we can choose a partition $\pi = (t_i)$ of a geodesic γ with respect to g^{ε} (and d^{ε}) so that

$$\begin{aligned} D^{\mathfrak{X}}(x_0, x_1) &\leqslant \eta + \sum d^{\mathfrak{X}}\big(\gamma(t_{i-1}), \gamma(t_i)\big) \\ &\leqslant 2\eta + \sum d^{\varepsilon}\big(\gamma(t_{i-1}), \gamma(t_i)\big) \\ &\leqslant 2\eta + d^{\varepsilon}(x_0, x_1) \leqslant 2\eta + d^{\mathfrak{X}}(x_0, x_1), \end{aligned}$$

by (3.1). Because that is true for all $\eta > 0$, we conclude that $D^{\mathfrak{X}} \leqslant d^{\mathfrak{X}}$. The opposite inequality is always true, so $d^{\mathfrak{X}}$ is a length metric on M.

Perhaps $(M, \mu_g, E^{\mathfrak{X}})$ is not an admissible Dirichlet space.

Example 3.6. Riemannian Lipschitz manifolds

A *Lipschitz n-manifold* is a separable topological manifold with a maximal atlas $\mathcal{U} = (\theta_i, U_i)$ such that all transition maps

$$\theta_{ji} = \theta_j \circ \theta_i^{-1} : \theta_i(U_i \cap U_j) \to \theta_j(U_i \cap U_j)$$

are locally Lipschitz between domains of $\mathbb{R}^n$. A map $\varphi : M \to N$ between Lipschitz manifolds is *locally Lipschitz* if $\chi_p \circ \varphi \circ \theta_i^{-1}$ is locally Lipschitz for any charts (θ_i, U_i) and (χ_p, V_p) with $\varphi(U_i) \subset V_p$.

The transformation of integral formula is valid for Lipschitz maps between domains of Euclidean spaces. Thus there is a notion of n-dimensional null set of a Lipschitz n-manifold, i.e., M has an *n-measure class* [Fe 1969, Chapter 3.2].

Rademacher's theorem [Fe 1969, Chapter 3.1] asserts that *the transition maps θ_{ji} above have almost everywhere differentials $d\theta_{ji}$, and they are locally bounded and measurable on their domains.* These define transition maps

$$\lambda_{ji} = d\theta_{ji} : U_i \cap U_j \to \mathrm{GL}(\mathbb{R}^n),$$

the group of linear automorphisms of $\mathbb{R}^n$, and so determine a bounded measurable analogue of the principal tangent bundle $\mathrm{GL}(M) \to M$. It has, of course, all the usual associated bundles.

Let g denote a locally bounded measurable section of the associated bundle $\odot^2 T^*(M)$ of symmetric 2-covectors on M which is positive definite a.e. If M is oriented, then (M, g) has the Hodge operator

$$* : \wedge^k T^*(M) \to \wedge^{n-k} T^*(M) \qquad (0 \leqslant k \leqslant n)$$

defined a.e. on M. In particular, $*1 = \mu_g$, *the volume element of* (M, g). As usual, $d\mu_g = \sqrt{\det g}\, dx$, and $\mathrm{vol}(M, g) = \int_M d\mu_g$. We shall also require that g satisfy the following uniformity condition [Te 1983]: Relative to an atlas $\mathcal{U} = (\theta_i, U_i)$ for the Lipschitz structure, there are constants $C_i \geqslant 1$ such that for any locally bounded measurable differential form ω with compact support in U_i, its L^2-norm $\|\omega\|$ satisfies

$$C_i^{-1} \|(\theta_i^{-1})^* \omega\| \leqslant \|\omega\| \leqslant C_i \|(\theta_i^{-1})^* \omega\|. \tag{3.2}$$

A *Riemannian Lipschitz manifold* (M, g) is a Lipschitz manifold with a metric tensor field g satisfying (3.2). In (3.2) we shall also assume that if $\mathrm{vol}(M, g) = \infty$, then $\inf C_i^{-1} > 0$. *There are constants $c_i \geqslant 1$ such that for a.e. $x \in U_i$ and all tangent vectors $v \in T_x(M)$*

$$c_i^{-2} |d\theta_i(x) v|^2_{\mathbb{R}^n} \leqslant g(x)(v, v) \leqslant c_i^2 |d\theta_i(x) v|^2_{\mathbb{R}^n}.$$

Consider the harmonic functions on open subsets of a Riemannian Lipschitz manifold (M, g). These are the distributional solutions u of Laplace's equation

$$\Delta u = \operatorname{div} \nabla u = 0.$$

In a chart of M let (g^{ij}) denote the contravariant representative of g, a symmetric positive definite matrix of locally bounded measurable functions. Then

$$\Delta u = (\det g)^{-\frac{1}{2}} \partial_i \big(g^{ij} (\det g)^{\frac{1}{2}} \partial_j u \big).$$

These harmonic functions are Hölder continuous (De Giorgi [Gi 1957] and Nash [Na 1958]). *They determine a Brelot harmonic sheaf on M*, and M has, at least locally, a symmetric Green function (R.–M. Hervé [Her 1964]).

If a certain Poincaré inequality holds (cf. Theorem 5.1) then the Dirichlet integral $\frac{1}{2} \int_M g(\nabla u, \nabla u)\, d\mu$ is a regular Dirichlet form of diffusion type on $L^2(M, \mu)$, and $\mathrm{Lip}_c(M)$ (the compactly supported Lip functions on M) is a core.

Following the method of De Cecco and Palmieri [CP 1988] we now construct an *intrinsic distance function* on (M, g). That is a delicate matter, for g is defined only a.e. (See also Chapter 4 below, The intrinsic distance d_X, now with M replacing X.)

Let $\mathrm{Lip}(I, M)$ denote the set of all Lipschitz paths $\gamma : I \to M$. If Z is a null subset of (M, g), let

$$\mathrm{Lip}^Z(I, M) = \{\gamma \in \mathrm{Lip}(I, M) : \ \mathrm{meas}_I\, \gamma^{-1}(Z) = 0\}.$$

The *Z-length function*

$$L^Z(\gamma) = \int_I |\gamma'(t)|\, dt$$

is defined on $\mathrm{Lip}^Z(I, M)$. For any two points $x_0, x_1 \in M$ set

$$d^Z(x_0, x_1) = \inf \left\{ L^Z(\gamma) : \gamma \in \mathrm{Lip}^Z(I, M) \text{ with } \gamma(0) = x_0, \gamma(1) = x_1 \right\};$$

and finally, let

$$d^g(x_0, x_1) = \sup_Z d^Z(x_0, x_1). \tag{3.3}$$

De Cecco and Palmieri [CP 1988; 1990] have used (3.3) to show that d^g *is an intrinsic metric compatible with the topology of M. Furthermore, (M, d^g) is a length space.*

Remark 3.4. Analytical expressions have been obtained for the geometric metric d^g [CP 1990; 1991]. For example, with the notation $\mathrm{Lip}_{x_0,x_1}(M, \mathbb{R}) =$

$\{u \in \mathrm{Lip}(M, \mathbb{R}) : u(x_0) = 0, u(x_1) = 1\}$, define $D^g(x_0, x_0) = 0$, and for $x_0 \neq x_1$ let

$$\begin{aligned} D^g(x_0, x_1) &= \left(\inf\{\|du\|_\infty : u \in \mathrm{Lip}_{x_0,x_1}(M, \mathbb{R})\}\right)^{-1} \\ &= \sup\{|u(x_0) - u(x_1)| : u \in \mathrm{Lip}(M, \mathbb{R}), \|du\|_\infty \leqslant 1\}. \end{aligned}$$

Then, under mild hypotheses, $d^g = D^g$. The latter expression for $D^g(x_0, x_1)$ is a particular case of (2.2).

Remark 3.5. A triangulable Riemannian Lipschitz manifold is an admissible Riemannian polyhedron. Consider a non-differentiable one. De Cecco and Palmieri [CP 1993] showed that certain of these Y are *not isometrically embeddable* in any Euclidean space (and therefore not in any smooth Riemannian manifold). Indeed, the differential of the function $y \mapsto d_Y^2(b, y)$ does not produce $|v|_b^2$ when taken in direction $v \in T_b(Y)$.

Remark 3.6. In striking contrast to smooth Riemannian manifolds, *unique continuation properties* of harmonic functions on Riemannian Lipschitz manifolds have substantial variety. See Proposition 5.4 and Remark 5.7 below.

Remark 3.7. The geometry/topology of Lipschitz manifolds has its origins in ideas of Whitney in the mid-forties. Sullivan [Su 1979] has shown, provided $n \neq 4$, that any topological n-manifold admits a Lipschitz structure, and that is unique. In contrast with this, Donaldson and Sullivan [DS 1989] proved that for $n = 4$, such structures may not exist, and are not necessarily unique.

There are compact topological (even Lipschitz) n-manifolds M with no piecewise linear structure, [KiSi 1969] and [Si 1971] for $n \geqslant 5$, and [Fr 1982] for $n = 4$. Some of these manifolds are not homeomorphic to a simplicial polyhedron [DJ 1991].

Example 3.7. The infinite dimensional torus $\mathbb{T}^\infty$

This is the simplest (and earliest) example of an infinite dimensional space carrying a harmonic structure, and is due independently to Bendikov [Ben 1974] and Berg [Berg 1975; 1976]. Given a sequence $a_n > 0$ and $\lambda \geqslant 0$, consider on $\mathbb{T}^\infty$ the formal second order differential operator in infinitely many variables

$$L = \sum_{k=1}^{\infty} a_k \frac{\partial^2}{\partial \theta_k^2} - \lambda.$$

Assign to any open set $U \subset \mathbb{T}^\infty$ the linear space $\mathcal{H}(U)$ of all continuous functions $u : U \to \mathbb{R}$ which satisfy $Lu = 0$ in U in the distributional sense, using as test functions those C^∞-functions on $\mathbb{T}^\infty$ which depend only on finitely many coordinates $\theta_n \in \mathbb{T}$ $(= \mathbb{R}/\mathbb{Z})$ of the argument $\theta \in \mathbb{T}^\infty$.

The translation invariant harmonic sheaf $\mathcal{H}$ thus defined is a Brelot sheaf if $\sum 1/a_k < \infty$, [Berg 1977]. In fact, $\mathcal{H}$ is a Brelot sheaf iff $t^{-1} \log N(t) \to 0$ as $t \to \infty$, whereby $N(t) := \#\{k : a_k \leqslant t\}$ [BeS–C 1997]. Furthermore, *every* translation invariant Brelot sheaf on $\mathbb{T}^\infty$ (more generally on $\mathbb{R}^n \times \mathbb{T}^\infty$) is formed by the germs of distributional solutions u of an equation

$$\sum_{i,j=1}^{\infty} a_{ij} \frac{\partial^2 u}{\partial\theta_i \partial\theta_j} + \sum_{i=1}^{\infty} b_i \frac{\partial u}{\partial\theta_i} = 0$$

with constant coefficients such that the infinite matrix a_{ij} is positive definite [Ben 1995].

The above results were derived from a study of the symmetric convolution semigroups on the group $\mathbb{T}^\infty$, say. Such a semigroup is the transition semigroup of a Hunt process whose excessive functions are precisely the superharmonic functions $\geqslant 0$ on X with respect to the Brelot sheaf $\mathcal{H}$ in question.

The stated results have been extended to the case where $\mathbb{T}^\infty$ is replaced by any connected and locally connected, locally compact group having a countable base for its topology [BeS–C 1999]. In case such a metrizable group G is even path connected and locally path connected, then G has a left invariant metric (bi-invariant if G is compact) making it a geodesic space with no terminals (see Example 3.2) and curvature bounded below (by 0 if G is compact) [P 1998, §13].

4

Riemannian polyhedra

This chapter presents in detail the necessary background concerning polyhedra, starting with countable locally finite simplicial complexes. It continues with locally bi-Lipschitz embeddings into Euclidean spaces. Then Riemannian polyhedra; these have a Riemannian metric tensor with bounded measurable components on each maximal simplex, locally equivalent to a Euclidean metric.

Lip continuous map. Lip homeomorphism

When passing from Riemannian manifolds to the more general Riemannian polyhedra, the role played by smooth functions and maps is taken over by Lip continuous functions and maps.

For metric spaces (X, d_X) and (Y, d_Y), $\mathrm{Lip}(X, Y)$ denotes the class of all (globally) Lipschitz maps $X \to Y$, i.e., maps f for which there is a constant c (Lip constant) such that

$$d_Y(f(x), f(x')) \leqslant c d_X(x, x'), \qquad x, x' \in X.$$

$\mathrm{Lip}_{\mathrm{loc}}(X, Y)$ denotes the class of all *Lipschitz continuous*, i.e., locally Lipschitz, maps $f : X \to Y$. Thus every point of X shall have a neighbourhood U such that $f_{|U} \in \mathrm{Lip}(U, Y)$. It easily follows that $f_{|U} \in \mathrm{Lip}(U, Y)$ holds for every relatively compact open set $U \subset X$. (If not, there would be sequences $(x_j), (x'_j)$ in U such that $d_Y(f(x_j), f(x'_j)) > j d_X(x_j, x'_j)$. We may assume that $x_j \to x$, $x'_j \to x'$ in X, but this is impossible both if $x \neq x'$ and if $x = x'$.)

By a *Lip homeomorphism*, or locally bi-Lip homeomorphism, we shall understand a bijection which along with its inverse is Lip continuous.

If $Y = \mathbb{R}$ with the standard metric, we may write $\mathrm{Lip}(X, \mathbb{R}) = \mathrm{Lip}(X)$ and $\mathrm{Lip}_{\mathrm{loc}}(X, \mathbb{R}) = \mathrm{Lip}_{\mathrm{loc}}(X)$.

Recall that every metric space (X, d_X) is *paracompact*. For every locally finite open cover (U_j) of X there even exists a *Lip continuous partition of unity* (φ_j) on X subordinated to (U_j), i.e., $\operatorname{supp}\varphi_j \subset U_j$. In fact, for disjoint closed sets A, B of X, the function $\varphi : X \to [0,1]$ defined by

$$\varphi(x) = \frac{d_X(x,A)}{d_X(x,A) + d_X(x,B)}$$

is Lipschitz, and $A = \varphi^{-1}(0)$, $B = \varphi^{-1}(1)$. The proof of the existence of a continuous partition of unity subordinated to (U_j) therefore carries over, cf. e.g., [Bou 1974, §4, Proposition 3].

Simplicial complex

A basic reference for simplicial complexes and their polyhedral realizations is [Sp 1966, Chapter 3, §§1–3]. We shall restrict our attention to a subclass, as follows: A *countable locally finite simplicial complex* K, or briefly: a *complex*, consists of a countable set $\{v\}$ of elements, called vertices, and a set $\{s\}$ of finite nonvoid subsets of vertices, called simplexes, such that

- any set consisting of exactly one vertex is a simplex,
- any nonvoid subset of a simplex is a simplex,
- every vertex belongs to only finitely many simplexes (the local finiteness condition).

A *q-simplex* is a simplex with exactly $q+1$ vertices; q is the *dimension* of s. A subset $s' \subset s$ is called a *face* of s. The *dimension* of K, written $\dim K$, is the supremum of the dimensions of its simplexes.

The *p-skeleton* $K^{(p)}$ of K is the complex consisting of all its simplexes of dimension $\leqslant p$. We denote by $\mathcal{S}^{(p)}(K)$ the collection of all p-simplexes of K; and $\mathcal{S}(K) := \bigcup_p \mathcal{S}^{(p)}(K)$.

The *space* $|K|$ of K is the set of all formal finite linear combinations $\alpha = \sum_{v\in K} \alpha(v)v$ of vertices of K such that $0 \leqslant \alpha(v) \leqslant 1$, $\sum_{v\in K} \alpha(v) = 1$ and $\{v : \alpha(v) > 0\}$ is a simplex of K.

The values $\alpha(v)$ are called the *barycentric coordinates* of $\alpha \in |K|$.

The space $|K|$ is a subset of the linear space $\operatorname{lin} K$ of all formal finite linear combinations of vertices of K. A vertex v_0 of K will be identified with the formal linear combination $1v_0$; thus formal linear combinations of vertices become true linear combinations in $\operatorname{lin} K$.

The space $|s|$ of a simplex s of K is convex as a subset of $\operatorname{lin} K$. A map $f : |K| \to V$ into a linear space V is called *simplexwise affine*, but henceforth we shall simply say *affine*, if it is the restriction to $|K|$ of an affine-linear map $\operatorname{lin} K \to V$; or equivalently, if the restriction of f to each simplex $|s|$ is an affine-linear map $|s| \to V$.

$|K|$ is made into a metric space with *barycentric distance* $d(\alpha, \beta)$ between two points $\alpha = \sum \alpha(v)v$ and $\beta = \sum \beta(v)v$ of $|K|$ given by the finite sum

$$d(\alpha, \beta)^2 = \sum_{v \in K} (\alpha(v) - \beta(v))^2.$$

Then $|K|$ is locally compact, for that is equivalent to K being locally finite. And $|K|$ is separable, K being countable.

A map $\varphi : K \to K'$ between complexes K, K' is called *simplicial* if φ maps each simplex of K onto a simplex of K'. Such a map extends uniquely to a map (likewise termed simplicial) of $|K|$ into $|K'|$ which is affine on $|s|$ for each simplex s of K.

Lemma 4.1. *Let K be a countable, locally finite simplicial complex of finite dimension n, and V a Euclidean space of dimension $2n+1$. There exists an affine Lip homeomorphism f of $|K|$ onto a closed subset of V.*

Proof. There exists an affine homeomorphism f of $|K|$ into V such that $f(\mathcal{S}^{(0)}(K))$ (the image of the vertices of K) is a set of points in general position in V (i.e., any $2n+2$ distinct points of $f(\mathcal{S}^{(0)}(K))$ are affinely independent); moreover, $f(|K|)$ is closed, or equivalently: f is proper. See e.g., [Sp 1966, p. 120]. We show that any such map $f : |K| \to V$ *is a Lip homeomorphism* (into V). Because $|K|$ and V are locally compact and f is proper, this reduces to the case where $|K|$ is compact, or where K is a *finite* complex, say with vertices $v_0, \cdots, v_N$. Writing $f(v_i) = e_i$ (supposed to be in general position), we have in terms of a norm $|\cdot|$ in V, taking e_0 as origin 0 in V,

$$f(\alpha) = \sum_0^N \alpha_i f(v_i) = \sum_1^N \alpha_i e_i$$

for any point $\alpha = \sum_0^N \alpha_i v_i \in |K|$. For any other point $\beta = \sum_0^N \beta_i v_i \in |K|$, write $\alpha_i - \beta_i = \xi_i$, $i = 0, 1, \cdots, N$, and insert $\xi_0 = -\sum_1^N \xi_i$. Then $|f(\alpha) - f(\beta)|^2 = \left|\sum_1^N \xi_i e_i\right|^2$ is bounded from above and below by positive constants times $\sum_1^N \xi_i^2$ because the $2n+1$ or fewer vectors e_i with $i \in \{1, \cdots, N\}$ and $\xi_i \neq 0$ (hence $\alpha_i > 0$ or $\beta_i > 0$) are linearly independent. This completes the proof since

$$\sum_1^N \xi_i^2 \leqslant d(\alpha, \beta)^2 = \sum_0^N \xi_i^2 = \Big(\sum_1^N \xi_i\Big)^2 + \sum_1^N \xi_i^2 \leqslant (N+1) \sum_1^N \xi_i^2.$$

□

Polyhedron

We shall use the term *polyhedron* to mean a connected locally compact separable Hausdorff space X for which there exists a simplicial complex K and a homeomorphism θ of $|K|$ onto X. Any such pair $T = (K, \theta)$ is called a *triangulation* of X. K is necessarily countable and locally finite, [Sp 1966, p. 120], and X is path connected and locally contractible. The *dimension* of X is defined as $\dim X = \dim K$; it is independent of the triangulation T.

A *subpolyhedron* of a polyhedron X with a specific triangulation $T = (K, \theta)$ is a polyhedron $X' \subset X$ having as a triangulation $T' = (K', \theta|_{|K'|})$ where K' is a subcomplex of K, i.e., K' is a complex whose vertices and simplexes are some of those of K, hence $|K'|$ may be viewed as a subset of $|K|$.

When X is a polyhedron with a specified triangulation $T = (K, \theta)$, we shall speak of vertices or simplexes of (X, T) as the images under θ of vertices or simplexes $|s|$ of K. Furthermore, we may simply write s for the space $|s|$ of a simplex of K or of (X, T). Thus our simplexes become compact subsets of X.

A triangulation $T' = (K', \theta')$ of X is said to be a *subdivision* of $T = (K, \theta)$ if every simplex s' of T' is a subset of some simplex of T.

If T' is a subdivision of T, and T'' a subdivision of T', then clearly T'' is a subdivision of T. It can be shown that any two subdivisions of a triangulation of X have a common subdivision.

The *interior* of a p-simplex s of (X, T) (i.e., the points of s not in any $(p-1)$-face of s) is denoted by s°. (A 0-simplex s is a vertex, and $s^\circ = s$.)

We denote by $\mathcal{S}^{(p)}(X) = \mathcal{S}^{(p)}(X, T)$ the collection of all p-simplexes of (X, T), and by $\mathcal{S}(X) = \mathcal{S}(X, T) = \bigcup_p \mathcal{S}^{(p)}(X)$ the collection of all simplexes of (X, T).

A simplex s of (X, T) will be termed *maximal* if s is not a proper face of any simplex of (X, T); and we denote by $\mathcal{S}^*(X) = \mathcal{S}^*(X, T)$ the collection of all maximal simplexes of (X, T).

For $0 \leqslant p \leqslant \dim X$, the *p-skeleton* $X^{(p)}$ of X (the union of all simplexes $|s|$ of dimension $\leqslant p$) is a subpolyhedron of X, independent of the triangulation of X, and closed in X.

The (open) *star* of an open simplex s° of $X = (X, T)$ is defined as $\mathrm{st}(s^\circ) = \mathrm{st}_X(s^\circ) = \bigcup\{t^\circ : t \in \mathcal{S}(X) \text{ with } t \supset s\}$. The star $\mathrm{st}(a)$ of a point $a \in X$ is defined as the star of its *carrier,* the unique open simplex s° containing a. Every star is a path connected open subset of X and contains the star of its points. In particular, X is *locally path connected.* The closure $\overline{\mathrm{st}}(s^\circ)$ is a finite subpolyhedron of (X, T).

A polyhedron X will be called *admissible* if in some (hence in any) triangulation,

(i) X is *dimensionally homogeneous*, i.e., all maximal simplexes have the same dimension n ($= \dim X$), or equivalently: every simplex is a face of some n-simplex; and

(ii) X is *locally* $(n-1)$-*chainable* (cf. [Chen 1995], [Wh 1986, §2]); i.e., for every connected open set $U \subset X$, the open set $U \setminus X^{n-2}$ is connected. Equivalently, for any simplex s, any two n-simplexes containing s can be joined by a chain of contiguous $(n-1)$- and n-simplexes containing s. (Simple examples show that it is not enough to require this just for vertices s.)

Every connected open subset U of a polyhedron X is a polyhedron (usually infinite), cf. [Sp 1966, p. 149]. If X is admissible then so is U.

The *boundary* bX of a polyhedron X is the union of all non-maximal simplexes contained in only one maximal simplex. bX is a closed subset of X and a subpolyhedron of (X, T), independent of the triangulation of X. For example, if X has just one maximal simplex s, then $bX = s \setminus s^\circ$.

Circuit

By an *n-circuit* we mean a polyhedron X of homogeneous dimension n such that in some, and hence any triangulation,

(a) every $(n-1)$-simplex is a face of at most two n-simplexes (exactly two iff $bX = \emptyset$); and

(b) X is $(n-1)$-chainable; i.e., $X \setminus X^{(n-2)}$ is connected; or equivalently: any two n-simplexes can be joined by a chain of contiguous $(n-1)$- and n-simplexes.

Let $S = S(X)$ denote the *singular set* of an n-circuit X, i.e., the complement of the set of all points of X having a neighbourhood which is a topological n-manifold (possibly with boundary). S is a closed triangulable subspace of X of codimension $\geqslant 2$, and $X \setminus S$ is a topological n-manifold dense in X, see [Hae 1984].

An n-circuit X is called *normal* if S does not locally separate X at any point; i.e., every point of S has a base of neighbourhoods U such that $U \setminus S$ is (necessarily path) connected. Equivalently (having decreed that 0- and 1-dimensional circuits are normal), for $n \geqslant 2$ a circuit is normal iff it has a triangulation such that the link of every vertex is an $(n-1)$-circuit [EdFi 1976], [GMacP 1980, §4].

A circuit is normal if and only if it is admissible (as a polyhedron). The "if part" follows from the fact that a topological space is connected if it contains a connected dense open subset. The "only if part" reduces to the fact that an $(n-2)$-disc embedded in an n-manifold V does not locally separate V.

Example 4.1. Every C^1-manifold (even every triangulable Lipschitz manifold) is a normal circuit, cf. Example 8.4. A normal 2-circuit without boundary is a 2-manifold. Indeed, the link of each vertex s of a triangulation is a polyhedral circle, so $\overline{\mathrm{st}}(s)$ is a 2-disc. For $n \geqslant 3$, the link of a vertex may not bound an n-disc.

Lip polyhedron

Replacing 'homeomorphism' by 'Lip homeomorphism' throughout the two preceding sections leads us to the notion (and some properties) of a *Lip polyhedron*; i.e., a *metric space* X which is the image of the metric space $|K|$ of some complex K under a Lip homeomorphism $\theta : |K| \to X$. The pair (K, θ) is then called a *Lip triangulation* (or briefly: a triangulation) of the Lip polyhedron X.

Thus any polyhedron X with a triangulation (K, θ) becomes a Lip polyhedron when X is given the metric corresponding via θ to the barycentric metric d on $|K|$ (or any locally equivalent metric).

For a *Lip circuit* X with singular set S, $X \setminus S$ is a Lipschitz manifold (cf. Example 3.6).

By a *null set* in a Lip polyhedron X is understood a set $Z \subset X$ such that Z meets every maximal simplex s (relative to some, and hence any triangulation $T = (K, \theta)$ of X) in a set whose pre-image under θ has p-dimensional Lebesgue measure 0, $p = \dim s$. Note at this point that every proper face of s is a null set, and that every Lip homeomorphism between domains of $\mathbb{R}^p$ preserves the sets of p-dimensional Lebesgue measure 0, cf. e.g., [Fe 1969, Chapter 3]; see also Example 3.6 above. As usual, 'almost everywhere' (a.e.) means: everywhere except in some null set.

Remark 4.1. From Lemma 4.1 follows that every Lip polyhedron (X, g, d_X) can be mapped Lip homeomorphically and (simplexwise) affinely onto a closed subset of a Euclidean space.

Analogously, let X denote a locally compact separable metric space, each point of which has a neighbourhood that can be bi-Lip embedded as a closed subset of some Euclidean space. Then there is a proper Lip homeomorphism of X onto a closed subset of some Euclidean space. [LV 1977, Remark 4.3 (3)]. For instance, every Riemannian Lipschitz manifold can be embedded in this way.

Example 4.2. Every domain X in $\mathbb{R}^n$, endowed with the metric induced from the Euclidean distance on $\mathbb{R}^n$, is an admissible Lip polyhedron of homogeneous dimension n (without boundary), and indeed a normal Lip circuit. To see this, consider a usual locally finite Euclidean triangulation T of X, i.e., a covering of X by a sequence of Euclidean n-simplexes s_j, $j \in \mathbb{N}$, such that the intersection (if nonvoid) of any two of these is a face of each, and that every compact in X meets only finitely many s_j. In particular, the s_j are pairwise non-overlapping. The set K of all vertices v_i of T is then a simplicial complex such that $\{v_{i_0}, \cdots, v_{i_n}\}$ is an n-simplex of K iff the convex hull $\operatorname{conv}\{v_{i_0}, \cdots, v_{i_n}\}$ equals some s_j. The map $f : |K| \to X$ taking a formal finite linear combination $\sum \alpha_i v_i \in |K|$ to $\sum \alpha_i v_i \in \mathbb{R}^n$ is a bijection, and f is a Lip homeomorphism, by the proof of Lemma 4.1. The Lip polyhedron X becomes a Riemannian polyhedron (see below) when given a Riemannian metric with measurable components and local elliptic bounds, cf. Example 3.6.

Subexample 4.2. The closed halfspace $X = \mathbb{R}^n_+ = \{(x_1, \cdots, x_n) \in \mathbb{R}^n : x_n \geqslant 0\}$ with the Euclidean metric is an admissible Lip polyhedron of homogeneous dimension n with boundary $bX = \{x_n = 0\}$, and indeed a normal Lip circuit. This is proved as in the preceding example, using a locally finite Euclidean triangulation of $\mathbb{R}^n_+$. Furthermore, X becomes a Riemannian polyhedron when endowed with a Riemannian metric (as a Lipschitz manifold with boundary).

Riemannian polyhedron

A *Riemannian polyhedron* $X = (X, g)$ will be defined as a Lip polyhedron X endowed with a covariant bounded measurable Riemannian metric tensor g_s on each maximal simplex s, satisfying the ellipticity condition (4.1) below.

Let $T = (K, \theta)$ be a specific (Lip) triangulation of the Lip polyhedron X. We shall view $|K|$ as *embedded in a Euclidean space V via an affine Lip homeomorphism*, cf. Remark 4.1.

Suppose, to begin with, that X has *homogeneous dimension n*. Choose a measurable Riemannian metric g_s on the open Euclidean n-simplex $\theta^{-1}(s^\circ)$ of $|K| \subset V$. In terms e.g., of Euclidean coordinates $x^1, \cdots, x^n$ of points $x = \theta^{-1}(p) \in \theta^{-1}(s^\circ)$ (in its affine span in V), g_s thus assigns to almost every point $p \in s^\circ$, or to $x \in \theta^{-1}(s^\circ)$, an $n \times n$ symmetric positive definite matrix

$$g_s(x) = (g^s_{ij}(x))_{i,j=1,\cdots,n}$$

with measurable real entries; and there is a constant $\Lambda_s > 0$ such that (with the usual summation convention)

$$\Lambda_s^{-2} \sum_{i=1}^{n} (\xi^i)^2 \leqslant g_{ij}^s(x)\xi^i\xi^j \leqslant \Lambda_s^2 \sum_{i=1}^{n} (\xi^i)^2 \tag{4.1}$$

for a.e. $x \in \theta^{-1}(s^\circ)$ and every $\xi = (\xi^1, \cdots, \xi^n) \in \mathbb{R}^n$. This condition is independent of the choice of Euclidean frame on $\theta^{-1}(s^\circ)$. The second inequality in (4.1) amounts to the components of g_s being bounded.

The condition (4.1) also becomes independent of the chosen triangulation when we define, for any other (Lip) triangulation $T' = (K', \theta')$ of X, the Riemannian metric $g_{s'}$ a.e. on ${\theta'}^{-1}({s'}^\circ)$ for each n-simplex s' of T' by covariance. Explicitly, if the open set $U = s^\circ \cap {s'}^\circ$ in X is non-empty, define

$$(g_{s'})_{|\theta'^{-1}(U)} = (\theta^{-1} \circ \theta')^* (g_s)_{|\theta^{-1}(U)},$$

noting that $\theta^{-1}(U) \subset \theta^{-1}(s^\circ)$, etc. As in the case of Riemannian Lipschitz manifolds (Example 3.6) we have here used Rademacher's theorem on the differentiability almost everywhere of Lipschitz continuous functions. When s ranges over $\mathcal{S}^{(n)}(X,T)$ (the n-simplexes of (X,T)), the open sets ${\theta'}^{-1}(U)$ with U as above cover ${\theta'}^{-1}({s'}^\circ)$ up to a null set, and so $g_{s'}(x')$ is altogether defined a.e. for $x' \in {\theta'}^{-1}({s'}^\circ)$ as an $n \times n$ symmetric positive definite matrix with bounded measurable entries $g_{ij}^{s'}(x')$. Furthermore,

$$\Lambda_{s'}^{-2} \sum_{i=1}^{n} (\xi^i)^2 \leqslant g_{ij}^{s'}(x')\xi^i\xi^j \leqslant \Lambda_{s'}^2 \sum_{i=1}^{n} (\xi^i)^2$$

a.e. for $x' \in {\theta'}^{-1}({s'}^\circ)$ with

$$\Lambda_{s'} = \max_s \{\lambda_{s,s'}\Lambda_s : s^\circ \cap {s'}^\circ \neq \emptyset\},$$

whereby $\lambda_{s,s'}$ denotes a bi-Lip constant of $\theta^{-1} \circ \theta'$ considered on $s^\circ \cap {s'}^\circ$ for those finitely many $s \in \mathcal{S}^{(n)}(X,T)$ for which $s^\circ \cap {s'}^\circ \neq \emptyset$.

If X is not necessarily dimensionally homogeneous, define g_s as above for each homogeneous subpolyhedron X_n, where X_n denotes the union of all maximal n-simplexes of X. Clearly, each X_n is closed in X and independent of the chosen triangulation. Furthermore, $X_m \cap X_n$ is a null set when $m \neq n$.

The covariantly defined map $g : s \mapsto g_s$ thus defined on the set $\mathcal{S}^*(X,T)$ of all maximal simplexes s of T, is called the *Riemannian (polyhedral) structure*, or *metric*, on X.

For simplicity of statements we shall sometimes require that, relative to a fixed triangulation T of a Riemannian polyhedron X,

$$\Lambda: = \sup\{\Lambda_s : s \in \mathcal{S}^*(X,T)\} < \infty. \tag{4.2}$$

This condition of uniform ellipticity and uniform boundedness will, however, mostly be used with (X,T) replaced by some finite subpolyhedron, such as the closed star of a point of X, and here the finiteness of Λ is of course a consequence of that of each Λ_s. The smallest constant Λ in (4.2) will be called the *ellipticity constant* of $X = (X,T,g)$.

Not every admissible finite Riemannian polyhedron (X,g) can be *isometrically* embedded in a Euclidean space (cf. Remark 3.5, in which (X,g) is a non-differentiable triangulable Riemannian Lipschitz manifold).

A continuous map $\varphi : X \to Y$ between Riemannian polyhedra is called *simplicial* if it can be realized as a simplicial map between (Lip) triangulations of X and Y. Then $\varphi \in \mathrm{Lip}_{\mathrm{loc}}(X,Y)$.

A Riemannian metric g on a polyhedron X is said to be *continuous* if, relative to some (hence any) triangulation, g_s is continuous up to the boundary on each maximal simplex s, and for any two maximal simplexes s and s' sharing a face t, g_s and $g_{s'}$ induce the same Riemannian metric on t. There is a similar notion of *Lip continuous* Riemannian metric.

In Part III in particular we shall often require that the Riemannian metric g on a Riemannian polyhedron X be *simplexwise smooth.* This means that, relative to some triangulation (K,θ) of X, the Riemannian metric g_s on the interior s° of each maximal simplex s of K is the restriction to s° of a smooth Riemannian metric on a neighbourhood of s in the linear span of s. This property is preserved under subdivision.

Returning to the case of a general (bounded measurable) Riemannian metric g on X, we shall often consider, in addition to g, the *Euclidean Riemannian metric* g^e on a Lip polyhedron X with a specified triangulation $T = (K,\theta)$. For each $s \in \mathcal{S}^*(X,T)$, g^e_s is defined in terms of a Euclidean frame on $\theta^{-1}(s^\circ)$ as above by the unit matrix (δ_{ij}). Thus g^e is by no means covariantly defined and should be regarded as a mere reference metric on (X,T).

The ellipticity constant Λ of (X,T,g) from (4.2) equals the bi-Lip constant of the identity map $(X,T,g) \to (X,T,g^e)$ in terms of the associated intrinsic distance $d_X = d^g_X$ and its analogue d^e_X, both of which are defined in the next section.

Every connected open subset U of a Riemannian polyhedron X is a Riemannian polyhedron with the induced Riemannian structure.

Example 4.3. The cyclic group $\mathbb{Z}_m$ of the m-th roots of 1 acts on $\mathbb{C}$ isometrically by multiplication; it has fixed point $0 \in \mathbb{C}$. Form the orbit map $\pi_m : \mathbb{C} \to \mathbb{C}/\mathbb{Z}_m = X$.

Anticipating Examples 8.9 (ii) and 8.11 (iv), X is a simple example of a complex orbifold; and π_m a branched covering map with branching locus $0 \in X$. With the quotient metric, X is isometric to a standard cone of revolution in $\mathbb{R}^3$ with the geodesic distance on the cone (which is a geodesic space).

Choose a locally finite Euclidean triangulation of $\mathbb{C}$, invariant under $\mathbb{Z}_m$. It induces a triangulation of X, which thus becomes an admissible 2-dimensional Lip polyhedron without boundary (cf. the argument in Example 4.2). Topologically, X is a 2-manifold, cf. Example 4.1.

The projection map π_m induces a Riemannian metric g on X (as in the above definition of a Riemannian polyhedron) from the Euclidean Riemannian metric g^e on $\mathbb{C}$. In local (polar) coordinate representation, g has the form

$$g = dr^2 + m^{-2} r^2 d\theta^2.$$

In terms of Cartesian coordinates $(x, y) = (r\cos\theta,\ r\sin\theta)$ we have $rdr = xdx + ydy$, $r^2 d\theta = xdy - ydx$, and g is therefore indeed a Riemannian metric with measurable components and elliptic bounds (with ellipticity constant m). Thus (X, g) becomes a Riemannian Lipschitz manifold with one conical singularity (cf. Examples 3.6, 6.1 and 8.8).

More generally, replacing $1/m$ in the above representation by any real number $\lambda > 0$, we still obtain a metric g on $\mathbb{C}$ with a single conical singularity. It can be shown that the curvature is $\geqslant 0$ if $\lambda \leqslant 1$ (as above), and $\leqslant 0$ if $\lambda \geqslant 1$ (as in the next example).

Example 4.4. For an integer $m > 1$ the Riemann surface X of $z^{1/m}$ is conceived in classical terms as consisting of m copies of $\mathbb{C}$ ("sheets") glued together so that there is a homeomorphism $f_m : \mathbb{C} \to X$ for which

$$(p_m \circ f_m)(re^{i\theta}) = re^{im\theta}, \qquad r \geqslant 0,\ \theta \in \mathbb{R}/(2\pi\mathbb{Z}),$$

where $p_m : X \to \mathbb{C}$ denotes the natural projection. Denote by $0 \in X$ the single point of $p_m^{-1}(0)$. In view of Example 4.1, X becomes a 2-dimensional normal Lip circuit without boundary when we take as an (auxiliary) distance function on X the image under f_m of the Euclidean distance e on $\mathbb{C}$. Then $X \setminus \{0\}$ is a smooth manifold, and $p_m : X \setminus \{0\} \to \mathbb{C} \setminus \{0\}$ is a local diffeomorphism.

The whole of X becomes an admissible Riemannian polyhedron (and hence a Riemannian normal circuit, cf. below) when equipped (off 0) with the Riemannian metric $p_m^* g^e$, where g^e denotes the Euclidean Riemannian metric on $\mathbb{C} \setminus \{0\}$. Then X is also a *Riemannian Lipschitz manifold,* with an atlas consisting of the single chart $f_m^{-1} : X \to \mathbb{C}$. In this representation the Riemannian metric $p_m^* g^e$ on $X \setminus \{0\}$ corresponds to the Riemannian metric g on $\mathbb{C} \setminus \{0\}$ given in polar coordinates (r, θ) by

$$g = |d(re^{im\theta})|^2 = dr^2 + m^2 r^2 d\theta^2,$$

and g has indeed bounded measurable components in terms of Cartesian coordinates on $\mathbb{C}$, cf. the end of the preceding example. Thus $(X, p_m^* g^e)$ is (isometric to) a Riemannian Lipschitz manifold with one conical singularity.

Globally, we may view $p_m : X \to \mathbb{C}$ as a branched covering map with branching locus $0 = p_m^{-1}(0)$ on X. Thus p_m corresponds via the isometry f_m to the map $\pi_m : \mathbb{C} \to \mathbb{C}/\mathbb{Z}_m$ from Example 4.3.

Example 4.5. Every polyhedron X is clearly path connected, locally path connected and locally simply connected (being locally contractible). It follows that a Riemannian polyhedron X has a theory of covering spaces; in particular, a universal covering space $p : \widetilde{X} \to X$ with $\tilde{X}$ a Riemannian polyhedron, and p a simplicial map which is a local isometry. (See [Sp 1966, pp. 144f.].)

A *Riemannian circuit* is of course a circuit which is also a Riemannian polyhedron. The polyhedra in Examples 4.2 – 4.4 above are all (Riemannian) normal circuits; and the Riemannian normal circuits are among the most important admissible Riemannian polyhedra, cf. Chapter 8.

Example 4.6. Attaching a flat n-ball B^n along an equatorial $(n-1)$-sphere of the standard Euclidean n-sphere S^n produces an admissible Riemannian n-polyhedron $X = S^n \cup B^n$ $(n \geqslant 1)$, without boundary. Clearly X is not a circuit.

The intrinsic distance d_X

Relative to a given (Lip) triangulation $T = (K, \theta)$ of an n-dimensional Riemannian polyhedron (X, g) (not required to be admissible), we have on X the distance function e induced by the Euclidean distance on a Euclidean space V in which $|K|$ is affinely Lip embedded, cf. Lemma 4.1. This distance e on X is not intrinsic and will play an auxiliary role in defining an equivalent intrinsic distance $d_X^g = d_X$ as follows, by a slight adaptation of the procedure used by De Cecco and Palmieri [CP 1988; 1990] for the case of a Riemannian Lipschitz manifold, cf. Example 3.6.

Let $\mathcal{Z}$ denote the collection of all null sets $Z \subset X$ (cf. the section Lip polyhedron). For a given triangulation $T = (K, \theta)$ consider in particular the set $Z_T \in \mathcal{Z}$ obtained from X by removing from each maximal simplex s in X those points of s° which are Lebesgue points for g_s (that is, for every component g_{ij}^s). For $x, y \in X$ and any $Z \in \mathcal{Z}$ such that $Z \supset Z_T$ we denote by $\mathrm{Lip}^Z(x, y; X)$ *the family of all Lip continuous paths* $\gamma : [a, b] \to (X, e)$ *with* $\gamma(a) = x$, $\gamma(b) = y$ *which are transversal to* Z *in the sense that* $\gamma^{-1}(Z)$ *is a null subset* of $[a, b]$. The length $L_T(\gamma)$ of such a path γ is well defined by

$$L_T(\gamma) = \sum_{s \in \mathcal{S}^*(X,T)} \int_{\gamma^{-1}(s^\circ)} \sqrt{(g_{ij}^s \circ \theta^{-1} \circ \gamma)\dot{\gamma}^i \dot{\gamma}^j} \quad (\leqslant \infty),$$

where $(\gamma^1, \cdots, \gamma^n) = \theta^{-1} \circ \gamma$ in terms of Euclidean coordinates on the open Euclidean simplex $\theta^{-1}(s^\circ)$, and the dot means differentiation. Write

$$\varrho_Z(x, y) = \inf_\gamma \{L_T(\gamma) : \gamma \in \mathrm{Lip}^Z(x, y; X)\}.$$

($\varrho_Z(x, y)$ depends on T as well.) Clearly, $Z_1 \supset Z_2$ ($\supset Z_T$) implies $\varrho_{Z_1}(x, y) \geqslant \varrho_{Z_2}(x, y)$. Finally set

$$d_X(x, y) = \sup_Z \{\varrho_Z(x, y) : Z \in \mathcal{Z},\ Z \supset Z_T\}.$$

Note at this point that $\mathrm{Lip}^Z(x, y; X) \neq \emptyset$ (when $Z \in \mathcal{Z}$, $Z \supset Z_T$), as shown by application of Fubini's theorem, cf. [CP 1988, p. 160]. (Because X is connected, it suffices to consider points x, y of the same simplex s. The Euclidean segment $[x, y] \subset s$ can then be slightly deformed to a path $\gamma = [x, z] \cup [z, y] \subset s$ such that $\gamma^{-1}(Z)$ is null.)

Clearly ϱ_Z satisfies the triangle inequality, and so therefore does d_X.

For the Euclidean Riemannian metric g^e on X (induced by the Euclidean distance e on V) we have similar notions $L_T^e(\gamma)$, $\varrho_Z^e(x, y)$, $d_X^e(x, y)$. In view of (4.2), each of the quotients $L_T(\gamma)/L_T^e(\gamma)$, $\varrho_Z(x, y)/\varrho_Z^e(x, y)$ and $d_X(x, y)/d_X^e(x, y)$ lies between Λ^{-1} and Λ. It is easily verified that

$$e(x, y) \leqslant d_X^e(x, y) \qquad \text{for } x, y \in X. \tag{4.3}$$

Here the sign of equality holds if x and y are in the same maximal simplex s. This is shown by deforming $[x, y]$ into $\gamma = [x, z] \cup [z, y]$ with $\gamma^{-1}(Z)$ null and $L_T^e(\gamma) \leqslant e(x, y) + \varepsilon$, whereby $\varrho_Z(x, y)$ and hence $d_X^e(x, y)$ are $\leqslant e(x, y) + \varepsilon$.

Since X is connected it follows that $d_X^e(x, y) < \infty$ and hence $d_X(x, y) < \infty$ for *any* pair $x, y \in X$. By (4.3), $d_X^e(x, y) > 0$ and hence $d_X(x, y) > 0$ when $x \neq y$. Altogether, d_X and d_X^e are *equivalent metrics* on X, depending a priori on the triangulation T. They are locally equivalent to the given metric on X as a Lip polyhedron according to Lemma 4.2 below (to be used more fully in Lemma 4.3), applied to the star of any point of X. In view of Proposition 4.1 below, d_X is called the *intrinsic distance* on (X, g).

Example 4.7. Let X be the Riemann surface of $z^{1/m}$ with the Euclidean Riemannian metric (see Example 4.4). We shall identify points $x \in X$ with $p_m(x) \in \mathbb{C}$, except that $\arg x$ shall be counted mod $2m\pi$. The *intrinsic distance* d_X is then given by $d_X(x, 0) = |x|$, while for $x, y \neq 0$,

$$d_X(x, y) = \begin{cases} |x - y| & \text{if } |\arg x - \arg y| < \pi \\ |x| + |y| & \text{otherwise.} \end{cases}$$

(The former case may be expressed by saying that x and y belong to the same "half-sheet" of X.) In the latter case, the geodesic segment joining x with y in X is the segment $[x, 0]$ followed by $[0, y]$ (in the same sheet as x, resp. y).

Lemma 4.2. *Let X be a finite Lip polyhedron, affinely and Lip homeomorphically embedded in a Euclidean space V. The induced Euclidean distance e on X is then equivalent to the distance d_X^e associated with the Euclidean Riemannian structure g^e on X, hence also equivalent to the intrinsic distance d_X associated with a given Riemannian metric g on X.*

Proof. As noted above, d_X and d_X^e are equivalent distances. Clearly, d_X^e is continuous and hence bounded as a function on $X \times X$ (compact). Indeed, for given $x_0, y_0 \in X$ and for any $x \in \mathrm{st}(x_0)$, $y \in \mathrm{st}(y_0)$, we have $d_X^e(x, x_0) = e(x, x_0)$, $d_X^e(y, y_0) = e(y, y_0)$, as noted after (4.3); and so

$$|d_X^e(x, y) - d_X^e(x_0, y_0)| \leqslant e(x, x_0) + e(y, y_0) \to 0$$

as $(x, y) \to (x_0, y_0)$ in $X \times X$.

In view of (4.3) it remains to show that d_X^e / e remains bounded on $X \times X$. Suppose there are sequences $(x_j), (y_j) \subset X$ such that

$$\frac{d_X^e(x_j, y_j)}{e(x_j, y_j)} \to \infty \qquad \text{as } j \to \infty. \tag{4.4}$$

Since d_X^e is bounded on $X \times X$, $e(x_j, y_j) \to 0$ as $j \to \infty$. Passing to subsequences we may assume that there exists $a \in X$ such that $x_j \to a$, $y_j \to a$. Let s_0 denote the carrier of a, i.e., the lowest dimensional simplex of (X, T) containing a (necessarily as an inner point). Let s, resp. t, denote the lowest dimensional simplex of (X, T) containing s_0 and also infinitely many x_j, resp. y_j; we may assume that $x_j \in s$, $y_j \in t$ for *all* j. Using a as origin, consider in the Euclidean space V the linear subspaces V_{s_0}, V_s, V_t spanned by s_0, s, t, respectively. Denoting by x' the orthogonal projection of a point $x \in V$ on V_{s_0}, we may further assume that $x'_j, y'_j \in s_0$. If $s \neq t$ we finally arrange that V_s and V_t are perpendicular to one another in the sense that for every $x \in V_s$, $y \in V_t$, the vectors $x - x'$ and $y - y'$ are orthogonal; for if this is not already the case, it can be achieved by applying a linear, hence bi-Lip, bijection of V onto itself. In terms of the Euclidean norm $|\cdot|$ on V we then have (see text following (4.3))

$$\begin{aligned} d_X^e(x_j, y_j) &\leqslant |x_j - x'_j| + |x'_j - y'_j| + |y'_j - y_j| \\ &\leqslant 3|x_j - y_j| = 3e(x_j, y_j), \end{aligned}$$

noting that the vectors $x_j - x'_j \in V_s \ominus V_{s_0}$, $x'_j - y'_j \in V_{s_0}$ and $y'_j - y_j \in V_t \ominus V_{s_0}$ are mutually orthogonal with the sum $x_j - y_j$. This contradiction with (4.4) completes the proof, the remaining case $s = t$ being trivial, again see text following (4.3). □

Proposition 4.1. (a) *The distance function $d_X = d_X^g$ on a Riemannian polyhedron (X, g) is intrinsic, in particular independent of the chosen triangulation.*

(b) *(X, d_X) is a length space (hence a geodesic space, if complete).*

(c) *d_X equals the Carathéodory distance*

$$d_X(x, y) = \max\{|u(x) - u(y)| : u \in \mathrm{Lip}(X),\ |\nabla u| \leqslant 1 \text{ a.e. in } X\}.$$

Here $\mathrm{Lip}(X)$ denotes the class of Lip continuous functions $u : X \to \mathbb{R}$ (using any of the locally equivalent distances on X considered above); and $|\nabla u|$ is the Riemannian gradient, defined a.e. in X (that is, a.e. in each $s \in \mathcal{S}^*(X, T)$), for a given triangulation $T = (K, \theta)$ of X, by Rademacher's theorem for Lip functions on Euclidean domains, applied to u expressed in Euclidean coordinates in the interior of $\theta^{-1}(s)$, see e.g., [Fe 1969, 3.1.6].

Proof. The proof is quite similar to that in [CP 1990, (3.10)] for the case where X is instead a Riemannian Lipschitz manifold. Note first that, for given $x, y \in X$, there exists a null set $Z_T^* \in \mathcal{Z}$ with $Z_T^* \supset Z_T$ and large enough so that

$$\varrho_{Z_T^*}(x,y) = d_X(x,y). \tag{4.5}$$

(It suffices to choose an increasing sequence $(Z_j) \subset \mathcal{Z}$ with $Z_j \supset Z_T$ so that $\varrho_{Z_j}(x,y) \to d_X(x,y)$, and to take $Z_T^* = \bigcup_j Z_j$, noting that $Z_T^* \supset Z_T$ and $\varrho_{Z_T^*}(x,y) \geqslant \varrho_{Z_j}(x,y)$.)

Ad (a). Fix $x, y \in X$, and consider any other triangulation $T' = (K', \theta')$. For any maximal simplex $s_\alpha \in \mathcal{S}^*(X,T)$ and any maximal simplex $s'_\beta \in \mathcal{S}^*(X,T')$ such that $U := s_\alpha \cap {s'_\beta}^\circ \neq \emptyset$ we have the Lip homeomorphism

$$\theta'^{-1} \circ \theta_{|\theta^{-1}(U)} : \theta^{-1}(U) \to \theta'^{-1}(U)$$

between the open subsets $\theta^{-1}(U)$ and $\theta'^{-1}(U)$ of the open Euclidean simplexes $\theta^{-1}(s_\alpha^\circ)$ and $\theta'^{-1}({s'_\beta}^\circ)$. This Lip homeomorphism is differentiable off some Lebesgue null set $Z_{\alpha\beta} \subset \theta^{-1}(U)$, mapped by θ onto a null set $\theta(Z_{\alpha\beta}) \subset U$. Writing

$$Z_0 = Z_T \cup Z_{T'} \cup \bigcup_{\alpha,\beta} \theta(Z_{\alpha\beta}) \quad (\in \mathcal{Z}),$$

we have for any $Z \in \mathcal{Z}$ with $Z \supset Z_0$ and any path $\gamma \in \mathrm{Lip}^{Z_0}(x,y;X)$

$$L_T(\gamma) = L_{T'}(\gamma)$$

because this holds by covariance for the contributions from $\gamma^{-1}(s_\alpha \cap s'_\beta)$ to $L_T(\gamma)$ and to $L_{T'}(\gamma)$.

For any $Z \in \mathcal{Z}$ with $Z \supset Z_{T'}$ we have the quantities $\varrho'_Z(x,y)$ and $d'_X(x,y)$ corresponding to $\varrho_Z(x,y)$ and $d_X(x,y)$, respectively, but relative to T' in place of T. For any path $\gamma \in \mathrm{Lip}^{Z \cup Z_0}(x,y;X)$ we obtain

$$\varrho'_Z(x,y) \leqslant \varrho'_{Z \cup Z_0}(x,y) \leqslant L_{T'}(\gamma) = L_T(\gamma).$$

By varying $\gamma \in \mathrm{Lip}^{Z \cup Z_0}(x,y;X) \;\; (\subset \mathrm{Lip}^{Z_0}(x,y;X))$ we conclude that

$$\varrho'_Z(x,y) \leqslant \varrho_{Z \cup Z_0}(x,y) \leqslant d_X(x,y),$$

and finally $d'_X(x,y) \leqslant d_X(x,y)$. Similarly, $d_X(x,y) \leqslant d'_X(x,y)$.

Ad (b). With the intrinsic distance d_X we associate in the standard way the *intrinsic length* $L(\gamma)$ of a Lip path $\gamma : [a,b] \to (X, d_X)$ (or equivalently

$\gamma : [a,b] \to (X,e)$, by Lemma 4.2, which is applicable because only finitely many maximal simplexes of X meet the compact image $\gamma([a,b])$):

$$L(\gamma) = \sup_{\pi} \sum_{i=1}^{k} d_X(x_{i-1}, x_i),$$

where π ranges over all subdivisions

$$a = x_0 < x_1 < \cdots < x_{k-1} < x_k = b$$

of $[a,b]$.

Returning to the null set Z_T^* from (4.5) (for given $x, y \in X$ and a given triangulation T of X) we may clearly choose $Z = Z_T^* \in \mathcal{Z}$ so as to satisfy (4.5) for all pairs of points x, y from a countable dense subset X^* of X. With that set Z we show that

$$L(\gamma) \leqslant L_T(\gamma) \tag{4.6}$$

for any Lip path $\gamma : [a,b] \to (X,e)$ such that $\gamma^{-1}(Z)$ is null. Via a subdivision of $[a,b]$ this in fact reduces to $d_X(x,y) \leqslant L_T(\gamma)$. For given $\varepsilon > 0$ choose $x^*, y^* \in X^*$ so that $d_X(x,x^*)$, $d_X(y,y^*) < \varepsilon$. By the definition of $\varrho_Z(x,y)$ ($\leqslant d_X(x,y)$), etc., there are paths $\alpha \in \mathrm{Lip}^Z(x^*, x; X)$ and $\beta \in \mathrm{Lip}^Z(y, y^*; X)$ such that $L_T(\alpha)$, $L_T(\beta) < \varepsilon$. The path $\gamma^* = \alpha \cup \gamma \cup \beta$ belongs to $\mathrm{Lip}^Z(x^*, y^*; X)$, and by the choice of Z above,

$$\begin{aligned} d_X(x,y) &\leqslant d_X(x^*,y^*) + 2\varepsilon = \varrho_Z(x^*,y^*) + 2\varepsilon \\ &\leqslant L_T(\gamma^*) + 2\varepsilon \leqslant L_T(\gamma) + 4\varepsilon. \end{aligned}$$

This shows that $d_X(x,y) \leqslant L_T(\gamma)$, thus establishing (4.6).

We are now prepared to show that (X, d_X) is a *length space* because

$$d_X(x,y) = \inf\{L(\gamma) : \gamma \in \mathrm{Lip}(x,y;X)\}, \tag{4.7}$$

for every $x, y \in X$, where $\mathrm{Lip}(x,y;X)$ denotes the family of all Lip paths $\gamma : [a,b] \to (X, d_X)$ joining x to y. Choosing $Z = Z_T^* \in \mathcal{Z}$ conforming with (4.5) we then have from (4.6) when γ ranges over $\mathrm{Lip}^Z(x,y;X)$

$$\inf_{\gamma} L(\gamma) \leqslant \inf_{\gamma} L_T(\gamma) = \varrho_Z(x,y) = d_X(x,y).$$

A fortiori, the inequality sign $\geqslant$ holds in (4.7). The opposite inequality in (4.7) is obvious.

Ad (c). Let $\delta(x,y)$ denote the supremum of $|u(x)-u(y)|$ for the stated functions u. For any such function u take for Z the union of Z_T and the null set of points x of $X \setminus Z_T$ at which u is not differentiable (as a function on $s^\circ \ni x$, $s \in \mathcal{S}^*(X,T)$) with $|\nabla u(x)| \leqslant 1$. For any $\gamma \in \mathrm{Lip}_Z(x,y;X)$ we then have

$$\begin{aligned}|u(x)-u(y)| &= \Big| \sum_{s\in\mathcal{S}^*(X,T)} \int_{\gamma^{-1}(s^\circ)} \frac{d}{dt}(u\circ\gamma)\Big| \\ &\leqslant \sum_{s\in\mathcal{S}^*(X,T)} \int_{\gamma^{-1}(s^\circ)} |\langle(\nabla u)\circ\gamma,\dot\gamma\rangle| \leqslant L_T^Z(\gamma)\end{aligned}$$

by Cauchy–Schwarz, $\langle\cdot,\cdot\rangle$ and $|\cdot|$ denoting the Riemannian metric and the norm on (the tangent bundle of) s° for each $s \in \mathcal{S}^*(X,T)$. It follows that $|u(x)-u(y)| \leqslant \varrho_Z(x,y) \leqslant d_X(x,y)$.

In the opposite direction, note that the Lip function $u := d_X(x,\cdot)$ on X competes in the definition of $\delta(x,y)$ because $|\nabla u| \leqslant 1$ a.e. in each s°, $s \in \mathcal{S}^*(X,T)$, by the triangle inequality, noting also that $d_X(x,y)$ equals the usual Riemannian distance between two points $x,y \in s^\circ$. (This is because the usual geodesic segment $[x,y]$ can be slightly modified to a path, e.g., $[x,z]\cup[z,y]$ of class $\mathrm{Lip}^Z(x,y;X)$ and of length only slightly bigger than that of $[x,y]$, by the usual Fubini argument, assuming y near to x.) □

It can be shown that, for a triangulable Lip manifold (X,g), cf. Examples 3.6 and 8.4, and in particular for a Riemannian manifold, the intrinsic distance $d_X(x,y)$ as defined above (viewing (X,g) as a Riemannian polyhedron) equals the usual Riemannian (Lip) distance between x and y.

Local structure in terms of cubes

For later use we introduce a certain Lip homeomorphism θ of the closed star $\overline{\mathrm{st}}(a)$ of a given point a of a Riemannian polyhedron X of homogeneous dimension n onto a certain cell complex $P \subset \mathbb{R}^k$, where k is the number of 1-simplexes of X containing a. The cells of P will be Euclidean *cubes* of unit edge-length; cubes appear sometimes to be technically more convenient than simplexes.

Step 1. We may assume that the given point a is a *vertex* of a triangulation T of X, and that (X,T) is affinely and Lip homemorphically embedded in a Euclidean space, cf. Remark 4.1. Let $a_1,\cdots,a_k$ denote the remaining vertices of $S=\overline{\mathrm{st}}(a)$, and let $e_1,\cdots,e_k$ be the natural base of $\mathbb{R}^k$. There is a unique map

$$\varphi : S \to \mathbb{R}^k$$

such that $\varphi(a) = 0$, $\varphi(a_j) = e_j$ $(j = 1, \cdots, k)$, and further that φ is affine-linear on each maximal simplex of S (hence on every simplex of S). Since $0, e_1, \cdots, e_k$ are affinely independent, it follows from Lemma 4.1 (for a finite complex) that φ is a Lip homeomorphism onto its image $\varphi(S) \subset \mathbb{R}^k$.

According to Lemma 4.2 we may here replace the Euclidean metric on S $(\subset V)$ by one of the equivalent metrics d_S^e or $d_S = d_S^g$ (see the preceding section), and similarly the Euclidean metric on $\varphi(S)$ $(\subset \mathbb{R}^k)$ by $d^e_{\varphi(S)}$.

Step 2. Define $\theta = \psi \circ \varphi : S \to [0,1]^k$, where $\psi : \mathrm{conv}\{0, e_1, \cdots, e_k\} \to [0,1]^k$ is the Lip homeomorphism from Sublemma 4.3 below. Then θ and $\theta(S) = P$ have the properties stated in the following lemma.

Lemma 4.3. *For any point $a \in X$ and any triangulation T of X having a as a vertex there exists a Lip homeomorphism θ of the closed star $S = \overline{\mathrm{st}}(a)$ onto an n-dimensional cell complex P, a subcomplex of $[0,1]^k$, such that θ induces a bijection of the collection of all simplexes of (X,T) containing a onto the collection of all cells (cubes) of P containing 0; in particular $\theta(a) = 0$.*

Remark 4.2. While topological notions for $S = \overline{\mathrm{st}}(a)$, such as boundary ∂A and interior A° of a subset A, are understood relative to X, those for $P = \theta(S)$ will be taken relative to the cone $\mathbb{R}_+ P$. It follows that

$$P^\circ = \bigcup_{0<r<1} rP, \tag{4.8}$$

and ∂P is the union of all $(n-1)$-faces of P not containing 0. Hence $\theta^{-1}(\partial P)$ is the union of all $(n-1)$-faces of S not containing a. In order that this latter union equal the topological boundary ∂S in X it is necessary and sufficient that $S^\circ = \mathrm{st}(a)$, or explicitly that every $(n-1)$-simplex t of S not containing a be a face of at least one n-simplex of (X,T) not containing a. The stated condition means that $t \not\subset bX$ (the boundary of X, cf. the above section Polyhedron), a condition which can always be fulfilled by passing to a suitable subdivision of T. In this way we achieve that

$$\theta(\partial S) = \partial P, \quad \theta(S^\circ) = P^\circ. \tag{4.9}$$

By abuse of notation we continue writing s° for the relative interior $s \setminus b(s)$ of a simplex s, even if $\dim s < n$.

Sublemma 4.3. *Let $Q = [0,1]^k$, $\Delta = \operatorname{conv}\{0, e_1, \cdots, e_k\}$, where $e_1, \cdots, e_k$ are the natural basis vectors for $\mathbb{R}^k$, $k = 1, 2, \cdots$. For $x = (x_1, \cdots, x_k) \in \mathbb{R}^k$ write $|x|_1 = \sum |x_i|$, $|x|_\infty = \max |x_i|$ and e.g., $|x| = |x|_2 = (\sum x_i^2)^{1/2}$. The map $\psi : \Delta \to \mathbb{R}^k$ given by $\psi(0) = 0$ and*

$$\psi(x) = \frac{|x|_1}{|x|_\infty} x, \qquad x \in \Delta \setminus \{0\},$$

is then a Lip homeomorphism of the simplex Δ onto the cube Q, and ψ induces a bijection of the collection of all faces of Δ containing 0 onto the corresponding collection for Q.

Proof. Writing $\mathbb{R}_+^k = [0, \infty[^k$, we clearly have

$$\Delta = \{x \in \mathbb{R}_+^k : |x|_1 \leqslant 1\}, \quad Q = \{x \in \mathbb{R}_+^k : |x|_\infty \leqslant 1\}.$$

Since $|\psi(x)|_\infty = |x|_1$, it follows that $\psi(\Delta) = Q$. The faces of Δ and Q containing 0 are

$$\Delta_J = \{x \in \Delta : x_j = 0 \text{ for } j \in J\}, \quad Q_J = \{x \in Q : x_j = 0 \text{ for } j \in J\},$$

respectively, where $J \subset \{1, \cdots, k\}$. Hence $\psi(\Delta_J) = Q_J$. For $x, y \in \Delta \setminus \{0\}$ we have

$$\begin{aligned}
&|x|_\infty |y|_\infty (\psi(x) - \psi(y)) \\
&\qquad = (|x|_1 |y|_\infty - |x|_\infty |y|_1) x + |x|_\infty |y|_1 (x - y) \\
&\qquad = (|x|_1 - |y|_1) |y|_\infty x + (|y|_\infty - |x|_\infty) |y|_1 x + |x|_\infty |y|_1 (x - y)
\end{aligned}$$

and hence

$$\begin{aligned}
|\psi(x) - \psi(y)| &\leqslant \left(|x - y|_1 + \frac{|y|_1}{|y|_\infty} |x - y|_\infty\right) \frac{|x|}{|x|_\infty} + \frac{|y|_1}{|y|_\infty} |x - y| \\
&\leqslant c_k |x - y|
\end{aligned}$$

by the equivalence of the norms $|\cdot|_1, |\cdot| = |\cdot|_2$ and $|\cdot|_\infty$ on $\mathbb{R}^k$, c_k being a dimensional constant. Clearly, ψ is bijective, and

$$\psi^{-1}(x) = \frac{|x|_\infty}{|x|_1} x, \qquad x \in Q \setminus \{0\}.$$

The above proof therefore carries over to show that ψ^{-1} likewise is Lipschitzian. □

Uniform estimate of ball volumes

Let (X, g) be a Riemannian polyhedron of homogeneous dimension n. The intrinsic Riemannian distance on X is denoted by d_X, and the Riemannian volume, resp. diameter, of a set A by $|A|$, resp. $\operatorname{diam} A$. We shall consider closed balls

$$B(x,r) = B_X(x,r) = \{y \in X : d_X(x,y) \leqslant r\}.$$

As in (4.2) let $\Lambda \geqslant 1$ denote the ellipticity constant of the Riemannian metric g on X, now restricted to the compact closure of a given set $X_0 \Subset X$. The Riemannian distance on X induced by the Euclidean Riemannian metric g^e on X is denoted by d_X^e, and corresponding balls are denoted by $B^e(x,r)$. The volume, resp. diameter, of a set $A \subset X$ with respect to g^e is denoted by $|A|_e$, resp $\operatorname{diam}_e A$. The gradient operator on X with respect to g^e is written ∇_e.

Lemma 4.4. *For any set $X_0 \Subset X$ there is a constant $c_0 \geqslant 1$, depending only on X as a mere polyhedron and on X_0, such that*

$$c_0^{-1}\Lambda^{-2n}r^n \leqslant |B(x,r)| \leqslant c_0\Lambda^{2n}r^n$$

whenever $B(x, 2\Lambda^2 r) \subset X_0$ and (in the former inequality) $r \leqslant \operatorname{diam} X_0$.

Proof. If $B(x, 2\Lambda^2 r) \subset X_0 \Subset X$ then $2\Lambda^2 \operatorname{diam} X_0 < \infty$, and

$$|B(x,r)| \leqslant \Lambda^n |B(x,r)|_e \leqslant \Lambda^n |B^e(x, \Lambda r)|_e \leqslant c_0 \Lambda^{2n} r^n,$$

$$|B(x,r)| \geqslant \Lambda^{-n} |B(x,r)|_e \geqslant \Lambda^{-n} |B^e(x, \Lambda^{-1} r)|_e \geqslant c_0^{-1} \Lambda^{-2n} r^n.$$

Here we use the fact that $B^e(x, 2\Lambda r) \subset B(x, 2\Lambda^2 r) \subset X_0$ and that there is a constant $c_0 = c_0(X, X_0) \geqslant 1$ such that

$$c_0^{-1} r^n \leqslant |B^e(x,r)|_e \leqslant c_0 r^n \quad \text{when} \quad B^e(x, 2r) \subset X_0 \tag{4.10}$$

(and when $r \leqslant \Lambda^{-1} \operatorname{diam} X_0$ in the former inequality). For if there were no such constant c_0 then there would be a sequence of balls $B^e(x_j, r_j)$ with $B^e(x_j, 2r_j) \subset X_0$ such that

$$\text{either} \quad |B^e(x_j, r_j)|_e < j^{-1} r_j^n \quad \text{or} \quad |B^e(x_j, r_j)|_e > j r_j^n \tag{4.11}$$

(and $r_j \leqslant \Lambda^{-1} \operatorname{diam} X_0 < \infty$ in the former alternative, while of course $j r_j^n \leqslant |X_0|_e < \infty$ in the latter). In either situation we may suppose that

$x_j \to x$, $r_j \to r < \infty$ with $x \in \overline{X}_0$, and even $B^e(x, 2r) \subset \overline{X}_0$ if $r > 0$ (possible only in the former alternative).

If $r > 0$ we may assume that $|r_j - r| < r/4$ and $d^e_X(x_j, x) < r/4$, and so $|B^e(x, 3r/2)|_e \geqslant |B^e(x_j, r_j)|_e \to \infty$, contradicting $|B^e(x, r)|_e \leqslant |X_0|_e < \infty$.

In the remaining case $r_j \to 0$ we may assume that all x_j, and hence x, lie in one and the same n-simplex s_0 of X. By homogeneity and compactness,

$$|B^e(x_j, r_j)|_e \geqslant |s_0 \cap B^e(x_j, r_j)|_e \geqslant \alpha r_j^n$$

for large j, with $\alpha = \alpha(X, X_0) > 0$. On the other hand, $B^e(x_j, r_j)$ ($\subset X_0$) meets any n-simplex of X in a set of diameter $\leqslant 2r_j$, hence of Euclidean volume $\leqslant \beta r_j^n$, and so

$$|B^e(x_j, r_j)|_e \leqslant N\beta r_j^n,$$

where $\beta = \beta(X, X_0)$, and N is the number of n-simplexes of X meeting $\overline{X}_0$. These two estimates contradict (4.11), and so prove (4.10). □

Corollary 4.1. (Ball doubling property.) *For any* $\kappa \geqslant 1$,

$$|B(x, \kappa r)| \leqslant \kappa^n c_0^2 \Lambda^{4n} |B(x, r)|$$

whenever $B(x, 2\kappa\Lambda^2 r) \subset X_0$.

This follows from Lemma 4.4 if $r \leqslant \operatorname{diam} X_0$, and is trivial otherwise, both balls being then equal to X_0.

Remark 4.3. Thorough studies of ball doubling, Hölder continuity and bi-Lipschitz embeddings of metric measure spaces have been made in [Gro 1999, Appendix B by S. Semmes]. Relationships to Poincaré inequalities are obtained in [Chee 1999] (see also the beginning of Chapter 6 below).

Part II

Potential theory on polyhedra

5

The Sobolev space $W^{1,2}(X)$. Weakly harmonic functions

The space of Lip continuous (i.e., locally Lipschitz) functions $u : X \to \mathbb{R}$ on an admissible Riemannian polyhedron X admits Sobolev $(1,p)$-norms, requiring that u and its first differential be p-th power integrable. We restrict our attention to the case $p = 2$. Weakly harmonic and weakly subharmonic functions on X are defined as weak solutions/subsolutions of class $W^{1,2}(X)$ to the Laplace equation on X.

For classical Sobolev space theory we refer to [Ad 1975].

The Sobolev space $W^{1,2}(X)$

In the present chapter $X = (X, g)$ will denote an *admissible* Riemannian polyhedron of dimension n, and d_X the intrinsic distance on X, cf. Chapter 4. We denote by $\mathrm{Lip}^{1,2}(X)$ the linear space of all Lip continuous functions $u : (X, d_X) \to \mathbb{R}$ for which the Sobolev $(1,2)$-norm $\|u\|$ defined by

$$\|u\|^2 = \int_X (u^2 + |\nabla u|^2) = \sum_{s \in \mathcal{S}^{(n)}(X)} \int_s (u^2 + |\nabla u|^2)$$

is finite, $\mathcal{S}^{(n)}(X)$ denoting the collection of all n-simplexes s of X (relative to a given triangulation), and ∇u the Riemannian length of the Riemannian gradient on each s. Here and elsewhere the integration is with respect to the Riemannian volume measure on X, resp. on each $s \in \mathcal{S}^{(n)}(X)$, in the absence of other indications. By Rademacher's theorem (cf. e.g., [Fe 1969]) ∇u exists a.e. in each $s \in \mathcal{S}^{(n)}(X)$ (hence a.e. in X).

The Lebesgue space $L^2(X) = \bigoplus_{s \in \mathcal{S}^{(n)}(X)} L^2(s)$ is likewise formed with respect to the volume measure.

The *Sobolev space* $W^{1,2}(X)$ is defined as the completion of $\mathrm{Lip}^{1,2}(X)$ with respect to the above Sobolev norm $\|\cdot\|$. We shall realize this completion as the linear subspace of $L^2(X)$ consisting of all $u \in L^2(X)$ for which there exists a Cauchy sequence (u_j) in $(\mathrm{Lip}^{1,2}(X), \|\cdot\|)$ such that $u_j \to u$ in $L^2(X)$.

The gradient $\nabla u \in L^2(X)$ is then defined as the L^2-limit of (∇u_j), and the Sobolev norm $\|u\| := \lim \|u_j\|$ can be expressed as above.

When (M, g) is a Riemannian manifold (with or without boundary), the above definition of $W^{1,2}(M)$ agrees with the usual definition as the completion of $C^\infty(M)$ in the Sobolev $(1,2)$-norm. (This is because every function of class $C^\infty(M)$ of finite Sobolev norm is of class $\mathrm{Lip}^{1,2}(M)$, hence in $W^{1,2}(M)$ as above; and on the other hand, $\mathrm{Lip}^{1,2}(M)$ is contained in the usual Sobolev space $W^{1,2}(M)$.)

Returning to a general admissible Riemannian polyhedron X, we have, for any $s \in \mathcal{S}^{(n-1)}(X)$, the continuous trace map $u \mapsto u_{|s} : \mathrm{Lip}^{1,2}(X) \to L^2(s)$ (this reduces to the known case where X is a simplex), and that map extends uniquely to a continuous map $W^{1,2}(X) \to L^2(s)$, likewise called the *trace map*.

$\mathrm{Lip}_c(X)$ denotes the linear subspace of $\mathrm{Lip}(X)$ (and of $\mathrm{Lip}^{1,2}(X)$) consisting of all Lip functions of compact support in X. Similarly for $W_c^{1,2}(X)$.

$W_0^{1,2}(X)$ denotes the closure of $\mathrm{Lip}_c(X)$ in $W^{1,2}(X)$.

$\mathrm{Lip}_c^+(X)$ denotes the set of all functions $\geqslant 0$ (a.e.) of class $\mathrm{Lip}_c(X)$. Similarly for other spaces, e.g., $W_0^{1,2}(X)^+$.

$W_{\mathrm{loc}}^{1,2}(X)$ denotes the linear space of all $u \in L^2_{\mathrm{loc}}(X)$ such that $u_{|U} \in W^{1,2}(U)$ for every relatively compact subdomain U of X. It suffices that X can be covered by such subdomains U; this follows from (a) in the following remark, applied to functions ψ from a Lip partition of unity on the metric space (X, d_X), as in the beginning of Chapter 4.

Remark 5.1. (a) For any $\psi \in \mathrm{Lip}_c(X)$ and $u \in W_{\mathrm{loc}}^{1,2}(X)$, $\psi u \in W_0^{1,2}(X)$.

(b) For such ψ, the map $u \mapsto \psi u$ of $W^{1,2}(X)$ into $W_0^{1,2}(X)$ is norm continuous.

(c) $W_c^{1,2}(X) \subset W_0^{1,2}(X)$; hence $W_0^{1,2}(X)$ is the closure of $W_c^{1,2}(X)$.

Ad (a). There are constants a, b such that

$$|\psi(x)| \leqslant a, \ |\nabla\psi(x)| \leqslant b \qquad \text{a.e. for } x \in X\,.$$

For $u \in \mathrm{Lip}^{1,2}(X)$ it follows that $\psi u \in \mathrm{Lip}_c(X)$ and $\nabla(\psi u) = \psi \nabla u + u \nabla \psi$ (a.e.), hence after elementary manipulations

$$\|\psi u\| \leqslant (2a^2 + 2b^2)^{1/2} \|u\|\,. \tag{5.1}$$

For $u \in W^{1,2}(X)$ choose $(u_j) \subset \mathrm{Lip}^{1,2}(X)$ so that $\|u_j - u\| \to 0$. Then $(\psi u_j) \subset \mathrm{Lip}_c(X)$ is Cauchy, hence converges in $W_0^{1,2}(X)$ to its L^2-limit

ψu, which therefore is in $W_0^{1,2}(X)$, and $\|\psi u\|$ satisfies the above inequality (5.1). Finally, for $u \in W_{\text{loc}}^{1,2}(X)$ choose a relatively compact subdomain U of X containing $\operatorname{supp}\psi$; this is possible because (X, d_X) is a locally compact length space (Proposition 4.1), and therefore locally connected (even locally path connected). Then $u_U \in W^{1,2}(U)$, and so $\psi u \in W_0^{1,2}(U) \hookrightarrow W_0^{1,2}(X)$.

Ad (b). If $(u_j) \subset W^{1,2}(X)$ converges to $u \in W^{1,2}(X)$ then (ψu_j) is a Cauchy sequence in $W_0^{1,2}(X)$, by (5.1) for functions in $W^{1,2}(X)$, as has been established. Hence (ψu_j) converges in $W_0^{1,2}(X)$ to its L^2-limit ψu.

Ad (c). If $u \in W^{1,2}(X)$ has compact support, we may choose $\psi \in \operatorname{Lip}_c(X)$ so that $\psi = 1$ on $\operatorname{supp} u$, and then $u = \psi u \in W_0^{1,2}(X)$, by (a).

Remark 5.2. If a function $u \in W^{1,2}(X)$ (or just $u \in W_{\text{loc}}^{1,2}(X)$) is constant (a.e.) in some measurable set $A \subset X$ then $\nabla u = 0$ a.e. in A. (This is well known; in fact it reduces to the case where X is a Euclidean n-simplex, and further to the easy case of an absolutely continuous function u of one real variable, cf. [DL 1954, pp. 313-316].) It follows that

$$\langle \nabla u_+, \nabla u_- \rangle = 0 \qquad \text{a.e. for any } u \in W_{\text{loc}}^{1,2}(X)$$

because $u_+ = 0$ in $\{u \leqslant 0\}$ and $u_- = 0$ in $\{u \geqslant 0\}$.

Lemma 5.1. *$W^{1,2}(X)$ is the class of functions $u \in W_{\text{loc}}^{1,2}(X)$ of finite Sobolev norm $\|u\|$.*

Proof. Suppose that $u \in W_{\text{loc}}^{1,2}(X)$ has finite Sobolev $(1,2)$-norm $\|u\|$. Consider a locally finite, hence countable, cover (U_j) of X by relatively compact subdomains U_j such that $u_{|U_j} \in W^{1,2}(U_j)$. Choose a Lip partition of unity (φ_j) on X subordinated to (U_j) (see the beginning of Chapter 4). For given $\varepsilon > 0$ there exist Lip continuous functions u_j on U_j satisfying

$$\|u_{|U_j} - u_j\| \leqslant \varepsilon 2^{-j} / (\|\varphi_j\|_{L^\infty} + \|\nabla \varphi_j\|_{L^\infty}).$$

The function $v = \sum_j \varphi_j u_j$ is well defined and Lip continuous on X, the cover (U_j) being locally finite. Since $\sum_j \varphi_j = 1$ we have a.e. in X

$$v - u = \sum \varphi_j (u_j - u),$$
$$\nabla(v - u) = \sum \varphi_j \nabla(u_j - u) + \sum (u_j - u) \nabla \varphi_j.$$

It follows that $\|v - u\| \leqslant 2\varepsilon \sum 2^{-j} = 2\varepsilon$. □

Lemma 5.2. (Chain rule.) *Let* $u = (u^1, \cdots, u^p) \in [W^{1,2}(X)]^p$ *for some* $p = 1, 2, \cdots$, *and let* $\varphi \in C^1(\mathbb{R}^p)$. *If* $|\nabla\varphi|$ *is bounded and* $\varphi(0) = 0$ *then* $\varphi \circ u \in W^{1,2}(X)$ *and (with summation over* $\alpha = 1, \cdots, p$*)*

$$\nabla(\varphi \circ u) = [(D_\alpha\varphi) \circ u]\nabla u^\alpha \qquad \text{a.e. in } X.$$

Proof. If $u \in [\mathrm{Lip}^{1,2}(X)]^p$ then $\varphi \circ u \in \mathrm{Lip}^{1,2}(X)$ $(\subset W^{1,2}(X))$, and the classical chain rule applies a.e. in X (viz., at every point of differentiability of u in the interior of any n-simplex). In the general case suppose $p = 1$, for simplicity of writing. Choose a sequence $(u_j) \subset \mathrm{Lip}^{1,2}(X)$ so that $\sum \|u_j - u\| < \infty$. The function

$$f := |\nabla u| + \sum |\nabla u_j - \nabla u| \in L^2(X)$$

majorizes each $|\nabla u_j|$. Clearly, $u_j \to u$ and $\nabla u_j \to \nabla u$ hold also pointwise a.e.; and $\varphi \circ u_j \to \varphi \circ u$ holds in L^2-norm. Finally,

$$\nabla(\varphi \circ u_j) = (\varphi' \circ u_j)\nabla u_j \to (\varphi' \circ u)\nabla u,$$

first pointwise a.e., and next in L^2-norm because $|\nabla(\varphi \circ u_j)|$ is majorized by $f \sup_t |\varphi'(t)| \in L^2(X)$. □

Proposition 5.1. (a) $W^{1,2}(X)$ *is a Dirichlet space in the Sobolev* $(1,2)$-*norm* $\|\cdot\|$.

(b) $|||u||| = \|u\|$ *for every* $u \in W^{1,2}(X)$.

(c) *The self-map* $u \to |u|$ *of* $W^{1,2}(X)$ *is norm continuous.*

(d) $\mathrm{Lip}^{1,2}(X)^+$ *is norm dense in* $W^{1,2}(X)^+$, *and* $\mathrm{Lip}_c^+(X)$ *in* $W_0^{1,2}(X)^+$.

Proof. Ad (a). Clearly $(W^{1,2}(X), \|\cdot\|)$ is a Hilbert function space relative to the volume measure on X, cf. [De 1970, Chapitre 1]. Thus $W^{1,2}(X)$ has a continuous dense inclusion in $L^1_{\mathrm{loc}}(X)$. To see that $W^{1,2}(X)$ is a Dirichlet space we shall prove that, for any normal contraction $T : \mathbb{R} \to \mathbb{R}$ and any $u \in W^{1,2}(X)$, $v := Tu$ $(= T \circ u)$ belongs to $W^{1,2}(X)$, and $\|v\| \leqslant \|u\|$, cf. [De 1970, Chapitre 2].

According to [Fuk 1980, Theorem 1.4.1] this is equivalent to the closed Dirichlet form $\int_X |\nabla u|^2$ (with domain $W^{1,2}(X) \subset L^2(X)$) being *Markovian* in the sense of [Fuk 1980, p. 5]; and that follows immediately from Lemma 5.2 above, applied for each $\varepsilon > 0$ to a smooth function $\varphi_\varepsilon : \mathbb{R} \to [-\varepsilon, 1 + \varepsilon]$ with $0 \leqslant \varphi_\varepsilon' \leqslant 1$ such that $\varphi_\varepsilon(t) = t$ for $t \in [0, 1]$, cf. [Fuk 1980, Problem 1.2.1].

Ad (b). That $|u| \in W^{1,2}(X)$ follows from (a) applied to $Ts = |s|$. And $\||u|\| = \|u\|$ follows from $|\nabla|u(x)|| = |\nabla u(x)|$ for a.e. x, which holds by Remark 5.2 because $|u| - u = 0$ in $\{u \geqslant 0\}$, hence $\nabla|u| - \nabla u = 0$ there, and similarly for $\{u \leqslant 0\}$.

Ad (c). Consider a convergent sequence $(u_j) \subset W^{1,2}(X)$ with limit u. In view of (b), $\||u_j|\| = \|u_j\| \to \|u\| = \||u|\|$, and this implies that indeed $\||u_j| - |u|\| \to 0$ according to a general result [De 1970, p. 140].

Ad (d). If e.g., $(\varphi_j) \subset \mathrm{Lip}^{1,2}(X)$ converges to $\varphi \geqslant 0$ in $W^{1,2}(X)$ then $|\varphi_j| \to |\varphi| = \varphi$ in $W^{1,2}(X)^+$, by (c). □

For $u, v \in W^{1,2}(X)$ we write for brevity

$$E(u,v) = \frac{1}{2}\int_X \langle \nabla u, \nabla v \rangle = \frac{1}{2} \sum_{s \in \mathcal{S}^{(n)}(X)} \int_s \langle \nabla u, \nabla v \rangle \tag{5.2}$$

(the mutual energy of u and v), and

$$E(u) = E(u,u) = \frac{1}{2}\int_X |\nabla u|^2 \tag{5.3}$$

(the *energy* of u). Clearly, $E(u) \geqslant 0$. And $E(u) = 0$ (if and) only if u is *constant* a.e., first in each $s \in \mathcal{S}^n(X)$, next in the star of any vertex p of X (by considering the traces of u on the $(n-1)$-simplexes of X containing p, and using the local $(n-1)$-chainability of X, see Chapter 4, Polyhedron), and finally in all of X (connected).

Lemma 5.3. *Let $u \in W^{1,2}(X)$. If X is non-compact, suppose that*

$$\operatorname*{ess\,lim}_{x \to \infty_X} u(x) = 0\,,$$

where ∞_X denotes the Alexandroff point. Then $u \in W_0^{1,2}(X)$.

Proof. If X is compact, $u \in W_c^{1,2}(X) = W_0^{1,2}(X)$, by Remark 5.1 (c). For $\varepsilon > 0$ let $T_\varepsilon : \mathbb{R} \to \mathbb{R}$ be odd, of compact support $[-2\varepsilon, 2\varepsilon]$, and defined for $s \geqslant 0$ by

$$T_\varepsilon(s) = \begin{cases} s & \text{for } 0 \leqslant s \leqslant \varepsilon \\ 2\varepsilon - s & \text{for } \varepsilon \leqslant s \leqslant 2\varepsilon\,. \end{cases}$$

Then T_ε is a normal contraction on $\mathbb{R}$ (because T_ε is Lipschitzian with $|T'_\varepsilon(s)| \leqslant 1$ a.e. and $T_\varepsilon(0) = 0$). We approximate any $u \in W^{1,2}(X)$ by

$$u_\varepsilon := u - T_\varepsilon u \in W_c^{1,2}(X)$$

in view of Proposition 5.1 (a) and the hypothesis that there exists a compact set $K \subset X$ such that $|u| < \varepsilon$ (hence $u_\varepsilon = 0$) a.e. in $X \setminus K$. By Remark 5.1 (c) it remains to prove that $T_\varepsilon u \to 0$ in $(W^{1,2}(X), \|\cdot\|)$. We obtain

$$\|T_\varepsilon u\|^2 \leqslant \int_{\{|u| \leqslant 2\varepsilon\}} (u^2 + |\nabla u|^2) \to 0 \qquad \text{as } \varepsilon \to 0.$$

In $\{|u| \leqslant \varepsilon\}$ we have in fact $T_\varepsilon u = u$, hence $\nabla(T_\varepsilon u) = \nabla u$ (cf. Remark 5.2). And in $\{\varepsilon \leqslant |u| \leqslant 2\varepsilon\}$ we have $|T_\varepsilon u| = 2\varepsilon - |u|$, hence $\nabla(T_\varepsilon u) = -\nabla u$. Finally, in $\{|u| \geqslant 2\varepsilon\}$ we have $T_\varepsilon u = 0$, hence $\nabla(T_\varepsilon u) = 0$. □

A Poincaré inequality

For any ball $B(x,r)$ in the admissible Riemannian polyhedron (X,g) and any function u integrable over $B(x,r)$ with respect to the Riemannian volume measure μ_g on X we denote by $\overline{u}$ the *meanvalue* of u:

$$\overline{u} = \frac{1}{\mu_g(B(x,r))} \int_{B(x,r)} u \, d\mu_g,$$

and similarly under analogous circumstances, e.g., when $B(x,r)$, μ_g, $|\nabla u|$ are replaced by $B^e(x,r)$, μ_e, $|\nabla_e u|$, referring to the Euclidean Riemannian structure g^e on X, see Chapter 4, Riemannian polyhedron.

As shown by Sturm [St 1996] the "weak" Poincaré inequality in the following theorem implies the corresponding "strong" inequality whereby the integral on the left is taken over the same ball $B(x,r)$ as that on the right, and the factor $c(X, X_0)\Lambda^{6n}$ is replaced by some $c(X, \Lambda, X_0)$. A much weaker version of the Poincaré inequality was given in [Wh 1986].

Theorem 5.1. *For any relatively compact subset X_0 of X there is a constant $c = c(X, X_0)$ such that, for any function $u \in W^{1,2}(X)$ and any ball $B(x,r) \subset X_0$,*

$$\int_{B(x,\frac{r}{2\Lambda^2})} (u - \overline{u})^2 d\mu_g \leqslant c\Lambda^{6n} r^2 \int_{B(x,r)} |\nabla u|^2 d\mu_g.$$

Proof. Note that $B(x, \frac{r}{2\Lambda^2}) \subset B^e(x, \frac{r}{2\Lambda})$ and $B^e(x, \frac{r}{\Lambda}) \subset B(x,r) \subset X_0$.

Writing $|A| = \int_A d\mu_g$ and $|A|_e = \int_A d\mu_e$ for measurable sets $A \subset X$, we get

$$\begin{aligned}
&\int_{B(x,\frac{r}{2\Lambda^2})} (u-\overline{u})^2 d\mu_g \\
&= \frac{1}{2|B(x,\frac{r}{2\Lambda^2})|} \int_{B(x,\frac{r}{2\Lambda^2})} \int_{B(x,\frac{r}{2\Lambda^2})} (u(x_1)-u(x_2))^2 d\mu_g(x_1)\, d\mu_g(x_2) \\
&\leqslant \frac{\Lambda^{2n}}{2|B(x,\frac{r}{2\Lambda^2})|} \int_{B^e(x,\frac{r}{2\Lambda})} \int_{B^e(x,\frac{r}{2\Lambda})} (u(x_1)-u(x_2))^2 d\mu_e(x_1)\, d\mu_e(x_2) \\
&= \frac{|B^e(x,\frac{r}{2\Lambda})|_e}{|B(x,\frac{r}{2\Lambda^2})|} \Lambda^{2n} \int_{B^e(x,\frac{r}{2\Lambda})} (u-\overline{u})^2 d\mu_e \\
&\leqslant c_0^2 c_1 \Lambda^{5n-2} r^2 \int_{B^e(x,\frac{r}{\Lambda})} |\nabla_e u|^2 d\mu_e \\
&\leqslant c\Lambda^{6n} r^2 \int_{B(x,r)} |\nabla u|^2 d\mu_g
\end{aligned}$$

with $c = c_0^2 c_1$. The first equation reduces to the easy case $\overline{u} = 0$, replacing u by $u - \overline{u}$. The second inequality is derived from Lemma 4.4 (assuming to begin with $\frac{r}{2\Lambda^2} \leqslant \operatorname{diam} X_0$), from (4.10), and from the following "Euclidean" weak Poincaré inequality (applied with r replaced by $\frac{r}{\Lambda}$):

$$\int_{B^e(x,\frac{1}{2}r)} (u-\overline{u})^2 d\mu_e \leqslant c_1 r^2 \int_{B^e(x,r)} |\nabla_e u|^2 d\mu_e, \tag{5.4}$$

valid when $B^e(x,r) \subset X_0$; here $c_1 = c_1(X, X_0)$. In the remaining case $\frac{r}{2\Lambda^2} > \operatorname{diam} X_0$ we have

$$B(x,r) \subset X_0 \subset B(x, \operatorname{diam} X_0) \subset B(x, \tfrac{r}{2\Lambda^2}) \subset B(x, \tfrac{1}{2}r),$$

and so $B(x,r) = X_0 = B(x, \frac{1}{2}r)$. By hypothesis there exists $y \in X \setminus X_0$, hence $d_X(x,y) > r$. Because X is connected there also exists $z \in X$ with $d_X(x,z) = \frac{3}{4}r$, contradicting $B(x,r) = B(x, \frac{1}{2}r)$.

The proof of (5.4) is by iteration, exploiting fully the polyhedral structure of X. The same iterative scheme will be used later (in the proofs of Proposition 6.2 and Theorem 7.4). We may suppose that $r \leqslant 2 \operatorname{diam}_e X_0$; otherwise

$B^e(x, \frac{1}{2}r) = X_0 \subset B^e(x, \operatorname{diam}_e X_0)$, and (5.4) follows from its validity for $r = 2\operatorname{diam}_e X_0$. Suppose there is no such constant c_1. There are then sequences of balls $B_j := B^e(x_j, r_j) \subset X_0$ and of functions $u_j \in W^{1,2}(X)$ (or just as well $u_j \in W^{1,2}(B_j^\circ)$) such that

$$\int\limits_{B^e(x_j,\frac{1}{2}r_j)} |u_j - \overline{u}_j|^2 d\mu_e > j{r_j}^2 \int\limits_{B^e(x_j,r_j)} |\nabla_e u_j|^2 d\mu_e. \tag{5.5}$$

The same inequality then holds a fortiori when $\overline{u}_j$ is replaced by *any* number. We may assume that $x_j \to x \in \overline{X}_0$, $r_j \to r \leqslant 2\operatorname{diam}_e X_0$ $(< \infty)$.

Step 1. The case $r > 0$. We may then further assume that $d^e_X(x_j, x) < \frac{1}{8}r$ and $|r_j - r| < \frac{1}{8}r$. It follows that

$$B^e(x_j, \tfrac{1}{2}r_j) \subset B^e(x, \tfrac{3}{4}r) \subset B^e(x_j, r_j),$$

and so (5.5) contradicts the following "Euclidean" *strong Poincaré inequality* (applied to the radius $\frac{3}{4}r$):

$$\int\limits_{B^e(x,r)} |u - \overline{u}|^2 d\mu_e \leqslant Cr^2 \int\limits_{B^e(x,r)} |\nabla_e u|^2 d\mu_e, \tag{5.6}$$

valid for *fixed* $B^e(x,r) \subset X_0$ with C independent of $u \in W^{1,2}(B^e(x,r)^\circ)$.

The proof of (5.6) is similar to the corresponding part of the proof of [Chen 1995, Proposition 2.2]: We may assume that $\overline{u} = 0$ and, writing $B^e(x,r)^\circ = B$, that there is a sequence $(u_j) \subset W^{1,2}(B)$ such that $\overline{u}_j = 0$ and

$$1 = \int_B u_j^2\, d\mu_e > jr^2 \int_B |\nabla_e u_j|^2 d\mu_e. \tag{5.7}$$

These Dirichlet integrals thus converge to 0, and the Sobolev norms $||u_j||$ over B remain bounded. In terms of a triangulation T of X let $s_1, \cdots, s_k$ denote the n-simplexes of T meeting B $(\subset X_0)$. By Rellich's compactness theorem, applied to each $s_i^\circ \cap B$, we may assume that u_j converges in $L^2(B)$ to some $u \in L^2(B)$; and since $\nabla_e u_j \to 0$ in $L^2(B)$ we have $u \in W^{1,2}(B)$ and $\nabla_e u = 0$. It follows that u is constant (a.e.) in each $s_i \cap B$, and these constants must be equal. Indeed, the Euclidean open ball B in the length space (X, d^e_X) is connected[1] and therefore an admissible Riemannian

[1]For any $y \in B$ there is (since $d^e_X(x,y) \leqslant r$) a path γ in X joining x to y and of Euclidean length $< r + d^e_X(y, X \setminus B)$, whence γ cannot meet $X \setminus B$.

subpolyhedron (B, T') with each n-simplex s' of (B, T') being contained in some n-simplex of T; in particular, u is constant (a.e.) in s'. Any two n-simplexes s', s'' of (B, T') can therefore be joined by consecutive $(n-1)$- and n-simplexes of (B, T'), and the trace of $u \in W^{1,2}(B)$ on any such $(n-1)$-simplex is the same from both sides. Consequently, u is constant (a.e.) on B, and in fact $u = 0$ on B because $\overline{u} = \lim_j \overline{u}_j = 0$. Thus $u_j \to 0$ in $L^2(B)$, contradicting the equation in (5.7). This proves (5.6), and so rules out the possibility $r > 0$.

Step 2. We denote by s the carrier of x (i.e., x is an inner point of the simplex s). After embedding the star $\mathrm{st}(s^\circ) = \mathrm{st}(x)$ in a Euclidean space V with origin at x and norm $|\cdot|$, we denote by p: $V \to \mathbb{R}s$ the orthogonal projection on $\mathbb{R}s$. Choose $\varrho > 0$ so that $B^e(x, 4\varrho) \Subset \mathrm{st}(x)$. Because $x_j \to x$ $(= 0)$ and $p(x_j) \to p(x) = x$, we may assume that $x_j \in B^e(x, \varrho)$ $(\subset \mathrm{st}(x))$ and $p(x_j) \in s^\circ \cap B^e(x, \varrho)$. Denote by $\psi_j : V \to V$ the homothety of V with centre $p(x_j)$ and factor

$$\alpha_j = \frac{1}{\varrho} \max\{|x_j - p(x_j)|, r_j\} > 0. \tag{5.8}$$

Writing $x'_j = \psi_j^{-1}(x_j)$ $(\in V)$, we have

$$x'_j - p(x_j) = \alpha_j^{-1}(x_j - p(x_j)), \tag{5.9}$$

and by (5.8)

$$|x'_j - p(x_j)| = \alpha_j^{-1}|x_j - p(x_j)| \leqslant \varrho. \tag{5.10}$$

Since $|p(x_j| \leqslant \varrho$ it follows from (5.10) that $|x'_j| \leqslant 2\varrho$, which in turn implies that x'_j lies in $K' := \{z \in V : |z| \leqslant 2\varrho\}$, a compact subset of $\mathrm{st}(x)$ (by the choice of ϱ), and so is $L' := \{z \in V : |z| \leqslant 3\varrho\}$.

Since the centre $p(x_j)$ of ψ_j is in s and $\alpha_j < 1$ we have $\psi_j(s) \subset s$, $\psi_j(\mathrm{st}(x)) \subset \mathrm{st}(x)$. Writing $r'_j := \alpha_j^{-1} r_j \leqslant \varrho$, by (5.8), and recalling the notation $B_j = B^e(x_j, r_j)$, we see that

$$\begin{aligned} B'_j := \psi_j^{-1}(B_j) &= \{z \in V : |z - x'_j| \leqslant r'_j\} \\ &\subset \{z \in V : |z| \leqslant 3\varrho\} = L' \subset \mathrm{st}(x), \end{aligned}$$

and so $B'_j = B^e(x'_j, r'_j)$ $(\subset \mathrm{st}(x))$.

Writing $u'_j = u_j \circ \psi_j \in W^{1,2}(B'_j)$, (5.5) transforms under ψ_j^{-1} into

$$\int_{B^e(x'_j, \frac{1}{2}r'_j)} |u'_j - \overline{u}'_j|^2 d\mu_e > j{r'_j}^2 \int_{B^e(x'_j, r'_j)} |\nabla_e u'_j|^2 d\mu_e. \tag{5.11}$$

Because $x'_j \in K'$ and $r'_j \leqslant \varrho$ we may assume that $x'_j \to x'$ $(\in K' \subset \mathrm{st}(x))$ and $r'_j \to r' \in [0, \varrho]$. Write $X' = \mathrm{st}(x)$. From Step 1, now with the "dashed" quantities, we infer from (5.11) that $r'_j \to 0$. We may therefore assume that $r'_j < \varrho$, and so r_j $(= \alpha_j r'_j) < \alpha_j \varrho = |x_j - p(x_j)|$, by (5.8). Since $p(x_j) \to 0$ it therefore follows from (5.10) that

$$|x'| = \lim |x'_j - p(x_j)| = \lim \alpha_j^{-1} |x_j - p(x_j)| = \varrho.$$

Moreover, $x' = \lim_j (x'_j - p(x_j))$ is orthogonal to $\mathbb{R}s$ because $x'_j - p(x_j)$ is so for each j, by (5.9). It follows that $x' \in \mathrm{st}(x) \setminus V_s$, and hence the carrier s' of x' in $X' = \mathrm{st}(x)$ has $\dim s' > \dim s$. Iterating the above procedure at most $n - \dim s$ times therefore leads to an absurdity. This proves (5.4), and thereby the stated Poincaré inequality. □

Weakly harmonic and weakly sub/superharmonic functions

Let (X, g) be an admissible Riemannian polyhedron of dimension n.

Definition 5.1. A function $u \in W^{1,2}_{\mathrm{loc}}(X)$ is said to be *weakly harmonic* if

$$\int_X \langle \nabla u, \nabla \varphi \rangle \, d\mu_g = 0 \qquad \text{for every } \varphi \in \mathrm{Lip}_c(X).$$

Definition 5.2. A function $u \in W^{1,2}_{\mathrm{loc}}(X)$ is said to be *weakly subharmonic*, resp. *weakly superharmonic*, if

$$\int_X \langle \nabla u, \nabla \varphi \rangle \, d\mu_g \leqslant 0, \text{ resp. } \geqslant 0, \qquad \text{for every } \varphi \in \mathrm{Lip}_c^+(X).$$

In these three definitions, $\mathrm{Lip}_c(X)$ and $\mathrm{Lip}_c^+(X)$ can be replaced equivalently by $W^{1,2}_c(X)$, resp. $W^{1,2}_c(X)^+$, the latter by Proposition 5.1 (d). If $u \in W^{1,2}(X)$ then $\mathrm{Lip}_c(X)$ and $\mathrm{Lip}_c^+(X)$ can similarly be replaced by $W^{1,2}_0(X)$, resp. $W^{1,2}_0(X)^+$.

Every constant function is weakly harmonic. If X is compact then every weakly harmonic function $u \in W^{1,2}(X)$ is constant (a.e.). Indeed, $u \in W^{1,2}_0(X)$, hence $E(u, u) = 0$, by Definition 5.1 and the above observation, and so u is constant, as noted in text following (5.3).

The concepts of weakly harmonic and weakly sub/superharmonic functions are *local* concepts. For example, the restriction $u_{|U}$ of a weakly harmonic function u on X to an open set U is clearly weakly harmonic (strictly speaking: on each component) in U. And if a function u is weakly harmonic in each member of a family of open sets then so it is in their union U; in particular, u is of class $W^{1,2}_{\text{loc}}(U)$. (This is obtained using a Lip partition of unity on U, subordinate to a covering as above, cf. the first section of Chapter 4.)

A classical variational argument (cf. e.g., [KS 1993, p. 622]) leads to the following characterization of weakly harmonic functions of class $W^{1,2}(X)$ as local E-minimizers:

Proposition 5.2. *A function $u \in W^{1,2}(X)$ is weakly harmonic if and only if u minimizes the energy $E(v)$ among all functions $v \in W^{1,2}(X)$ such that $v - u \in W_0^{1,2}(X)$.*

Remark 5.3. Under suitable smoothness assumptions, weak harmonicity and subharmonicity can be characterized in more classical terms. Consider a domain $U \subset X$ which meets exactly one $(n-1)$-simplex t. Let $s_1, \cdots, s_k$ denote the n-simplexes of X containing t, and suppose that the Riemannian metric g is continuous and has smooth restrictions to each $s_j^\circ \cap U$ up to the boundary $t^\circ \cap U$ (i.e., g_U is continuous and simplewise smooth, cf. Chapter 4, Riemannian polyhedron). Consider a continuous function u on U of class C^2 in each $s_j^\circ \cap U$ and of class C^1 in each $(s_j^\circ \cup t^\circ) \cap U$. Then u is weakly harmonic (resp. subharmonic) in U if and only if (i) u is weakly harmonic (resp. subharmonic) in each $s_j^\circ \cap U$, $j = 1, \cdots, k$, and (ii)

$$\sum_{j=1}^{k} D_j u(x) = 0 \quad (\text{resp. } \geqslant 0)$$

at almost every point x of $t \cap U$. Here $D_j u(x)$ denotes the inner normal derivative of $u_{|s_j \cap U}$ at x. (Proof by Green's formula.) Cf. Example 8.1 for the 1-dimensional case.

The following maximum principles, including Theorem 5.3 below, are due to R.-M. Hervé [Her 1964, §1] for the particular case where X is a bounded domain in $\mathbb{R}^n$ carrying a Riemannian metric with measurable components and elliptic bounds.

Proposition 5.3. *Let $u \in W^{1,2}(X)$ be weakly subharmonic on X. In either of these two cases it follows that $u \leqslant 0$ (a.e. in X):*

(i) *$u \in W_0^{1,2}(X)$ and $1 \notin W_0^{1,2}(X)$, or*

(ii) *X is non-compact and $\operatorname{ess\,lim\,sup}_{x \to \infty_X} u(x) \leqslant 0$.*

Proof. If u is in $W_0^{1,2}(X)$, as in case (i), then in view of Proposition 5.1 (c) so is $|u|$, $\mathrm{Lip}_c(X)$ being dense in $W_0^{1,2}(X)$. In case (ii), u_+ is in $W^{1,2}(X)$ by Proposition 5.1 (a), and in $W_0^{1,2}(X)$ by Lemma 5.3. Thus, in either case, $u_+ \in W_0^{1,2}(X)$. Taking $\varphi = u_+$ in Definition 5.2 we obtain $E(u, u_+) \leqslant 0$, and so $E(u_+, u_+) = 0$ because $E(u_-, u_+) = 0$ by (5.2) and Remark 5.2. Consequently u_+ is constant (see text following (5.3)), and indeed $u_+ = 0$ in either case (i) or (ii). □

It follows from this proposition that a weakly *harmonic* function u of class $W^{1,2}(X)$ is $\equiv 0$ if $\operatorname{ess\,lim}_{x \to \infty_X} u(x) = 0$ (assuming that X is noncompact).

Remark 5.4. Proposition 5.3 extends, as in [Her 1964], to weakly subharmonic functions u of class $W_{\mathrm{loc}}^{1,2}(X)$. In case (ii) there is, for any $\varepsilon > 0$, a compact set $K \subset X$ such that $u \leqslant \varepsilon$ a.e. in $X \setminus K$. With $U \supset K$ open connected relatively compact in X and containing a given point $x_0 \in X$, it follows that $u - \varepsilon \leqslant 0$ (a.e.) in $U \setminus K$, hence in U because $u - \varepsilon$ is subharmonic and of class $W^{1,2}(U)$; and so, in particular, $u(x_0) \leqslant \varepsilon$. As to case (i), see Theorem 5.3 below, applied to $v = u - \varepsilon$ and $f = 0$.

Theorem 5.2. (The variational Dirichlet problem.) *Suppose the following Poincaré inequality holds:*

$$\int_X u^2 \leqslant c \int_X |\nabla u|^2 \qquad \text{for all } u \in W_0^{1,2}(X) \tag{5.12}$$

(or equivalently for all $u \in \mathrm{Lip}_c(X)$), with c depending only on the admissible Riemannian polyhedron X. Then $1 \notin W_0^{1,2}(X)$ and so X is non-compact.

For any $f \in W^{1,2}(X)$ the class of competing maps

$$W_f^{1,2}(X) = \{v \in W^{1,2}(X) : v - f \in W_0^{1,2}(X)\}$$

contains a unique weakly harmonic function u. That function is the unique solution u of the equation $E(u) = E_0$, where

$$\begin{aligned} E_0 &:= \inf\{E(v) : v \in W^{1,2}(X),\ v - f \in \mathrm{Lip}_c(X)\} \\ &= \min\{E(v) : v \in W_f^{1,2}(X)\}. \end{aligned} \tag{5.13}$$

If $f \geqslant 0$ off some compact subset of X, then $u \geqslant 0$. If instead f is weakly subharmonic, then $u \geqslant f$.

Proof. First note that $1 \notin W_0^{1,2}(X)$, by (5.12). In particular, X is not compact (for then $1 \in W^{1,2}(X) = W_0^{1,2}(X)$). Consider a sequence (u_j) in $W^{1,2}(X)$ such that

$$u_j - f \in \mathrm{Lip}_c(X), \quad E(u_j) \to E_0.$$

with E_0 defined by the former equation (5.13). Then $\frac{1}{2}(u_i+u_j)$ also competes for E_0, and by the parallelogram identity

$$E(u_i - u_j) \leqslant 2E(u_i) + 2E(u_j) - 4E_0 \to 0 \qquad \text{as } i, j \to \infty,$$

(∇u_j) is a Cauchy sequence in $L^2(X)$ and therefore convergent. Since

$$u_i - u_j = (u_i - f) - (u_j - f) \in \mathrm{Lip}_c(X),$$

the Poincaré inequality (5.12) shows that (u_j) is itself a Cauchy sequence in $L^2(X)$, hence also in $W^{1,2}(X)$, and so has a limit $u \in W^{1,2}(X)$. Moreover, $u - f = \lim_j (u_j - f) \in W_0^{1,2}(X)$, and $E(u) = \lim E(u_j) = E_0$, so that u is *minimizing* for the variational problem (5.13), and so u is *weakly harmonic*, by Proposition 5.2.

For any two weakly harmonic functions u, u' of class $W_f^{1,2}(X)$ we have $E(u - u') = 0$ by Definition 5.1 because $u - u'$ is weakly harmonic and of class $W_0^{1,2}(X)$. Consequently, $u - u'$ is constant, as mentioned earlier (after (5.3)), and $u - u' = 0$ because $1 \notin W_0^{1,2}(X)$. This establishes the uniqueness assertions.

Next, suppose that $\mathrm{supp}\, f_-$ is compact, and again let u denote the minimizing function. Then $|u| \in W^{1,2}(X)$ and $\|\,|u|\,\| = \|u\|$, by Proposition 5.1 (a), (b) and hence

$$E(|u|) = E(u) \ \ (= E_0).$$

It remains to show that $f - |u| \in W_0^{1,2}(X)$, from which it follows that $u = |u|$ $(\geqslant 0)$ by the latter uniqueness property above.

Write $f - u = \varphi$ $(\in W_0^{1,2}(X))$, and choose functions $(\varphi_j) \subset \mathrm{Lip}_c(X)$ so that $\|\varphi_j - \varphi\| \to 0$. Then

$$u_j := f - \varphi_j \to f - \varphi = u \qquad \text{in } W^{1,2}(X),$$

and consequently $|u_j| \to |u|$ in $W^{1,2}(X)$ by Proposition 5.1 (c). It follows that

$$f - |u| = \lim_j \,(f - |u_j|) \in W_0^{1,2}(X)$$

because $f - |u_j| = f - |f - \varphi_j|$ has compact support ($\subset \operatorname{supp}\varphi_j \cup \operatorname{supp} f_-$).

Finally, if instead f is weakly subharmonic, then so is $f - u \in W_0^{1,2}(X)$, and hence $f - u \leqslant 0$ by case (i) of Proposition 5.3. □

Remark 5.5. Note that this type of Poincaré inequality (5.12) is different from that of Theorem 5.1. After a suitable subdivision of a given triangulation of X, the inequality (5.12) holds for the (open) star $\operatorname{st}(a)$ of any prescribed point $a \in X$. In fact, with $S := \overline{\operatorname{st}}(a)$, consider the Lip homeomorphism $\theta : S \to P$ from Lemma 4.3. Every n-cell Q of P has an $(n-1)$-face F not containing 0, that is, $F \subset \partial P = P \setminus P^\circ$, cf. Remark 4.2. This establishes the Poincaré inequality (5.12) for P°, and hence for $\operatorname{st}(a) = \theta^{-1}(P^\circ)$, because it is known that (5.12) holds with X replaced by the above n-cube Q (e.g., endowed with the Euclidean metric) for any $u \in W^{1,2}(Q)$ vanishing near the above $(n-1)$-face F; cf. the proof of the Poincaré inequality for a strip in $\mathbb{R}^n$ in [DL 1954, p. 318].

By Theorem 5.2, writing $Lf = u$, the operator $L : W^{1,2}(X) \to W^{1,2}(X)$ is *linear* and *positive*, hence increasing, with $\ker L = W_0^{1,2}(X)$; and we have $f \leqslant Lf$ (a.e.) if $f \in W^{1,2}(X)$ is weakly subharmonic. Furthermore, L is *continuous* in the Sobolev norm $\|\cdot\|$ because $f \in W_f^{1,2}(X)$ and hence $E(u) \leqslant E(f)$. Indeed, (5.12) implies

$$\|Lf\|^2 = 2E(u) + \int_X u^2 \leqslant (1+c)2E(u) \leqslant (1+c)2E(f) \leqslant (1+c)\|f\|^2.$$

The following *maximum principle* for weakly subharmonic functions is carried over from [Her 1964, Théorème 1], with a simplified proof:

Theorem 5.3. *Let X be an admissible Riemannian polyhedron which satisfies the Poincaré inequality (5.12). If a weakly subharmonic function $v \in W_{\mathrm{loc}}^{1,2}(X)$ is majorized off some compact set $K \subset X$ by a function $f \in W_0^{1,2}(X)$, then $v \leqslant 0$ a.e. in X.*

Proof. If $v \in W_0^{1,2}(X)$ we have in fact from the above:

$$v \leqslant Lv \leqslant Lf = 0 \quad \text{a.e. in } X$$

because $f - v \geqslant 0$ in $X \setminus K$ and $0 \in W_f(X)$.

In the general case, note first that $E(\cdot)^{\frac{1}{2}}$ is a norm on $W_0^{1,2}(X)$ equivalent to the Sobolev norm $\|\cdot\|$ according to the Poincaré inequality (5.12). For any connected open set $U \subset X$ we view $W_0^{1,2}(U)$ as a closed linear subspace of the Hilbert space $W_0^{1,2}(X)$ with the inner product $E(\cdot,\cdot)$, noting that the extension of any $v \in W_0^{1,2}(U)$ by $v = 0$ in $X \setminus U$ is of class $W_0^{1,2}(X)$. The orthogonal complement $H(U)$ of $W_0^{1,2}(U)$ within $W_0^{1,2}(X)$ then consists of all $u \in W_0^{1,2}(X)$ for which $u_{|U}$ is *weakly harmonic* in U, as seen from Definition 5.1. Let P_U denote the operator projecting $W_0^{1,2}(X)$ onto $H(U)$ with respect to $E(\cdot,\cdot)$. For any $f \in W_0^{1,2}(X)$ we have $f - P_U f \in H(U)^\perp = W_0^{1,2}(U)$, and so

$$(P_U f)_{|U} = L_U f_{|U}, \tag{5.14}$$

where L_U is the operator on $W_0^{1,2}(U)$ corresponding to L above, now with U in place of X.

Given a subharmonic function $v \in W_{\mathrm{loc}}^{1,2}(X)$ and a function $f \in W_0^{1,2}(X)$ such that $v \leqslant f$ off a compact set $K \subset X$, we exhaust X by an increasing sequence of relatively compact subdomains $U_j \supset K$ as in Remark 5.4; then

$$\bigcap_j H(U_j) = 0.$$

Indeed, if a function $u \in W_0^{1,2}(X)$ is orthogonal to all $W_0^{1,2}(U_j)$ then u is weakly harmonic in all U_j, and hence in X, again by Definition 5.1; and so $u = 0$ because $E(u,u) = 0$ and $u \in W_0^{1,2}(X)$, while $1 \notin W_0^{1,2}(X)$, by (5.12). It follows that $P_{U_j} f \to 0$, i.e., $E(P_{U_j} f) \to 0$ and hence

$$\int (P_{U_j} f)^2 \to 0$$

by (5.12). After extracting a subsequence we may assume that $P_{U_j} f \to 0$ pointwise a.e. in X. Applying Theorem 5.2 (in U_j) and the identity (5.14), we now obtain a.e. for $x \in U_i$ (i fixed) and for $j > i$

$$v(x) = v_{|U_j}(x) \leqslant (L_{U_j} f_{|U_j})(x) = (P_{U_j} f)(x) \to 0 \qquad \text{as } j \to \infty,$$

and so $v \leqslant 0$ a.e. in each U_i, hence a.e. in X. □

Unique continuation of harmonic functions

We close this chapter with an immediate extension to Riemannian *circuits* of the *unique continuation theorem* for harmonic functions on a manifold with smooth or just Lipschitz continuous, Riemannian metric, see [AKS 1962] and [K 1988]. With reference to Theorem 6.2 we call a function $u : X \to \mathbb{R}$ *harmonic* if u is *continuous* and weakly harmonic.

Recall from Chapter 4, Circuit, that if X is a Lip circuit with singular set S then $X \setminus S$ is a Lip manifold (cf. Example 3.6). Furthermore, a continuous Riemannian structure g on X is said to be *Lip continuous* if, for every topdimensional simplex s in some triangulation of X, the components of g_s are Lipschitz continuous in s up to the boundary.

Proposition 5.4. *Let X denote a normal Riemannian n-circuit, and suppose that the Riemannian structure g on X is Lipschitz continuous. Then every harmonic function u which vanishes identically in a non-void open set $U \subset X$ vanishes in all of X.*

Proof. This follows immediately from the unique continuation theorem for manifolds, quoted above, when applied to the dense n-dimensional submanifold $X \setminus S$ of X. □

Remark 5.6. In the above proposition one cannot replace a normal circuit by an admissible polyhedron X (not even in dimension 1): Let X denote the admissible Riemannian polyhedron consisting of three n-dimensional Euclidean half-spaces X_1, X_2, X_3 meeting in a common $(n-1)$-space X_0, and consider the harmonic function u on X (cf. Remark 5.3) defined by $u = 0$ in X_3, and $u(x) = (-1)^j$ times the Euclidean distance between x and X_0 for $x \in X_j$, $j = 1, 2$. For $n = 1$ this example is due to A. Cornea (unpublished).

Remark 5.7. For $n \leqslant 2$ the hypothesis in the above proposition that g be Lip continuous can be dropped; this is elementary for $n = 1$, and for $n = 2$ it follows from the unique continuation of complex solutions to the Beltrami equation with bounded measurable coefficients [BNir 1955], see also [BJS 1964]. When $n > 2$, that hypothesis cannot be removed, not even in the case of a triangulable Riemannian Lipschitz manifold X; this appears from an example of non-uniqueness given (for $n = 3$) in [Mi 1974], whereby the Riemannian metric on the manifold X can be Hölder continuous with any given precise exponent < 1.

6

Harnack inequality and Hölder continuity for weakly harmonic functions

As mentioned in Chapter 1, Theorems 6.1, 6.2 and 6.3, based primarily on ideas of Moser and De Giorgi, could alternatively be deduced from corresponding results obtained by Biroli and Mosco [BM 1995]. This is because the Dirichlet space $W^{1,2}(X)$ on an admissible Riemannian polyhedron X has locally the properties required in [BM 1995], in particular the ball doubling property and the weak Poincaré inequality (Corollary 4.1 and Theorem 5.1).

Theorem 6.1. *If $u \geqslant 0$ is weakly harmonic on an admissible Riemannian polyhedron $X = (X, g)$, then*

$$\operatorname*{ess\,sup}_K u \leqslant c \operatorname*{ess\,inf}_K u$$

for every nonempty compact set $K \subset X$, whereby the constant c depends on (X, g) and K only.

More precisely: For any admissible polyhedron X, for any compact set $K \subset X$, and for any number $\Lambda > 0$, there exists a constant $c = c(X, K, \Lambda)$ such that the stated Harnack inequality holds for any function $u \geqslant 0$ on X which is weakly harmonic with respect to some Riemannian structure g on X with ellipticity constant $\leqslant \Lambda$.

The proof of the theorem in this amplified form will be adapted from [Mos 1961], covering the particular case of a domain X in $\mathbb{R}^n$ carrying a Riemannian metric with bounded measurable components and elliptic bounds. Clearly, we may suppose that $u \in W^{1,2}(X)$ (rather than just $W^{1,2}_{\mathrm{loc}}(X)$).

Proof of Theorem 6.1 in the locally bounded case

Lemma 6.1. *Let $u \in W^{1,2}_{\mathrm{loc}}(X)$ be weakly harmonic, and suppose that $u > 0$ a.e. and that u and $1/u$ are essentially bounded. Writing*

$$v = u^{p/2}, \qquad p \in \mathbb{R},\ p \neq 1,$$

we have

$$\int_X \eta^2|\nabla v|^2 \leqslant \Big(\frac{p}{p-1}\Big)^2 \int_X |\nabla\eta|^2 v^2, \qquad \eta \in \mathrm{Lip}_c(X).$$

Remark 6.1. The essential boundedness of $u \in W^{1,2}_{\mathrm{loc}}(X)$ and $1/u$ insures that $u^{p/2} \in W^{1,2}_{\mathrm{loc}}(X)$. In [Mos 1961] no such boundedness assumption is mentioned, but the tools needed in order to remove this extra assumption, as we do in the next section, are all present in combination with the earlier paper [Mos 1960]. For $p = 2$ the essential boundedness of u and $1/u$ can of course be dropped because $v = u$; and in that case the lemma holds with a similar proof for a weakly harmonic *signed* function $u : X \to \mathbb{R}$ (this is known in the case where X is a domain in $\mathbb{R}^n$).

Remark 6.2. The proof of Lemma 6.1 given below shows that the lemma extends to the case of a weakly *subharmonic* function u of class $W^{1,2}_{\mathrm{loc}}(X)$ if $p > 1$, resp. weakly *superharmonic* if $p < 1$ (keeping the assumptions that $u > 0$ and that u and $1/u$ be essentially bounded in either case).

Proof of Lemma 6.1. We may assune that $p \neq 0$. In Definition 5.1 take

$$\varphi = u^{p-1}\eta^2 = v^{2-2/p}\,\eta^2,$$

which is of class $W^{1,2}_0(X)$ by Remark 5.1 (a). We obtain

$$\nabla\varphi = (2 - 2/p)v^{1-2/p}\,\eta^2\nabla v + 2v^{2-2/p}\,\eta\nabla\eta,$$

$$\nabla u = \tfrac{2}{p}v^{2/p-1}\nabla v,$$

$$0 = \int_X \langle\nabla u, \nabla\varphi\rangle = \frac{4}{p}\Big(1 - \frac{1}{p}\Big)\int_X \eta^2|\nabla v|^2 + \frac{4}{p}\int_X \langle\eta v\nabla\eta, \nabla v\rangle,$$

$$\int_X \eta^2|\nabla v|^2 = -\frac{p}{p-1}\int_X \langle\eta\nabla v, v\nabla\eta\rangle,$$

whence the asserted inequality by Schwarz' inequality. □

Now to the proof of Theorem 6.1 for locally bounded u. Fix a point $a \in X$. By a subdivision of the given triangulation of X we arrange that a is a vertex and that $\mathrm{st}(a)$ is the interior of its compact closure S, cf. Remark

4.2. Recall from Lemma 4.3 the bi-Lip homeomorphism θ of (S, g) onto (the set underlying) the n-dimensional cell complex

$$P = \theta(S) \subset [0, 1]^k,$$

whereby $\theta(a) = 0$, and k is the number of 1-simplexes of X containing a. Here P is endowed with the Euclidean Riemannian structure (the Euclidean metric on each n-cell), and the n-cells are Euclidean n-cubes of unit edge-length.

We denote by λ the bi-Lip constant of $\theta : S \to P$ (the larger of the Lip constants of θ and θ^{-1}), whereby P is endowed as above with the Euclidean Riemannian structure and the associated distance function d_P^e, while S is equipped with the intrinsic distance induced by the given Riemannian structure g on X (restricted to S). In terms of the ellipticity constant Λ for S, cf. (4.1) and (4.2), we clearly have

$$1 \leqslant \lambda \leqslant |\theta| \Lambda, \tag{6.1}$$

where $|\theta|$ denotes the bi-Lip constant of θ, now with the Euclidean Riemannian structure on both S and P. Thus $|\theta|$ depends only on X as a mere polyhedron, and on the point a. Write

$$S_r = \theta^{-1}(rP), \quad \text{hence } S_r^\circ = \theta^{-1}(rP^\circ), \qquad 0 < r \leqslant 1, \tag{6.2}$$

where S_r° denotes the interior of S_r in X, and $P^\circ = \theta(S^\circ)$ that of P in $\mathbb{R}_+ P$ (cf. Remark 4.2). In particular, $S_1 = S$.

Let now r, $0 < r \leqslant 1$, be given, and consider a weakly harmonic function

$$u \in W^{1,2}(S_r^\circ), \quad u > 0 \text{ a.e. in } S_r^\circ. \tag{6.3}$$

Equivalently, $u \circ \theta^{-1}$ is weakly harmonic and > 0 a.e. in rP° ($\subset P$) when P (after simplicial subdivision of its cubic n-cells) is endowed with the Riemannian structure $(\theta^{-1})^* g$, g being the structure on $S \subset X$. For $0 < h < r$ and $p \in \mathbb{R} \setminus \{0\}$ we shall write

$$\Phi(p, h) = \left(\fint_{S_h} u^p \right)^{1/p} = \left(\fint_{hP} (u \circ \theta^{-1})^p \right)^{1/p},$$

the integration being with respect to the Riemannian volume measure on S_h, resp. its image measure on hP under θ. Here $\fint_E$ denotes the *meanvalue* over a measurable set E, that is, the integral divided by the (Riemannian) measure of E. Defining $\Phi(\infty, h) = \lim_{p\to\infty} \Phi(p, h)$ and $\Phi(-\infty, h) = \lim_{p\to-\infty} \Phi(p, h)$, we have

$$\Phi(\infty, h) = \operatorname*{ess\,sup}_{S_h} u, \quad \Phi(-\infty, h) = \operatorname*{ess\,inf}_{S_h} u. \tag{6.4}$$

The image under θ of the Riemannian volume measure on S has a density between λ^{-n} and λ^n with respect to n-dimensional Lebesgue measure on each n-cell of P. From Hölder's inequality it is clear that $\Phi(p, h)$ is increasing in $p \in \mathbb{R}$. Moreover, for $0 < h' < h < r$,

$$\begin{aligned} \Phi(p, h') &\leqslant \lambda^{2n/p}(h/h')^{n/p}\Phi(p, h) && \text{if } p > 0 \\ \Phi(p, h') &\geqslant \lambda^{-2n/|p|}(h/h')^{-n/|p|}\Phi(p, h) && \text{if } p < 0. \end{aligned} \tag{6.5}$$

The key to Theorem 6.1 (and subsequent comments) in the case where u is locally bounded is now the following local version, in which we keep the notation $S = \overline{\text{st}}(a)$, $\theta : S \to P$ from Lemma 4.3, and $S_r = \theta^{-1}(rP)$ $(0 < r \leqslant 1)$ from (6.2):

Proposition 6.1. *For any $a \in X$, $0 < r \leqslant 1$, and any essentially bounded weakly harmonic function $u \in W^{1,2}(S_r^\circ)$ with $u \geqslant 0$ a.e. in S_r° (the interior of $\theta^{-1}(rP)$) we have*

$$\operatorname*{ess\,sup}_{S_{r/5}} u \leqslant c_0 \operatorname*{ess\,inf}_{S_{r/5}} u.$$

Here c_0 (as well as subsequent constants $c_1, c_2, \cdots$) only depends on (X, g) and a, and more precisely on $n = \dim X$, $N =$ the number of n-simplexes containing a, and $\lambda =$ the bi-Lip constant of $\theta : S \to P$ with the given Riemannian structure on S $(\subset X)$ and the Euclidean Riemannian structure on P.

The dependence of c_0, say, on λ, and similarly on N, should be understood in analogy with the precise formulation given after Theorem 6.1. The uniformity not just in u, but also in r, and the confinement in (6.3) of the inequality $u > 0$ to S_r° will be crucial in the proof of Theorem 6.2 below. Since $S_{r/5}$ is a neighbourhood of a, Theorem 6.1 for locally essentially bounded u follows from the above proposition for fixed r (e.g., $r = 1$) by a standard compactness argument (not quite as in [Mos 1961, p. 582], which seems to be erroneous), cf. e.g., [Hel 1969, 2.16]. The comment relating to Theorem 6.1 about the way c depends on (X, g) likewise follows from Proposition 6.1 in view of (6.1) above. The choice of the denominator 5 (in $\frac{r}{5}$) in the proposition will be explained later. ([Mos 1961] uses 4 in place of 5, but this reflects an insignificant error, viz. the interchange of 2 and 3 in (4.5), p. 586 of the quoted paper.) Actually, an easy modification of the proof of Proposition 6.1 will show that $S_{r/5}$ could be replaced by $S_{\varrho r}$ for an arbitrary constant ϱ with $0 < \varrho < 1$; with this change, c_0 of course also depends on ϱ.

We begin by reducing Proposition 6.1 to the case $r = 1$. Using $\theta : S \to P$ we transform the problem to that of reducing the inequality

$$\operatorname*{ess\,sup}_{(r/5)P} v \leqslant c_0 \operatorname*{ess\,inf}_{(r/5)P} v, \qquad 0 < r \leqslant 1, \tag{6.6}$$

to the case $r = 1$, $v = u \circ \theta^{-1}$ denoting any essentially bounded function of class $W^{1,2}(rP^\circ)$ which is $\geqslant 0$ a.e. and weakly harmonic in rP° with respect to the image $g' := (\theta^{-1})^* g$ of the Riemannian structure g on S ($\subset X$). Note that the bi-Lip constant λ of $\theta : (S, g) \to (P, g'^e)$ (g'^e denoting the Euclidean Riemannian structure on P) equals the ellipticity constant of g' because $\theta : (S, g) \to (P, g')$ is an isometry. (See the paragraph containing (4.1), and Chapter 4, The intrinsic distance.)

For given r, $0 < r < 1$, use the homothety $\psi(\xi) = \frac{1}{r}\xi$ of rP on P to transport the structure $g'_{|rP}$ to $(\psi^{-1})^* g'$ on P. Then $v \circ \psi^{-1} \geqslant 0$ is weakly harmonic in P° with respect to the structure $(\psi^{-1})^* g'$, hence also with respect to $g'' := r^{-2}(\psi^{-1})^* g'$. Moreover, the ellipticity constant λ of g' majorizes that of g'' because $rP \subset P$, and (6.6) will therefore follow from the inequality

$$\operatorname*{ess\,sup}_{(1/5)P} (v \circ \psi^{-1}) \leqslant c_0 \operatorname*{ess\,inf}_{(1/5)P} (v \circ \psi^{-1})$$

because $\psi^{-1}(\frac{1}{5}P) = \frac{r}{5}P$.

In proving Proposition 6.1 in the remaining case $r = 1$, and thereby Theorem 6.1 (for locally bounded u), we assume to begin with *that* $u > 0$ *in* S° *and also that* $1/u$ *is essentially bounded* in S°, so that Lemma 6.1 applies. Of course, this additional assumption can be removed by applying the proposition to $u + \varepsilon$ and letting $\varepsilon \to 0$.

Fix a constant κ, $1 < \kappa < \frac{n}{n-2}$. (For $n > 2$ one might equally well take $\kappa = \frac{n}{n-2}$.) Following [Mos 1961] we shall establish the following analogues (6.7) and (6.8) of his Theorems 2 and 3, respectively, for the present case of an admissible Riemannian polyhedron X:

Lemma 6.2. *Suppose that $0 < r \leqslant 1$ and that the weakly harmonic function $u \in W^{1,2}(S_r^\circ)$ is positive, bounded and bounded away from 0. Then*

$$\Phi(\infty, r/5) \leqslant c_1\Big(\frac{p}{p-1}\Big)^2 \Phi(p, 2r/5), \qquad p > 1, \tag{6.7}$$

$$\Phi(p, 2r/5) \leqslant c_2 \frac{1}{(\kappa - p)^2} \Phi(-\infty, r/5), \qquad 0 < p < \kappa. \tag{6.8}$$

Applying (6.7), (6.8) with $r = 1$, e.g., to $p = \frac{1+\kappa}{2}$, leads to Proposition 6.1 in the remaining case $r = 1$, and hence to Theorem 6.1 under the temporary extra hypothesis that u be locally bounded. Lemma 6.2 reduces to the case $r = 1$ in the same manner as described above for Proposition 6.1.

Remark 6.3. Corresponding to what was established in [Mos 1961], the inequality (6.7) holds for any bounded weakly *subharmonic* function $u \geqslant 0$ on S_r°, and (6.8) for any bounded weakly *superharmonic* function $u \geqslant 0$ (the latter will be used for $p = 1$ in the proof of Lemma 10.1). In view of Remark 6.2 above, this will appear from the proof of the remaining case $r = 1$ in Lemma 6.2, for which we now prepare the following lemma, cf. [Mos 1961, (4.2), (4.3)]:

Lemma 6.3. *Suppose that the weakly harmonic function $u \in W^{1,2}(S^\circ)$ is positive, bounded, and bounded away from 0. For $0 < h' < h \leqslant 2h'$ and $h < 1$ we then have*

$$\Phi(\kappa p, h') \leqslant c_3^{1/p}\Big(\frac{h}{h'} - 1\Big)^{-2/p}\Big|\frac{p}{p-1}\Big|^{2/p} \Phi(p, h) \tag{6.9}$$

when $p > 0$ and $p \neq 1$, while for $p < 0$ we have

$$\Phi(\kappa p, h') \geqslant c_3^{-1/|p|}\Big(\frac{h}{h'} - 1\Big)^{2/|p|} \Phi(p, h). \tag{6.10}$$

Proof. We apply Lemma 6.1 with X replaced by $S^\circ = \theta^{-1}(P^\circ)$, taking $\eta = \widetilde{\eta} \circ \theta_{|S^\circ}$, where

$$\widetilde{\eta}(x) = \max\big\{0, 1 - \tfrac{1}{h-h'} d_P^e(x, h'P)\big\}, \quad x \in P^\circ.$$

As explained earlier, d_P^e denotes the distance function on the length space P endowed with the Euclidean Riemannian metric on each n-cell of P (cf.

Chapter 4, The intrinsic distance). Clearly, $\widetilde{\eta} \in \mathrm{Lip}(P^\circ)$ has support in hP, and $|\nabla\widetilde{\eta}| \leqslant \frac{1}{h-h'}$. Hence $\eta \in \mathrm{Lip}_c(S^\circ)$ has support in S_h ($\subset S^\circ$), and

$$|\nabla\eta| \leqslant \frac{\lambda}{h-h'}.$$

Lemma 6.1, applied within S°, therefore yields for $p \neq 1$

$$\int_{S_{h'}} |\nabla v|^2 \leqslant \Big(\frac{\lambda}{h-h'}\Big)^2 \Big(\frac{p}{p-1}\Big)^2 \int_{S_h} v^2,$$

where $v = u^{p/2}$. Writing $v \circ \theta^{-1} = \widetilde{v}$, we infer that

$$\int_{h'P} |\nabla\widetilde{v}|^2 \leqslant \lambda^{2n+2} \Big(\frac{1}{h-h'}\Big)^2 \Big(\frac{p}{p-1}\Big)^2 \int_{hP} \widetilde{v}^2.$$

By application of Sobolev's inequality in each n-cell of $h'P$, as in [Mos 1961, p. 584], this leads to (6.9) after summation over these cells. Similarly, one obtains (6.10), noting that $0 < \frac{p}{p-1} < 1$ when $p < 0$. □

As in [Mos 1961, p. 585], (6.9) leads to (6.7) by iteration in the remaining case $r = 1$. (Replace h_ν, h'_ν by $\frac{1}{5}h_\nu$, $\frac{1}{5}h'_\nu$, $\nu = 0, 1, \cdots$.) Similarly, (6.10) leads to the following first step towards (6.8) (with $r = 1$):

$$c_4\Phi(-\infty, \tfrac{1}{5}) \geqslant \Phi(-q, \tfrac{2}{5}) \tag{6.11}$$

for a suitable $q > 0$, depending on n, N, λ only. Next, for $q < p < \kappa$, apply (6.9) successively to $p_\nu = p/\kappa^{\mu-\nu+1}$ ($\nu = 0, \cdots, \mu$), where the integer $\mu \geqslant 0$ is defined by $p/\kappa^\mu > q \geqslant p/\kappa^{\mu+1}$. This leads to

$$\Phi(p, \tfrac{2}{5}) \leqslant \frac{c_5}{(\kappa-p)^2}\Phi(p/\kappa^{\mu+1}, \tfrac{3}{5}) \leqslant \frac{c_5}{(\kappa-p)^2}\Phi(q, \tfrac{3}{5}), \quad 0 < p < \kappa, \tag{6.12}$$

noting that we may assume $q < p$, the resulting inequality (6.12) being trivial for $q \geqslant p$, in view of the former inequality (6.5) (with q in place of p).

As in [Mos 1961, p. 586] one completes the proof of (6.8), hence altogether of Lemma 6.2, by establishing

$$\Phi(q, \tfrac{3}{5}) \leqslant c_6\Phi(-q, \tfrac{3}{5}) \leqslant c_7\Phi(-q, \tfrac{2}{5}); \tag{6.13}$$

and finally by combining (6.12), (6.13), and (6.11). The latter inequality (6.13) follows from the second inequality (6.5). For the former inequality (6.13) we set

$$v = \log u, \quad \varphi = \frac{1}{u}\eta^2 \qquad \text{in } S^\circ$$

with $\eta \in \mathrm{Lip}_c(S^\circ)$. Then $v \in W^{1,2}(S^\circ)$, $\varphi \in W_c^{1,2}(S^\circ)$ and

$$\langle \nabla u, \nabla \varphi \rangle = -\eta^2 |\nabla v|^2 + 2\eta \langle \nabla \eta, \nabla v \rangle.$$

Since u is weakly harmonic on S° it follows that

$$\int_{S^\circ} \eta^2 |\nabla v|^2 = 2 \int_{S^\circ} \eta \langle \nabla v, \nabla \eta \rangle,$$

or, using Schwarz' inequality,

$$\int_{S^\circ} \eta^2 |\nabla v|^2 \leqslant 4 \int_{S^\circ} |\nabla \eta|^2.$$

Inserting $\widetilde{\eta} = \eta \circ (\theta^{-1})_{|P^\circ}$ $(\in \mathrm{Lip}_c(P^\circ))$ and $\widetilde{v} := v \circ (\theta^{-1})_{|P^\circ}$ $(\in W^{1,2}(P^\circ))$, we have

$$\int_{(P,g^e)} \widetilde{\eta}^2 |\nabla \widetilde{v}|^2 \leqslant 4\lambda^{2n+4} \int_{(P,g^e)} |\nabla \widetilde{\eta}|^2, \tag{6.14}$$

where (P, g^e) means P endowed with its Euclidean Riemannian structure g^e and corresponding volume measure and gradient.

Fix for a while an n-cell F of P, and consider any "parallel" cube Q within the cube $\frac{3}{5}F$. For the above function $\eta \in \mathrm{Lip}_c(S^\circ)$ we choose $\eta = \widetilde{\eta} \circ \theta_{|S^\circ}$, where

$$\widetilde{\eta}(x) = \max \left\{0, 1 - \tfrac{3}{h} d_P^e(x, Q)\right\}, \qquad x \in P^\circ,$$

h $(< \frac{3}{5})$ denoting the edge-length of Q. Using that $\frac{3}{5} + \frac{h}{3} \leqslant \frac{4}{5}$ one finds that $\operatorname{supp} \widetilde{\eta} \subset \frac{4}{5}P$, [2] and so $\widetilde{\eta} \in \mathrm{Lip}_c(P^\circ)$, hence $\eta \in \mathrm{Lip}_c(S^\circ)$. For $x, y \in \operatorname{supp} \widetilde{\eta}$ we have

$$\begin{aligned} d_P^e(x, y) &\leqslant d_P^e(x, Q) + d_P^e(y, Q) + \operatorname{diam} Q \\ &\leqslant 2 \cdot \tfrac{h}{3} + h\sqrt{n} = O(h), \end{aligned}$$

[2] To see this, let $x \in \operatorname{supp} \widetilde{\eta}$. Then $d_P^e(x, \frac{3}{5}P) \leqslant d_P^e(x, \frac{3}{5}F) \leqslant d_P^e(x, Q) \leqslant \frac{h}{3} \leqslant \frac{1}{5}$ because $Q \subset \frac{3}{5}F$. Being compact, the length space (P, d_P^e) is a geodesic space. Let γ denote a shortest path in P from x to $\frac{3}{5}P$, and let $y \in \frac{3}{5}P$ be the terminal point of γ, so that $d_P^e(x, y) \leqslant \frac{1}{5}$. The star σ of y in $\frac{4}{5}P$ is a neighbourhood of y, hence contains some $z \in \gamma$, $z \neq y$. Let $\frac{4}{5}F'$ be an n-cell of σ (hence of $\frac{4}{5}P$) containing z (and of course y). Then the straight line segment $[z, y] \subset \frac{4}{5}F'$ is a final segment of γ. Since the assertion to be proved is easy to verify in the case where P is a single cube (in this case F'), we find that x belongs to the straight half-line in F' from y through z, and hence indeed $x \in \frac{4}{5}F' \subset \frac{4}{5}P$.

and so $\operatorname{supp}\widetilde{\eta}$ meets each n-cell of $\frac{4}{5}P$ in a set of diameter $O(h)$, hence of volume $O(h^n)$. Thus

$$|\operatorname{supp}\widetilde{\eta}| = O(h^n), \quad |\nabla\widetilde{\eta}| \leqslant \frac{3}{h},$$

and we obtain from (6.14), since $|\widetilde{\eta}| \leqslant 1$, $\widetilde{\eta} = 1$ on Q:

$$\fint_{(Q,g^e)} |\nabla\widetilde{v}|^2 \leqslant c_8 h^{-2},$$

where (Q, g^e) means the cube Q with its Euclidean gradient and volume measure. Writing $w_Q = \fint_{(Q,g^e)} w$ for any integrable function w on (Q, g^e), we infer by Poincaré's inequality (cf. e.g., [Mos 1961, Lemma 1]) that

$$\fint_{(Q,g^e)} (\widetilde{v} - \widetilde{v}_Q)^2 \leqslant c_9 \tag{6.15}$$

for every parallel cube $Q \subset \frac{3}{5}F$.

The John–Nirenberg inequality extends to functions defined in any finite union of non-overlapping n-cubes in $\mathbb{R}^n$ of equal size. (In the proof given in [JoNi 1961] just replace 2^n throughout by $N \cdot 2^n$, N being the number of cubes.) Applying the version of the inequality given in [Mos 1961, Lemma 3], thus extended, to the union of the n-cells $\frac{3}{5}F$ of $\frac{3}{5}P$ and to the function

$$w = c_9^{-1/2}\widetilde{v} = c_9^{-1/2} \log \widetilde{u}, \qquad \widetilde{u} = u \circ \theta^{-1},$$

leads in view of (6.15) to

$$\int_{(\frac{3}{5}P,g^e)} \widetilde{u}^q \int_{(\frac{3}{5}P,g^e)} \widetilde{u}^{-q} \leqslant \beta^2,$$

where $q = \alpha c_9^{-1/2}$, the positive constants α, β depending on n, N only.

Returning to integration with respect to the Riemannian measure on S, we infer (after pulling back to S) that

$$\int_{S_{3/5}} u^q \cdot \int_{S_{3/5}} u^{-q} \leqslant \beta^2 \lambda^{2n},$$

and by raising to the power $1/q$:

$$\Phi(q, \tfrac{3}{5}) \leqslant (\beta\lambda^n)^{2/q}\, \Phi(-q, \tfrac{3}{5}),$$

thus establishing the remaining former inequality (6.13). This proves Lemma 6.2 (as well as its extension in Remark 6.3), hence also Proposition 6.1, and consequently Theorem 6.1 for the case where $u \geqslant 0$ is locally essentially bounded in X. □

Completion of the proof of Theorem 6.1

We complete the proof by showing that *every weakly subharmonic function u on X is locally essentially bounded from above.* We may suppose that $u \in W^{1,2}(X)$, and also that $u \geqslant 0$ (otherwise replace u by $|u|$). It will follow that every weakly *harmonic* function u on X is locally essentially bounded (since $|u|$ is weakly subharmonic).

Let $2 \leqslant p < \infty$ be given. Inspired by [Mos 1960] we approximate as follows the convex even function $f(t) = |t|^{p/2}$, $t \in \mathbb{R}$, from below by an increasing family of convex even functions $f_m(t) \geqslant 0$, $m > 0$, having bounded continuous derivative $f'_m(t)$:

$$f_m(t) = \begin{cases} t^{p/2} & \text{for } 0 \leqslant t \leqslant m \\ m^{p/2} + \frac{p}{2} m^{p/2-1}(t-m) & \text{for } t \geqslant m, \end{cases}$$

i.e., by replacing for $t \geqslant m$ the graph of f by its tangent at m. Thus each f'_m is a bounded Lip function.

Given a weakly subharmonic function $u \geqslant 0$ of class $W^{1,2}(X)$ we write

$$\begin{aligned} v_m &= f_m \circ u && \in W^{1,2}(X), \\ \varphi_m &= [(f_m f'_m) \circ u]\, \eta^2 && \in W^{1,2}_c(X), \end{aligned}$$

for given $\eta \in \mathrm{Lip}_c(X)$. Each of the functions v_m and φ_m is indeed of class $W^{1,2}(X)^+$ because u is so and each of the functions $F = f_m$ and $F = f_m f'_m$ is uniformly Lipschitz with $F(0) = 0$, hence $F(t) \leqslant ct$ for $t \geqslant 0$, $c = \max F'$. For any function $u \in \mathrm{Lip}^{1,2}(X)^+$ it follows that $F \circ u \in \mathrm{Lip}^{1,2}(X)^+$ and $\|F \circ u\| \leqslant c\|u\|$; and in this statement we may clearly replace $\mathrm{Lip}^{1,2}(X)^+$ by $W^{1,2}(X)^+$, in view of Proposition 5.1 (d).

Write $v = u^{p/2}$, and for $0 < h < 1$, cf. the paragraph containing (6.3),

$$\Phi(p,h) = \left(\fint_{S_h} v^2\right)^{1/p} = \left(\fint_{S_h} u^p\right)^{1/p}, \tag{6.16}$$

$$\Phi_m(p,h) = \left(\fint_{S_h} v_m^2\right)^{1/p}. \tag{6.17}$$

For $m \nearrow \infty$ we have $f_m \nearrow f$ (pointwise), hence $v_m \nearrow v$, and so

$$\Phi_m(p,h) \nearrow \Phi(p,h) \qquad \text{as } m \nearrow \infty. \tag{6.18}$$

Proceeding as in the proof of Lemma 6.1 we obtain, since $\varphi_m \geqslant 0$,

$$\begin{aligned} 0 \geqslant \int_X \langle \nabla\varphi_m, \nabla u\rangle &= \int_X \left[(f_m f_m'' + (f_m')^2)\circ u\right] \eta^2 |\nabla u|^2 \\ &\quad + \int_X \left[(f_m f_m')\circ u\right] 2\eta\langle\nabla\eta, \nabla u\rangle. \end{aligned}$$

Inserting $f_m \circ u = v_m$ and $(f_m' \circ u)\nabla u = \nabla v_m$ in the second integral gives

$$\int_X \left[\left(\frac{f_m f_m''}{(f_m')^2} + 1\right)\circ u\right] \eta^2 |\nabla v_m|^2 \leqslant -2\int_X \eta v_m \langle\nabla\eta, \nabla v_m\rangle. \tag{6.19}$$

Here

$$\frac{f_m(t) f_m''(t)}{(f_m'(t))^2} + 1 = \begin{cases} 2 - 2/p \geqslant 1 & \text{for } t < m, \\ 1 & \text{for } t > m. \end{cases}$$

Applying Schwarz' inequality to the right hand side of (6.19) therefore leads to the following conclusion, similar to that of Lemma 6.1:

$$\int_X \eta^2 |\nabla v_m|^2 \leqslant 4\int_X v_m^2 |\nabla\eta|^2, \qquad \eta \in \mathrm{Lip}_c(X). \tag{6.20}$$

For the given weakly subharmonic function $u \geqslant 0$ of class $W^{1,2}(X)$ we now proceed exactly as in the proof of (6.7) in Lemma 6.2, replacing Lemma 6.1 by (6.20), and using the notation in (6.17). As in (6.9) this yields

$$\Phi_m(\kappa p, h') \leqslant c_{10}^{1/p}\left(\frac{h}{h'} - 1\right)^{-2/p} \Phi_m(p,h)$$

for $0 < h' < h \leqslant 2h'$ and $h < 1$. As in [Mos 1961, p. 585] it follows by iteration that

$$\Phi_m(\kappa^{\nu+1}p,\ h_{\nu+1}) \leqslant c_{11}\, \Phi_m(p, \tfrac{2}{5})$$

for $\nu = 0, 1, 2, \cdots$, whereby $h_\nu = \frac{1}{5}(1 + 2^{-\nu})$, hence $h_0 = \frac{2}{5}$. For $m \to \infty$ this leads to the same inequality with Φ_m replaced by Φ, cf. (6.17). Next, for $\nu \to \infty$, it follows as in [Mos 1961], in view of (6.16), (6.17) and the former inequality (6.5), that

$$\begin{aligned} \operatorname*{ess\,sup}_{S_{1/5}} u = \lim_{\nu\to\infty} \Phi(\kappa^{\nu+1}p, \tfrac{1}{5}) &\leqslant \liminf_{\nu\to\infty} \lambda^{2n/(\kappa^{\nu+1}p)} \Phi(\kappa^{\nu+1}p, h_{\nu+1}) \\ &\leqslant c_{11}\Phi(p, \tfrac{2}{5}). \end{aligned} \tag{6.21}$$

Because $u \in W^{1,2}(X) \subset L^2(X)$, $\Phi(p, \frac{2}{5})$ is finite e.g., for $p = 2$, and we have thus proved that *every weakly subharmonic function* $u : X \to \mathbb{R}$ *is locally essentially bounded from above,* viz. essentially bounded from above in a neighbourhood (denoted $S_{1/5} = \theta^{-1}(\frac{1}{5}P)$) of any point $a \in X$. □

Anticipating Definition 7.1 of *harmonic* functions as *continuous* weakly harmonic functions (with reference to Theorem 6.2 below) we derive in the following proposition a Harnack-style inequality for harmonic functions in the complement of a variable point y. The result will be used in the study of the Green function on X in Chapter 7. As in Chapter 4 we denote by $B^e(y,r) = B^e_X(y,r)$ the closed ball centred at y and of radius r for the distance function d^e_X associated with the Euclidean Riemannian structure g^e on X (induced by the metric on a Euclidean space in which X is affinely embedded).

Proposition 6.2. *Let (X, g) be an admissible Riemannian polyhedron of dimension $n > 1$ and ellipticity constant $\leqslant \Lambda$. For any compact proper subset K of X, any number $\varrho > 0$ such that $L := \{x \in X : d^e_X(x, K) \leqslant 2\varrho\}$ is a proper subset of X, any $y \in K$ and any harmonic function $u \geqslant 0$ in $X \setminus \{y\}$, we have*

$$\max_{\partial B^e(y,r)} u \leqslant c \min_{\partial B^e(y,r)} u \qquad \text{for } r \leqslant \varrho,$$

where $c = c(X, K, \Lambda, \varrho)$ is independent of y, r and u.

Proof. Clearly, $B^e(y,r) \neq X$, and so $\partial B^e(y,r) \neq \emptyset$. Suppose there is no such constant c. Then there is, for every natural number j, a ball $B_j = B^e(y_j, r_j)$ in X, $y_j \in K$, $r_j \leqslant \varrho$, and a harmonic function $u_j \geqslant 0$ in $X \setminus \{y_j\}$ with respect to some Riemannian structure g_j on X with ellipticity constant $\leqslant \Lambda$, such that

$$\max_{\partial B_j} u_j > j \min_{\partial B_j} u_j. \tag{6.22}$$

We may assume that $y_j \to y \in K$ and $r_j \to r \in [0, \varrho]$.

Step 1. The case $r > 0$. We may then further assume that $d^e_X(y_j, y) < r/4$ and $|r_j - r| < r/4$. It follows that $\partial B_j \subset A := L \setminus B^e(y, \frac{1}{2}r)^\circ$, and u_j is harmonic with respect to g_j and $\geqslant 0$ in the connected open set Y obtained from X by removing y and all the y_j. (Since $n > 1$, a countable set does not divide an n-simplex, and therefore not X.) This leads to a contradiction with Theorem 6.1, applied with X replaced by the open subpolyhedron Y, and K by A. We are thus left with the case $r_j \to 0$.

Step 2. Let s denote the carrier of y. After embedding $\mathrm{st}(s^\circ) = \mathrm{st}(y)$ in a Euclidean space V with origin at y and norm $|\cdot|$ we consider the homothety $\psi_j : V \to V$ from Step 2 of the proof of Theorem 5.1 with x, x_j, x'_j replaced by y, y_j, y'_j. We also consider the balls $B'_j := B^e(y'_j, r'_j) \subset \mathrm{st}(y)$ for which $\psi_j(B'_j) = B_j$ $(= B^e(y_j, r_j))$. Writing $u_j \circ \psi_j = u'_j$, we have from (6.22) its analogue

$$\max_{\partial B'_j} u'_j > j \min_{\partial B'_j} u'_j \,. \tag{6.23}$$

Here u'_j is harmonic and $\geqslant 0$ in $\mathrm{st}(y) \setminus \{y'_j\}$ with respect to the Riemannian structure $\alpha_j^{-2}(\psi_j^{-1})^* g_j$ with ellipticity constant $\leqslant \Lambda$. Continuing as in Step 2 of the proof of Theorem 5.1 we find from (6.23) that $r'_j \to 0$ (by Step 1 above), and we may assume that $y'_j \to y' \in \mathrm{st}(y) \setminus s$. The carrier s' of y' in $X' := \mathrm{st}(y)$ contains s properly, and so $\dim s < \dim s' \leqslant n$ $(= \dim X')$. Iterating the above procedure at most $n - \dim s$ times therefore leads to an absurdity. □

The hypothesis $n > 1$ cannot be dropped, not even for fixed y, r, u (take e.g., $X = \mathbb{R}$, $y = 0$ and $u(x) = \max\{x, 0\}$).

Hölder continuity

Theorem 6.2. *Every weakly harmonic function on X is Hölder continuous (after correction on a null set).*

Remark 6.4. In the particular case where X is a domain in $\mathbb{R}^n$ carrying a Riemannian metric with bounded measurable components, this is due to Morrey [Mor 1938] for $n = 2$ and to De Giorgi [Gi 1957] for arbitrary dimension n. The proof of Hölder continuity of *harmonic maps* of an admissible Riemannian polyhedron X into a compact subset of a nonpositively curved Riemannian polyhedron Y, given in [Chen 1995], is faulty, even in the present case $Y = \mathbb{R}$, in particular because the conditions needed to apply Morrey's Dirichlet growth lemma at the end of the proof are not fully established.

In our proof of Theorem 6.2 we continue to follow Moser [Mos 1961], who combined the Harnack inequality with an iterative procedure of De Giorgi. In this way one also obtains local essential *equicontinuity* of any locally uniformly bounded family of weakly harmonic functions, see Theorem 6.3 below.

Proof of Theorem 6.2. It suffices to consider a weakly harmonic function u of class $W^{1,2}(X)$. For a given point $a \in X$ and a suitable triangulation of

X having a as a vertex we continue to use the notation $S_h = \theta^{-1}(hP)$ for $0 < h \leqslant 1$, cf. (4.2) and Remark 4.2. Because u is locally essentially bounded (see the preceding section), Proposition 6.1 asserts that

$$u \geqslant 0 \text{ a.e. in } S_h^\circ \quad \text{implies} \quad \operatorname*{ess\,sup}_{S_{h/5}} u \leqslant c_0 \operatorname*{ess\,inf}_{S_{h/5}} u, \tag{6.24}$$

where c_0 is independent of u and of $h \leqslant 1$; in fact, c_0 depends only on n, N, λ. Write

$$M(h) = \operatorname*{ess\,sup}_{S_h} u, \quad \mu(h) = \operatorname*{ess\,inf}_{S_h} u,$$

so that $\omega(h) = M(h) - \mu(h)$ is the essential oscillation of u over S_h. Then $M(h) - u$ and $u - \mu(h)$ are of class $W^{1,2}(S_1^\circ)$ and weakly harmonic there, and $\geqslant 0$ a.e. in S_h, whence by (6.24)

$$M(h) - \mu(h/5) \leqslant c_0 \left(M(h) - M(h/5)\right),$$
$$M(h/5) - \mu(h) \leqslant c_0 \left(\mu(h/5) - \mu(h)\right).$$

Adding these inequalities gives, after simplification,

$$\omega(h/5) \leqslant q\omega(h), \qquad q = \frac{c_0 - 1}{c_0 + 1}.$$

By iteration we obtain

$$\omega(h/5^j) \leqslant q^j \omega(h), \qquad j = 1, 2, \cdots, \tag{6.25}$$

which leads to

$$\omega(h) \leqslant (5h/h')^\alpha \omega(h') \qquad \text{for } q = 5^{-\alpha},\ 0 < 5h \leqslant h' \leqslant 1. \tag{6.26}$$

In fact, choose j so that $h'/5^{j+1} < h \leqslant h'/5^j$. Then, by (6.25),

$$\omega(h) \leqslant \omega(h'/5^j) \leqslant q^j \omega(h') < (5h/h')^\alpha \omega(h').$$

Taking $h' = 1$ leads to

$$\omega(h) < (5h)^\alpha \omega(1) \qquad \text{for } 0 < h \leqslant \tfrac{1}{5}. \tag{6.27}$$

From this we derive estimates of the essential oscillations over Riemannian balls $B(a, r) = B_X(a, r)$ centred at the given point $a \in X$. By λ we denote, as before, the bi-Lip constant of $\theta : S \to P$ with the Euclidean Riemannian structure on P (see the paragraph containing (6.1)). Fix $\varrho > 0$ so that $B(a,\varrho) \subset \mathrm{st}(a)$ $(= S^\circ)$ and $5\lambda\varrho \leqslant 1$. For $r \leqslant \varrho$ we obtain

$$x \in B(a,r) \iff d_X(a,x) \leqslant r \implies d_P^e(0,\theta(x)) \leqslant \lambda r \implies x \in \lambda r P,$$

and so from (6.27)

$$\operatorname*{ess\,osc}_{B(a,r)} u \leqslant \omega(\lambda r) \leqslant (5\lambda r)^\alpha \omega(1) \qquad \text{for } r \leqslant \varrho. \tag{6.28}$$

Recall that α, from (6.26), only depends on n, N, λ (like c_0 and q).

Finally we show that u is equal a.e. to a continuous function u^*. Note first that

$$\operatorname*{ess\,lim\,sup}_{x\to a} u(x) = \operatorname*{ess\,lim\,inf}_{x\to a} u(x) \tag{6.29}$$

for every $a \in X$. In fact, for $r \leqslant \varrho$,

$$\operatorname*{ess\,sup}_{B(a,r)} u - \operatorname*{ess\,inf}_{B(a,r)} u = \operatorname*{ess\,osc}_{B(a,r)} u \leqslant (5\lambda r)^\alpha \omega(1)$$

by (6.28), and each member of (6.29) is squeezed between the above ess sup and ess inf. We may therefore define

$$u^*(a) = \operatorname*{ess\,lim}_{x\to a} u(x), \qquad a \in X.$$

Select u within its equivalence class in $L^2(X)$. For any compact $K \subset X$ there exists, by Lusin's theorem, a closed set $F \subset K$ such that $u_{|F}$ is continuous and the volume measure $|K \setminus F| < \varepsilon$. For any point $a \in F$ such that $|F \cap B(a,r)| > 0$ for every $r > 0$ we have

$$u^*(a) = \operatorname*{ess\,lim}_{x\to a,\, x\in F} u(x) = u(a).$$

The remaining points of F form a null set, and we have thus found that $u^* = u$ a.e. in F, hence a.e. in K, and therefore in X. For any ball $B(a,r)$, $a \in K$, and any $x \in B(a,r)^\circ$,

$$\operatorname*{ess\,inf}_{B(a,r)} u \leqslant u^*(x) \leqslant \operatorname*{ess\,sup}_{B(a,r)} u.$$

This implies

$$\operatorname*{osc}_{B(a,r)^\circ} u^* \leqslant \operatorname*{ess\,osc}_{B(a,r)} u \leqslant (5\lambda r)^\alpha \omega(1) \qquad \text{for } r \leqslant \varrho,$$

by (6.28), and so u equals a.e. the Hölder continuous function u^*. □

Example 6.1. (Cf. [Chen 1995].) The function $z \mapsto z^{1/m}$ on its Riemann surface X (containing 0), with the Euclidean Riemannian metric g^e (see Examples 4.4 and 4.7) is (complex) harmonic on all of X because $z^{1/m}$ is bounded near 0 and holomorphic on $X \setminus \{0\}$, and because $\{0\}$ is a polar set by Corollary 7.2 below, hence a removable singularity for locally bounded harmonic functions (cf. Chapter 2 with reference to [Br 1959]). (Explicitly, the continuous function s defined on X by $s(z) = -\log|z|$ for $z \in X \setminus \{0\}$ and $s(0) = \infty$ is superharmonic on X, being harmonic off 0, and equal to ∞ at 0.)

In view of Theorem 6.2 above we shall establish (with reference to Definition 7.1) the following result on local *uniform Hölder equicontinuity* of suitable families of *harmonic* (i.e., continuous and weakly harmonic) functions on an admissible Riemannian polyhedron X, cf. [Gi 1957] for the case where X is a domain in $\mathbb{R}^n$ carrying a Riemannain metric with bounded measurable components.

Theorem 6.3. *Every locally uniformly bounded family $\mathcal{F}$ of harmonic functions on X is locally uniformly Hölder equicontinuous, that is, for every compact set $K \subset X$ there are constants A, α, δ, depending on (X, g) and K only, such that*

$$\operatorname*{osc}_{B(x,r)} u \leqslant r^\alpha \qquad \text{for any } u \in \mathcal{F},\ x \in K,\ r \leqslant \delta. \tag{6.30}$$

As in Theorem 6.1, we show that A, α, δ can be chosen so as to depend only on X as a mere polyhedron, on K, and on the ellipticity constant Λ of the Riemannian structure g on X. Thus each $u \in \mathcal{F}$ is supposed to be harmonic on X with respect to some Riemannian structure g_u on X with ellipticity constant $\leqslant \Lambda$.

Proof. For a given point $a \in X$ and its closed star $S = \overline{\mathrm{st}}(a)$ (in a suitable triangulation of X having a as a vertex) write

$$M = 2 \sup_{u \in \mathcal{F}} \max_S |u| \quad (< \infty).$$

Recall from Lemma 4.3 the bi-Lip homeomorphism θ of S onto an n-dimensional cell complex $P \subset [0,1]^k$; in particular, $\theta(a) = 0$. The distance function

d_P^e is induced by the Euclidean Riemannian structure on P, and $P^\circ = \theta(S^\circ)$ is the interior of P relative to $\mathbb{R}_+P$, cf. Remark 4.2.

By a standard covering argument the proof of the theorem reduces to the case where

$$K = \theta^{-1}(\tfrac{1}{2}P)$$

(a neighbourhood of a in S° and hence in X), and in that case we show that A, α, δ depend only on n, N, λ (cf. Proposition 6.1), hence only on n, N, Λ in view of (6.1). For any $\xi \in \frac{1}{2}P$ we denote by $Q(\xi)$ the lowest dimensional face of P containing ξ, necessarily as an inner point, and write

$$m(\xi) = \dim Q(\xi).$$

If $m(\xi) > 0$ we denote by $\frac{1}{2}\delta(\xi)$ the shortest Euclidean distance between ξ and any $(m(\xi) - 1)$-face of $Q(\xi)$. If $m(\xi) = 0$ (i.e., $\xi = 0$) put $\delta(\xi) = 1$. Thus $0 < \delta(\xi) \leqslant 1$ in any case.

For $0 < h \leqslant \delta(\xi)$ denote by $Q(\xi, h)$ the (parallel) square (or $m(\xi)$-cube) in $Q(\xi)$ centred at ξ and of edge-length h. Finally, $P(\xi, h)$ will denote the union of all n-cubes contained in P and having $Q(\xi, h)$ as a face; there is precisely one such cube in each n-cell of P which contains ξ and hence has $Q(\xi)$ as a face. We denote by $P(\xi)$ the union of these cells (the star of ξ, i.e., of $Q(\xi)$ in P). In particular (for $\xi = 0$), $P(0) = P$ and

$$P(0, h) = hP, \quad 0 < h \leqslant 1.$$

If $\xi \neq 0$ there is among the $(m(\xi) - 1)$-faces of $Q(\xi)$ of minimal distance $\frac{1}{2}\delta(\xi)$ from ξ at least one containing 0 (the above factor $\frac{1}{2}$ secures that). We then denote by $p(\xi)$ the projection of ξ on one of these last faces. Note that $p(\xi) \in \frac{1}{2}P$ since $\xi \in \frac{1}{2}P$, and that

$$Q\left(p(\xi)\right) \subset Q(\xi), \quad m\left(p(\xi)\right) = m(\xi) - 1, \quad \delta\left(p(\xi)\right) \geqslant \delta(\xi). \tag{6.31}$$

The last relation shows that $P(p(\xi), \delta(\xi))$ is defined. Moreover,

$$P(\xi, h) \subset P(p(\xi), \delta(\xi)). \tag{6.32}$$

Consider now any function $u : X \to \mathbb{R}$ which is harmonic with respect to some Riemannian structure g_u on X with ellipticity constant $\leqslant \Lambda$; furthermore, let $\xi \in \frac{1}{2}P$ and $0 < h \leqslant \delta(\xi)$. If $u \geqslant 0$ in $\theta^{-1}(P(\xi, h))$ then we deduce from Theorem 6.1 the following Harnack inequality (which for $\xi = 0$ reduces to Proposition 6.1):

$$\max_{P(\xi, h/5)} (u \circ \theta^{-1}) \leqslant c_0 \min_{P(\xi, h/5)} (u \circ \theta^{-1}), \tag{6.33}$$

where c_0 depends only on n, N, λ (not on u, ξ, h). For this we proceed as in the reduction of Proposition 6.1 to the case $r = 1$: Let ξ' denote the centre of $Q(\xi)$, and ψ the homothety of $\mathbb{R}^k$ for which $\psi(Q(\xi,h)) = Q(\xi)$ $(= Q(\xi',1) \supset Q(\xi,h))$, hence

$$\psi(P(\xi,h)) = P(\xi) = P(\xi',1) \supset P(\xi,h).$$

In particular, $\psi(\xi) = \xi'$ (ξ' and ξ being the centres of $Q(\xi)$ and $Q(\xi,h)$). The factor of ψ is $1/h$, and the centre of ψ lies in $Q(\xi)$. The function $u \circ \theta^{-1} \circ \psi^{-1}$ is harmonic and $\geqslant 0$ in $P(\xi)^\circ$ with respect to the Riemannian structure $(\theta^{-1} \circ \psi^{-1})^* g_u$ on $P(\xi)$, hence also with respect to the structure $h^{-2}(\theta^{-1} \circ \psi^{-1})^* g_u$, whose ellipticity constant is no bigger than that of $\theta^* g_u$, which in turn equals λ from Proposition 6.1. Applying Theorem 6.1 with X replaced by $P(\xi)^\circ = P(\xi',1)^\circ$, g by $h^{-2}(\theta^{-1} \circ \psi^{-1})^* g_u$, and K by $P(\xi', \frac{1}{5})$, we therefore obtain

$$\max_{P(\xi',1/5)} (u \circ \theta^{-1} \circ \psi^{-1}) \leqslant c_0 \min_{P(\xi',1/5)} (u \circ \theta^{-1} \circ \psi^{-1}),$$

with c_0 depending only on n, N, λ because there are at most $2^n N$ different stars $P(\xi)$ in P, $\xi \in \frac{1}{2}P$. And this inequality implies (6.33) because $\psi(Q(\xi, \frac{h}{5})) = Q(\xi', \frac{1}{5})$, hence $\psi(P(\xi, \frac{h}{5})) = P(\xi', \frac{1}{5})$.

For u, ξ, h as above, in particular $h \leqslant \delta(\xi)$, define

$$\omega(\xi,h) = \operatorname*{osc}_{P(\xi,h)} (u \circ \theta^{-1}).$$

In analogy with the definition of α in (6.26) define a constant $\beta > 0$ by $1/5^\beta = (c_0 - 1)/(c_0 + 1)$, which depends on n, N, λ only. Following De Giorgi and Moser as in the proof of Theorem 6.2, which was based on (6.24), we similarly obtain from (6.33):

$$\omega(\xi,h) \leqslant (5h/h')^\beta \omega(\xi,h'), \qquad 0 < 5h \leqslant h' \leqslant \delta(\xi). \tag{6.34}$$

From this we shall derive the crucial inequality:

$$\omega(\xi,h) \leqslant 5^\beta M h^\alpha, \tag{6.35}$$

valid for any $u \in \mathcal{F}$, $\xi \in \frac{1}{2}P$ and $0 < h \leqslant \min\{\eta, \delta(\xi)\}$, where we define $\alpha = \alpha(n,N,\lambda)$ and $\eta = \eta(n,N,\lambda)$ by

$$\alpha = \beta/2^n, \quad \eta = 1/5^{2^n}. \tag{6.36}$$

In the case $\xi = 0$ this is clear from (6.34) with $h' = \delta(\xi)$ $(= 1)$, for then $\omega(\xi, h') \leqslant M$ and $h^\beta \leqslant h^\alpha$. Suppose therefore that $r := m(\xi) > 0$, and define recursively (with $\xi_0 := \xi$)

$$\xi_j = p(\xi_{j-1}), \quad h_j = \delta(\xi_{j-1}), \qquad j = 1, 2, \cdots, r.$$

By (6.31), $h_{j-1} \leqslant h_j$ (with $h_0 := h$), and $m(\xi_j) = r - j > 0$ for $j = 1, \cdots, r-1$, while $m(\xi_r) = 0$, i.e., $\xi_r = 0$. By (6.32),

$$P(\xi_{j-1}, h_{j-1}) \subset P(\xi_j, h_j), \qquad j = 1, \cdots, r. \tag{6.37}$$

Consider first the case where, for some $j = 1, \cdots, r$, we have

$$h_{j-1} \leqslant h_j^2 \text{ and } h_{j-1} \leqslant h^{2^{1-j}}. \tag{6.38}$$

Then

$$h_{j-1} = h_{j-1}^{1/2} h_{j-1}^{1/2} \leqslant h_j h^{2^{-j}} \leqslant h_j \eta^{2^{-j}} \leqslant h_j \eta^{2^{-n}} = h_j/5$$

for $h \leqslant \eta$, cf. (6.36), noting that $j \leqslant r \leqslant n$. Thus (6.34) applies with ξ, h, h' replaced by ξ_{j-1}, h_{j-1}, h_j, noting that $h_j = \delta(\xi_{j-1})$. When combined with (6.37), (6.34) and (6.38), this leads to (6.35) as follows, noting that $\omega(\xi_{j-1}, h_{j-1}) \leqslant M$:

$$\begin{aligned} \omega(\xi, h) &\leqslant \omega(\xi_{j-1}, h_{j-1}) \leqslant 5^\beta M \left(h_{j-1}/h_j\right)^\beta \\ &\leqslant 5^\beta M h_{j-1}^{\beta/2} \leqslant 5^\beta M h^{\beta/2^j} \leqslant 5^\beta M h^\alpha \end{aligned}$$

because $\beta/2^j \geqslant \beta/2^n = \alpha$, by (6.36).

In the remaining case, where (6.38) fails for every $j = 1, \cdots, r$, we see by induction that the latter inequality (6.38) does hold (trivially for $j = 1$), and hence $h_{j-1} > h_j^2$, that is,

$$h > h_1^2 > h_2^4 > \cdots > h_r^{2^r}. \tag{6.39}$$

Now applying (6.34) with ξ, h, h' replaced by $\xi_r = 0$, h_r, $\delta(\xi_r) = 1$ we therefore obtain (6.35) from (6.37) and (6.39):

$$\omega(\xi, h) \leqslant \omega\left(\xi_r, h_r\right) \leqslant (5h_r)^\beta \omega(0, 1) \leqslant 5^\beta M h^{\beta/2^r} \leqslant 5^\beta M h^\alpha.$$

This application of (6.34) is permissible because $5h_r \leqslant 1 = \delta(\xi_r)$ $(= h')$. When $h \leqslant \eta$ it follows from (6.39) and (6.36) that

$$h_r < h^{2^{-r}} \leqslant \eta^{2^{-r}} \leqslant \eta^{2^{-n}} = 1/5.$$

In terms of the "Euclidean" distance d_P^e on P, and the associated bi-Lip constant λ of $\theta : S \to P$ (see the paragraph containing (6.1)), the ball $B_P(\xi, \frac{h}{2})$ is clearly contained in $P(\xi, h)$ for $\xi \in \frac{1}{2}P$, $h \leqslant \delta(\xi)$.

We proceed to estimate the oscillation $\widetilde{\omega}(\xi, h)$ of $u \circ \theta^{-1}$ over $B_P(\xi, \frac{h}{2})$ without assuming $h \leqslant \delta(\xi)$. Since $\frac{1}{2}\delta(\xi)$ is the distance between ξ and $p(\xi)$ we have $B_P(\xi, \frac{1}{2}h) \subset B_P(p(\xi), \frac{1}{2}h + \frac{1}{2}\delta(\xi)) \subset B_P(\xi_1, \frac{1}{2}h_0 + \frac{1}{2}h_1)$, hence $\widetilde{\omega}(\xi, h) \leqslant \widetilde{\omega}(\xi_1, h_0 + h_1)$, and by repeated application

$$\widetilde{\omega}(\xi, h) \leqslant \widetilde{\omega}(\xi_k, h_0 + \cdots + h_k), \qquad k = 1, \cdots, r. \tag{6.40}$$

Denote by k the smallest among those numbers $k \in \{1, \cdots, r-1\}$ for which

$$h_0 + \cdots + h_k \leqslant h_{k+1} \ \ (= \delta(\xi_k)); \tag{6.41}$$

if there is no such k, write $k = r$. It follows by recursion that

$$h_j < 2^{j-1}h \qquad \text{for } j = 1, \cdots, k. \tag{6.42}$$

If $k < r$ and $h \leqslant \eta/2^n$ we obtain from (6.40), (6.41) and (6.35)

$$\begin{aligned} \widetilde{\omega}(\xi, h) &\leqslant \widetilde{\omega}(\xi_k, h_0 + \cdots + h_k) \\ &\leqslant \omega(\xi_k, h_0 + \cdots + h_k) \leqslant 5^\beta 2^{n\alpha} M h^\alpha. \end{aligned} \tag{6.43}$$

If $k = r$ we have $\xi_k = 0$, and (6.43) remains valid because $h_0 + \cdots + h_k < 2^k h \leqslant 2^n h \leqslant \eta < 1 = \delta(\xi_k)$, by (6.42), (6.35) and (6.36).

Having thus shown that $u \circ \theta^{-1}$ is uniformly Hölder equicontinuous on $\frac{1}{2}P$, we conclude (θ being bi-Lip) that u is itself uniformly Hölder equicontinuous on $\theta^{-1}(\frac{1}{2}P) = S_{\frac{1}{2}}$, a neighbourhood of the given point $a \in X$. □

Remark 6.5. It follows from (6.21) (in the proof of Theorem 6.1, carried over to the setting of the above proof), taking $p = 2$, that every family $\mathcal{F}$ of harmonic functions on X such that $\mathcal{F}$ is bounded in $L^2_{\text{loc}}(X)$, is also locally uniformly bounded, and so Theorem 6.3 applies to such $\mathcal{F}$. (This, too, is inspired by [Gi 1957].)

Remark 6.6. If $\dim X = 1$, 'Hölder' can be replaced by 'Lipschitz' in Theorem 6.3, as can be easily shown directly, see Example 8.1.

7

Potential theory on Riemannian polyhedra

We shall consider an arbitrary n-dimensional admissible Riemannian polyhedron $X = (X, g)$ and show that it carries a natural harmonic space structure in the sense of Brelot, cf. Chapter 2. We also study the Dirichlet space $L_0^{1,2}(X)$ of functions on X having a finite Dirichlet integral, and interrelate these two potential theoretic aspects in the light of the general study thereof by Feyel and de La Pradelle [FP 1978]. The Green function on X is studied in detail.

Harmonic space structure

Definition 7.1. A function $u : U \to \mathbb{R}$, defined in an open subset U of X, is said to be *harmonic* (in U) if u is continuous and weakly harmonic, in particular of class $W^{1,2}_{\text{loc}}$, in each connectivity component of U (cf. Definition 5.1).

According to Theorem 6.2 every weakly harmonic function in a subdomain U of X can be redefined uniquely in a set of zero (volume) measure so as to become continuous and therefore harmonic in U.

Denote by $\mathcal{H}(U)$ the linear space of functions harmonic in an open set $U \subset X$, and by $\mathcal{H} = \mathcal{H}_X$ the map $U \mapsto \mathcal{H}(U)$. $\mathcal{H}$ is a *sheaf*, as a consequence of the sheaf property of weakly harmonic functions (see the text preceding Proposition 5.2). We proceed to show that $\mathcal{H}$ is a *harmonic sheaf* in the sense of Brelot (cf. Chapter 2).

To see that $(X, \mathcal{H})$ has a base of regular domains we show that the classical-style Dirichlet problem with a given continuous boundary function is uniquely and positively solvable in the star $\text{st}(a)$ of any point a relative to a triangulation T of X such that a is a vertex and $\text{st}(a)$ is the interior of its closure S; this latter requirement can always be met by subdivision of T, cf. Remark 4.2. Recall that the Poincaré inequality (5.12) then holds in $S^\circ = \text{st}(a)$ (Remark 5.5).

Proposition 7.1. *Every continuous function $f : \partial S \to \mathbb{R}$ extends uniquely to a continuous function $\overline{u}$ on S whose restriction u to S° is harmonic. If $f \geqslant 0$ then $u \geqslant 0$.*

Proof. The uniqueness follows from Proposition 5.3 and Remark 5.4, applied (in case (ii)) to the noncompact admissible Riemannian polyhedron S° and the difference between any two solutions of the stated Dirichlet problem. As to the final assertion, replace u by $-u$ in Proposition 5.3.

As to the existence, consider first (in Step 1 and Step 2 below) the case where f extends to a Lip function $S \to \mathbb{R}$, likewise denoted f. Applying Theorem 5.2 to S° leads to a unique function $u \in W^{1,2}(S^\circ)$ with $u - f_{|S^\circ} \in W_0^{1,2}(S^\circ)$ such that u is weakly harmonic, hence continuous (after correction on a null set) by Theorem 6.2, and therefore harmonic (Definition 7.1). Write $\overline{u} = f$ on ∂S and $\overline{u} = u$ on S°.

We shall prove that $\overline{u}$ is continuous at any point $y_0 \in \partial S$:

$$u(x) \to f(y_0) \qquad \text{as } x \to y_0,\ x \in S^\circ.$$

Step 1. Suppose in addition that $f = 0$ in an open neighbourhood Y of y_0 in ∂S. With S embedded in $\mathbb{R}^k$ (as in Step 1 leading to Lemma 4.3, whereby $a \in S$ corresponds to $0 \in \mathbb{R}^k$), consider the Lip homeomorphism $\zeta : (\partial S) \times [0,1] \to S \setminus \frac{1}{2}S^\circ$ given by

$$\zeta(y,t) = (1 - \tfrac{1}{2}t)y, \qquad y \in \partial S,\ t \in [0,1];$$

or rather its restriction

$$\zeta : Y \times [0,1[\to U,$$

this image U being an open collar around $Y \subset \partial S$ in S. For brevity write

$$\begin{aligned} Z &= Y \times]-1,1[, \qquad Z_0 = Y \times \{0\}, \\ Z_+ &= Y \times]0,1[, \qquad Z_- = Y \times]-1,0[, \end{aligned}$$

so that $Y \times [0,1[= Z_+ \cup Z_0$ and $Z = Z_+ \cup Z_- \cup Z_0$ (pairwise disjoint). Extend ζ to Z by $\zeta(y,-t) = \zeta(y,t)$ for $t \in [0,1[$, and use this Lip map $\zeta : Z \to U$ to transport the Riemannian structure from $U \subset S$ to Z, which then becomes an admissible Riemannian polyhedron with a Riemannian structure symmetric about Z_0. Writing u in place of $u \circ \zeta_{|Z_+}$ we extend u to Z as a function of (y,t) (likewise denoted u) which is odd in t. Similarly, write f in place of $f \circ \zeta_{|Z_+ \cup Z_0}$, and extend this function to an odd function $f \in \mathrm{Lip}(Z)$ (using that $f_{|Z_0} = 0$ by hypothesis in the present Step 1).

There is a sequence $(\psi_j) \subset \mathrm{Lip}(Z_+)$ such that the support of each ψ_j has positive distance from Z_0 and that $\psi_j \to (u-f)_{|Z_+}$ in $W^{1,2}(Z_+)$. The odd extensions of the ψ_j are of class $\mathrm{Lip}(Z)$ and form a Cauchy sequence in $W^{1,2}(Z)$ converging to $u - f$ in $L^2(Z)$, and so $u - f \in W^{1,2}(Z)$, hence $u \in W^{1,2}(Z)$.

Because $u_{|Z_+}$ is weakly harmonic, so is $u_{|Z_-}$, and therefore also the odd function $u \in W^{1,2}(Z)$. Indeed, for any function $\varphi \in \mathrm{Lip}_c(Z)$ define $\check{\varphi} \in \mathrm{Lip}_c(Z)$ by

$$\check{\varphi}(y,t) = \varphi(y,-t), \qquad (y,t) \in Z.$$

The restriction of $\varphi - \check{\varphi}$ to Z_+ is in $W_0^{1,2}(Z_+)$ according to Lemma 5.3 since $\varphi - \check{\varphi} = 0$ on Z_0 and $\mathrm{supp}(\varphi - \check{\varphi})$ is compact in Z. Because $u_{|Z_+}$ is harmonic we therefore obtain

$$\int_Z \langle \nabla u, \nabla \varphi \rangle = \int_{Z_+} + \int_{Z_-} = \int_{Z_+} \langle \nabla u, \nabla(\varphi - \check{\varphi}) \rangle = 0,$$

which means that u is weakly harmonic (on Z). According to Theorem 6.2, u equals a.e. a harmonic, in particular continuous function, and being odd in t, u equals 0 on Z_0; in particular $u(y_0, 0) = 0 = f(y_0, 0)$.

Step 2. We merely assume that f extends to a Lip function $f : S \to \mathbb{R}$. For a given point $y_0 \in \partial S$ we may assume that $f(y_0) = 0$ (otherwise replace f by $f - f(y_0)$, hence u by $u - f(y_0)$). Define

$$f_j = f - T_{1/j} \circ f, \qquad j = 1, 2, \cdots,$$

on S in terms of the normal contraction T_ε from the proof of Lemma 5.3. Because f is Lipschitz on S, so is f_j, and $f_j = 0$ in the open neighbourhood $f^{-1}(]-1/j, 1/j[)$ of y_0 in S. Let $u_j : S \to \mathbb{R}$ denote the variational solution (as in Theorem 5.2) corresponding to the boundary function f_j. It follows from $-1/j \leqslant f - f_j \leqslant 1/j$ that $-1/j \leqslant u - u_j \leqslant 1/j$ according to one of the last assertions of Theorem 5.2 applied e.g., to $(f - f_j)_{|S^\circ} \in \mathrm{Lip}^{1,2}(S^\circ)$, with $u - u_j$ the corresponding solution to the variational Dirichlet problem in the admissible Riemannian polyhedron S°. Now, $f_j \to f$ and $u_j \to u$ uniformly on ∂S and on S°, respectively, and it follows from Step 1 applied to f_j (with solution u_j) that indeed $u(x) \to f(y_0)$ as $x \to y_0$, $x \in S^\circ$.

Step 3. This is the general case. The given continuous function f on ∂S extends to a continuous function f on X supported by a compact neighbourhood of S. And f can be approximated uniformly by a sequence of piecewise linear, hence Lip functions f_j associated with sufficiently fine subdivisions of

X and such that $\operatorname{supp} f_j \subset \{x \in X : f(x) \neq 0\}$. Let u_j denote the solution corresponding to $f_{j|S}$ as in Step 2 above. Because $-\varepsilon \leqslant f_i - f_j \leqslant \varepsilon$ for large enough i, j, we have $-\varepsilon \leqslant u_i - u_j \leqslant \varepsilon$ for the same i, j, and so u_j converges uniformly to some continuous function u in S°, whence $u(x) \to f(y_0)$ as $x \to y_0$, $x \in S^\circ$. It remains to prove that u is (weakly) harmonic (in S°). For any open set V of compact closure $\overline{V}$ in S° we may apply Lemma 6.1 for $p = 2$ to $u_i - u_j$ in place of u (cf. end of Remark 6.1), taking $\eta \in \mathrm{Lip}_c(S^\circ)$ with $\eta = 1$ on $\overline{V}$. It follows that the restrictions of u_j to V form a Cauchy sequence in $W^{1,2}(V)$ and hence converge in $W^{1,2}(V)$ to some function of class $W^{1,2}(V)$, necessarily to $u_{|V}$, the uniform, hence L^2, limit of $(u_{j|V})$. For given $\varphi \in \mathrm{Lip}_c(S^\circ)$ we may take $V \supset \operatorname{supp} \varphi$, and we conclude that

$$\int_{S^\circ} \langle \nabla u, \nabla \varphi \rangle = \int_V \langle \nabla u, \nabla \varphi \rangle = \lim_j \int_{S^\circ} \langle \nabla u_j, \nabla \varphi \rangle = 0$$

because u_j is weakly harmonic in S°. Consequently, u is likewise weakly harmonic in S°. □

Remark 7.1. An obvious extension of the above proposition is obtained when $S = \overline{\mathrm{st}}(a)$ is replaced by its image under any Lip homeomorphism σ of S into X. Indeed, a continuous function $\overline{u}$ on $\sigma(S)$ is harmonic in $\sigma(S)^\circ$ with the Riemannian structure $(\sigma^{-1})^* g$ on $\sigma(S)$ if and only if $\overline{u} \circ \sigma$ is harmonic in S° with the given structure g on S $(\subset X)$. For example, relative to a triangulation of X for which a is a vertex and $S^\circ = \mathrm{st}(a)$, every open ball $B^e(a, \varrho)^\circ$ with respect to the Euclidean Riemannian metric g^e on S such that $B^e(a, \varrho) \subset S^\circ$, is regular. (This follows from the above, e.g., by modifying Lemma 4.3, replacing $|\cdot|_\infty$ by $|\cdot|_2$ in Sublemma 4.3. Alternatively, the above proof of Proposition 7.1 carries over.)

We do not know whether every sufficiently small metric ball $B(a, \varrho)$ in X with respect to the intrinsic distance d_X associated with the *Riemannian* metric g is Lip homeomorphic to $S = \overline{\mathrm{st}}(a)$ (not even in the case of a Riemannian Lip *manifold* X, nor in the case of an admissible Riemannian polyhedron with simplexwise smooth Riemannian structure).

Every open star S° as above is relatively compact, and in the terminology of harmonic spaces (see Chapter 2), Proposition 7.1 asserts that S° is a *regular set* (for the Dirichlet problem). Thus *the regular domains in X form a base of the topology on X.*

For any regular set V in X and any continuous function $f : \partial V \to \mathbb{R}$ one denotes by H_f^V the unique harmonic function in V whose extension by f is continuous on $\overline{V}$, cf. Chapter 2. For any point $x \in V$ the map $f \mapsto H_f^V(x)$ is clearly linear, and it is positive: $f \geqslant 0 \Rightarrow H_f^V \geqslant 0$, by the definition of a regular set. By the F. Riesz representation theorem there is therefore a unique positive Radon measure ω_x^V on ∂V such that

$$H_f^V(x) = \int_{\partial V} f \, d\omega_x^V, \qquad f \in C(\partial V),\ x \in V. \tag{7.1}$$

This measure ω_x^V is called the *harmonic measure* for the regular set V and the point $x \in V$. It is a *probability measure* because $H_1^V = 1$. We may of course view ω_x^V as a measure on X supported by ∂V.

For any harmonic function u on X and any regular set V in X, the above applies to $f := u_{|\partial V}$, and hence (writing simply H_u^V in place of $H_{u|\partial V}^V$)

$$u(x) = H_u^V(x) = \int_{\partial V} u \, d\omega_x^V. \tag{7.2}$$

Conversely, if $u \in C(X)$, and if the above meanvalue property holds for every regular domain $V \subset X$ and every point $x \in V$, then u is harmonic in X because u coincides in V with the harmonic function H_u^V in V.

From the Harnack inequality (Theorem 6.1) together with the equicontinuity result in Theorem 6.3 we derive by a classical argument *Harnack's monotone convergence theorem*:

Proposition 7.2. *Let (u_j) be an increasing sequence of harmonic functions on a domain $U \subset X$, and write $u(x) = \sup_j u_j(x)$, $x \in U$. Then either $u \equiv \infty$, or u is harmonic.*

Proof. We may assume that each u_j is nonnegative (otherwise replace u_j by $u_j - u_1$). For any point $a \in U$ there is a regular domain $V \ni a$ in U and, by Theorem 6.1 applied to $K = \overline{V}$, a constant c such that

$$c^{-1}u_j(a) \leqslant u_j(x) \leqslant c u_j(a), \qquad x \in V, \text{ all } j.$$

Hence $c^{-1}u(a) \leqslant u(x) \leqslant cu(a)$ for $x \in V$, and so the sets $\{u < \infty\}$ and $\{u = \infty\}$ are both open. We may therefore assume that $u(x) < \infty$ for every $x \in U$. Being uniformly bounded in V, the sequence (u_j) is equicontinuous at a, by Theorem 6.3. Consequently, u is continuous at a, hence everywhere in U. For every $x \in V$ we finally have

$$u(x) = \lim_{j\to\infty} u_j(x) = \lim_{j\to\infty} \int_{\partial V} u_j \, d\omega_x^V = \int_{\partial V} u \, d\omega_x^V = H_u^V(x),$$

which shows that u is harmonic in each V, hence in U. □

The main results obtained so far (in particular Theorems 6.1, 6.2 and Propositions 7.1, 7.2) can be summed up as follows, with reference to [Br 1960; 1969], [Her 1962], [CC 1972] and [Bau 1984], see Chapter 2:

Theorem 7.1. *An admissible Riemannian polyhedron X becomes a Brelot harmonic space $(X, \mathcal{H})$ when we assign to each open set $U \subset X$ the linear space $\mathcal{H}(U)$ of all harmonic functions on U in the sense of Definition 7.1. For any sufficiently fine triangulation of X, the open star of any point $a \in X$ is regular for $(X, \mathcal{H})$. The function 1 is harmonic. Every harmonic function is Hölder continuous.*

For the particular case where X is a domain in $\mathbb{R}^n$ carrying a Riemannian metric with bounded measurable components, the corresponding result is contained in [Her 1962; 1964].

The Dirichlet space $L_0^{1,2}(X)$

We proceed to study the appropriate Dirichlet space structure on X, cf. Chapter 2, and relate it to the above harmonic space structure, invoking the general study of such a relationship by Feyel and de La Pradelle [FP 1978], see also Lassoued [La 1980]. In the present setting, the Dirichlet space on the admissible Riemannian polyhedron X will be the following Hilbert space $L_0^{1,2}(X)$, a subspace of $W_{\mathrm{loc}}^{1,2}(X)$, and with the energy norm $E(u)^{1/2}$, cf. (5.3).

Proposition 7.3. *Suppose that, for every compact set $K \subset X$,*

$$\left(\int_K |u|\right)^2 \leqslant c(K)E(u) \qquad \text{for all } u \in \mathrm{Lip}_c(X), \tag{7.3}$$

with $c(K)$ depending on X and K only; in particular, X is noncompact.

(a) *The completion $L_0^{1,2}(X)$ of $\mathrm{Lip}_c(X)$ within $L_{\mathrm{loc}}^1(X)$ with respect to the norm $E(u)^{1/2}$ is then a regular Dirichlet space of strongly local type (i.e., diffusion type). $L_0^{1,2}(X)$ is a subset of $W_{\mathrm{loc}}^{1,2}(X)$.*

(b) *$E(|u|) = E(u)$ for every $u \in L_0^{1,2}(X)$.*

(c) *The self-map $u \mapsto |u|$ of $L_0^{1,2}(X)$ is norm continuous.*

(d) *$\mathrm{Lip}_c^+(X)$ is norm dense in $L_0^{1,2}(X)^+$.*

If (X, d_X) is complete then $L_0^{1,2}(X)$ is an admissible Dirichlet space.

Proof. If X were compact, we could take $u = 1$ in (7.3) to obtain a contradiction. The inequality (7.3) extends immediately to functions $u \in L_0^{1,2}(X)$ and expresses then that $L_0^{1,2}(X)$ is a functional Hilbert space in the sense of [De 1970, Chapitre 1] with the norm $E(u)^{1/2}$ and the inner product $E(u,v)$ from (5.2) and (5.3). The proof of Proposition 5.1 (for the space $W^{1,2}(X)$) carries over mutatis mutandis to show that also $L_0^{1,2}(X)$ is a Dirichlet space, and to establish (b), (c) and (d).

As to the rest of (a), $L_0^{1,2}(X)$ is *regular* because $\mathrm{Lip}_c(X)$ is dense both in $L_0^{1,2}(X)$ (by definition) and in $C_c(X)$ (the continuous functions $X \to \mathbb{R}$ of compact support), as shown in the beginning of Step 3 in the proof of Proposition 7.1 above.

For any function $u \in \mathrm{Lip}_c(X)$ the meanvalue $\overline{u}$ of u over a sufficiently small (hence compact) ball $B = B(x,r)$ satisfies $\overline{u}^2 \leqslant c'(B)E(u)$ for some constant $c'(B)$, as a consequence of (7.3). When combined with Theorem 5.1 this shows that, for some constant $c''(B)$,

$$\int_B u^2 d\mu_g \leqslant 2\int_B (u-\overline{u})^2 d\mu_g + 2\int_B \overline{u}^2 d\mu_g \leqslant c''(B)E(u),$$

again by (7.3). Since any compact set $K \subset X$ can be covered by finitely many balls B as above, it follows that indeed $L_0^{1,2}(X) \subset W_{\mathrm{loc}}^{1,2}(X)$.

$L_0^{1,2}(X)$ has *strongly local type*, for if $u, v \in L^{1,2}(X)$ have compact supports, and if u is constant in a neighbourhood of $\operatorname{supp} v$, then $\nabla u = 0$ there, and so $\langle \nabla u, \nabla v\rangle = 0$ a.e. in X. Here we have applied Remark 5.2 to $u \in L_0^{1,2}(X) \subset W_{\mathrm{loc}}^{1,2}(X)$. By the same remark,

$$E(u_+, u_-) = 0 \quad \text{for every } u \in L_0^{1,2}(X).$$

Finally, if (X, d_X) is *complete* then the closed balls in X are compact, by the Hopf-Rinow theorem (cf. Chapter 2). We show that $\mathrm{Lip}_c(X)$ is μ-*separating*, and $L_0^{1,2}(X)$ is therefore an *admissible* Dirichlet space, in view of the ball doubling property (Corollary 4.1). Indeed, for fixed $x \in X$, the function $d_X(x,\cdot)$ is Lipschitz, by the triangle inequality; and $|\nabla d_X(x,\cdot)| \leqslant 1$ a.e. in X, as shown in the last paragraph of the proof of Proposition 4.1. For given $y \in X$ with $d_X(x,y) = r > 0$, the Lip function $u = (r - d_X(x,\cdot))_+$ has compact support $B_X(x,r)$, and $u(x) = r \neq 0 = u(y)$. Finally, $|\nabla u| \leqslant |\nabla d_X(x,\cdot)| \leqslant 1$ a.e. in X, by Remark 5.2. □

Remark 7.2. For local questions, (7.3) is not required because it is automatically satisfied with X replaced by the open star $\mathrm{st}(a)$ of a point a of X relatively to a sufficiently fine triangulation; this follows from Remark 5.5.

Remark 7.3. In the classical case, where X is a domain in $\mathbb{R}^n$, (7.3) fails to hold iff either $n = 1$ and $X = \mathbb{R}$, or $n = 2$ and $\mathbb{R}^2 \setminus X$ is locally of zero logarithmic capacity. Cf. [DL 1954, pp. 350ff.], where the functions of class $L_0^{1,2}$ are called *Beppo Levi functions.*

Theorem 7.2. *For any admissible Riemannian polyhedron X satisfying* (7.3), *the Dirichlet space $L_0^{1,2}(X)$ determines a Brelot harmonic structure on X as constructed by Feyel and de La Pradelle* [FP 1978], *and this structure is the same as that of Theorem 7.1.*

Proof. From the general theory of Dirichlet spaces we have the notion of an *E-potential*, that is, a function $p \in L_0^{1,2}(X)$ such that

$$E(p, \varphi) \geqslant 0 \qquad \text{for every } \varphi \in \mathrm{Lip}_c(X),\ \varphi \geqslant 0,$$

hence also for every $\varphi \in L_0^{1,2}(X)$, $\varphi \geqslant 0$, according to Proposition 7.3 (d). (Cf. [De 1970], where such a potential is called a pure potential.) Equivalently, p should be a weakly superharmonic function (in the sense of Definition 5.2) of class $L_0^{1,2}(X)$. It follows that $p \geqslant 0$ a.e. because $E(p_-, p_-) = -E(p, p_-) \leqslant 0$ (take $\varphi = p_-$ above).

As mentioned in Chapter 2, Dirichlet structures on a space (applied here to $\mathcal{D} = L_0^{1,2}(X)$), we also have the notion of *capacity*, the term *quasieverywhere* (q.e., that is, everywhere outside some set of zero capacity), and the existence of *quasicontinuous versions* of each element of $L_0^{1,2}(X)$, [De 1970, Chapitre 4], [Fuk 1980, §3]. By the spectral synthesis theorem discussed there we have, for any open set $U \subset X$,

$$L_0^{1,2}(U) = \{u \in L_0^{1,2}(X) : u = 0 \text{ q.e. in } X \setminus U\},$$

whereby u is taken to be quasicontinuous. (We identify a function of class $L_0^{1,2}(U)$ with its extension to X by 0 in $X \setminus U$.)

For any E-potential p on X, the *reduced function* (réduite) $R_p^{\complement U} \in L_0^{1,2}(X)$ (U open in X), as defined in [FP 1978, §2], can also be characterized by minimizing $E(u)$ among all $u \in L_0^{1,2}(X)$ such that $p - u \in L_0^{1,2}(U)$ (in analogy with Theorem 5.2). Hence $R_p^{\complement U}$ is the projection of p on

$$L_0^{1,2}(U)^{\perp} = \{u \in L_0^{1,2}(X) : u_{|U} \text{ is weakly harmonic}\}.$$

In particular, $R_p^{\complement U}$ is continuous in U after correction on a null set, by Theorem 6.2.

This establishes axiom (a) in [FP 1978, §III]. As to axiom (b), the equilibrium E-potential $p \in L_0^{1,2}(X)$ of a nonempty open, relatively compact set $U \subset X$, see [De 1970, p. 162], [Fuk 1980, §3.3], satisfies $p = 1$ a.e. in U, and so $p_{|U}$ is weakly harmonic. It follows that $p = R_p^{\complement U}$ a.e., and so $R_p^{\complement U} \neq 0$ a.e. In the terminology of [FP 1978] we have thus found that $L_0^{1,2}(X)$ is a *hypoelliptic Dirichlet space.*

For every quasicontinuous function $f \in L_0^{1,2}(X)$ which is weakly superharmonic in each component of a relatively compact open set $U \subset X$, the lower semicontinuous regularization

$$\hat{f}(x) := \operatorname{ess}\liminf_{y \to x} f(y), \qquad x \in X,$$

satisfies $\hat{f} = f$ q.e. in U, as shown in [FP 1978, 28°]. Let $\hat{\mathcal{F}}(U)$ denote the class of all restrictions $\hat{f}_{|U}$ obtained in this way. According to [FP 1978, p. 121], X carries a certain Brelot harmonic sheaf $\mathcal{H}'$ with the property that, for every relatively compact open set $U \subset X$, every function $\hat{f}_{|U} \in \hat{\mathcal{F}}(U)$ is hyperharmonic (in fact superharmonic) in U with respect to $\mathcal{H}'$, cf. [FP 1978, 30°].

Every weakly superharmonic function u on a domain $U \subset X$ is of class $W_{\mathrm{loc}}^{1,2}(U)$, and hence agrees in each subdomain $V \Subset U$ with some function $f \in W_0^{1,2}(X) \subset L_{\mathrm{loc}}^{1,2}(X)$, which we may take to be quasicontinuous. As mentioned above, $f_{|V}$ and hence $u_{|V}$ equal a.e. a function $\hat{u}$ of class $\hat{\mathcal{F}}(V)$, and $\hat{u}$ is superharmonic with respect to $\mathcal{H}'$. It follows that every weakly superharmonic function on U equals a.e. a function which is superharmonic with respect to $\mathcal{H}'$. When applied to $\pm u$ this shows that every function $u \in \mathcal{H}(U)$ is of class $\mathcal{H}'(U)$. We conclude that $\mathcal{H}(U) = \mathcal{H}'(U)$, by the maximality (with respect to inclusion) of every Brelot harmonic sheaf, established by Constantinescu and Cornea [CC 1965, p. 23] (see also Corollary 13.1 (ii) below). □

Remark 7.4. Even without assuming (7.3) we have seen in the last paragraph that *every weakly superharmonic function u on X can be redefined uniquely on a null set so as to become superharmonic* in the sense of the harmonic space $(X, \mathcal{H})$. Conversely, every superharmonic function u on $(X, \mathcal{H})$, majorized by a function of class $L_{\mathrm{loc}}^{1,2}(X)$, is weakly superharmonic, according to [La 1980, p. 141], who also proved that, in the presence of (7.3), every nonnegative superharmonic function on $(X, \mathcal{H})$, majorized by a function of class $L_0^{1,2}(X)$, is itself in that class (and is an E-potential, in particular weakly superharmonic).

As a prelude to Theorem 7.3 below we show that, if (7.3) holds, *there exists on X an $\mathcal{H}$-potential $p \not\equiv 0$* (hence $p > 0$). In fact, the equilibrium potential $p \in L_0^{1,2}(X)$ considered above in the verification of axiom (b) from [FP 1978] can be redefined on a null set so as to become $\mathcal{H}$-superharmonic, by Remark 7.4, and indeed an $\mathcal{H}$-potential, every harmonic minorant $u \geqslant 0$ of p being of class $L_0^{1,2}(X)$ and weakly harmonic (because u and $-u$ are majorized by the functions p and 0 of that class). Consequently, $E(u, \varphi) = 0$ for every $\varphi \in \mathrm{Lip}_c(X)$, or just $\varphi \in L_0^{1,2}(X)$, in particular for $\varphi = u$, and so $u = 0$.

In the rest of the present chapter we require that (X, g) is an admissible Riemannian polyhedron such that the inequality (7.3) *in Proposition 7.3 holds.* In particular, X is *noncompact.* See, however, Remark 7.8.

Remark 7.5. See [HeKo 1998] for a more general framework for potential theory. There (X, d_X, μ) is a metric measure space, regular with respect to Hausdorff dimension, and satisfying a Poincaré inequality.

The Green kernel

Using Theorem 7.2 we shall first deduce from results of Hervé [Her 1962], in combination with [FP 1978], that X has a *symmetric Green function*, and so $(X, \mathcal{H})$ is a selfadjoint harmonic space in the sense of Maeda [Mae 1980].

Theorem 7.3. *(X, g) has a unique symmetric, lower semicontinuous Green kernel*

$$G : X \times X \to]0, \infty],$$

finite and continuous off the diagonal in $X \times X$, such that

(1) *For every positive measure μ on X the function $G\mu : X \to [0, \infty]$ defined by*

$$G\mu(x) = \int_X G(x, y)\, d\mu(y), \qquad x \in X,$$

is a potential on X (in the harmonic space sense, Definition 2.3), *unless $G\mu \equiv \infty$.*

(2) *Every potential p on X has the form $p = G\mu$ for a unique positive measure μ on X. The support of μ is the complement of the biggest open set in which p is harmonic.*

(3) *The E-potentials on X are those functions on X which equal* a.e. *a (unique) potential $p = G\mu$ such that the energy $\frac{1}{2}\int G\mu\, d\mu$ is finite, hence equal to $E(p)$. For any $p \in L_0^{1,2}(X)$ and any measure $\mu \geqslant 0$ on X of finite energy we have $p = G\mu$* a.e. *iff*

$$E(p,\varphi) = \frac{1}{2}\int_X \varphi\, d\mu \qquad \text{for } \varphi \in L_0^{1,2}(X) \ \ (\text{or } \varphi \in \mathrm{Lip}_c(X)).$$

Proof. According to [FP 1978, 37°], any two potentials on $(X,\mathcal{H})$, harmonic off the same point of X, are proportional. Alternatively, this follows from Proposition 6.2 as in [Her 1965]. According to [Her 1962], this proportionality property implies that $(X,\mathcal{H})$ has a Green kernel G with the properties (1), (2) above (see also [CC 1972, Theorem 11.5.2]). According to [FP 1978, 37°], G also has the property (3) and is thereby uniquely determined. Finally, the symmetry of G derives from that of the Dirichlet form $E(u,v) = \frac{1}{2}\int_X \langle \nabla u, \nabla v\rangle$ as follows. For any $\varphi, \psi \in C_c^+(X)$, viewed as densities of positive measures μ, ν of finite energy on X,

$$\int_X \int_X G(x,y)\varphi(x)\psi(y)\, dx\, dy = \int_X G\mu\, d\nu = 2E(G\mu, G\nu)$$

is symmetric in φ, ψ, and so we have $G(x,y) = G(y,x)$ by the continuity of $G(x,y)$ for $x \neq y$. □

We next deduce the following three further properties, well known in classical potential theory:

(4) A set $e \subset X$ has zero capacity iff e is *polar*, that is, there should exist a superharmonic function u on X such that $u \equiv \infty$ on e.

In fact, if $\operatorname{cap} e = 0$ then there even exists a potential u of finite energy on X such that $u \equiv \infty$ on e, cf. e.g., [FP 1978, 21°]. Conversely, if this relation holds for some superharmonic function u on X, then likewise for some potential u of finite energy, by [FP 1978, 32°], and hence $\operatorname{cap} e = 0$ by a classical argument: For any number $\alpha > 0$, the set $\{u > \alpha\}$ is open and contains e, and so, as in [De 1970, p. 163], [Fuk 1980, §3],

$$\operatorname{cap} e \leqslant \operatorname{cap}\{u/\alpha > 1\} \leqslant E(u/\alpha) = E(u)/\alpha^2,$$

which tends to 0 as $\alpha \to \infty$.

(5) According to [FP 1978, p. 121] and [La 1980], the Brelot harmonic space $(X,\mathcal{H})$ satisfies the *axiom of domination*, introduced by Brelot: If a

locally bounded potential $p = G\mu$ is majorized quasieverywhere on $\operatorname{supp}\mu$ by a superharmonic function $u \geqslant 0$ on X then $p \leqslant u$ everywhere in X. The same holds for any E-potential $p = G\mu$; this is *Cartan's maximum principle*, cf. [Ca 1946]. In particular, if a potential $G\mu$ is $\leqslant 1$ q.e. on $\operatorname{supp}\mu$, then $G\mu \leqslant 1$ everywhere; this is *Frostman's maximum principle.*

(6) According to [JaNg 1975], the axiom of domination, together with Theorem 7.3, implies the *continuity principle* of Evans and Vasilesco [Va 1935]: If a potential $p = G\mu$ has a finite and continuous restriction to $\operatorname{supp}\mu$, then p is finite and continuous in all of X.

In view of Theorem 7.3 and the above properties (4), (5), (6), it is well known from the theory of Brelot harmonic spaces that *most qualitative results of classical potential theory carry over to the Brelot space* $(X, \mathcal{H})$ as defined in Theorem 7.1 for an admissible Riemannian polyhedron X. The inequality (7.3) from Proposition 7.3 should be imposed when it comes to global properties; and for unique continuation of harmonic functions, X is required to be a circuit with Lip continuous Riemannian structure (Proposition 5.4).

Proposition 7.4. *The Green kernel $G : X \times X \to]0, \infty]$ is Hölder continuous off the diagonal.*

Proof. Given $x, y \in X$, $x \neq y$, choose disjoint compact neighbourhoods U and V of x and y, respectively. The compact set $U \times V$ does not meet the diagonal in $X \times X$, and so G is finite and continuous, hence bounded, on $U \times V$. Choose compact neighbourhoods $K \subset U^\circ$ and $L \subset V^\circ$ of x and y, respectively. Applying Theorem 6.3 to the uniformly bounded family of harmonic functions $G(\cdot, y)$ in U°, $y \in L$, we obtain constants $A, \alpha, \delta > 0$ such that

$$|G(x_1, y_1) - G(x_2, y_1)| \leqslant A\, d_X(x_1, x_2)^\alpha \tag{7.4}$$

for $y_1 \in L$ and $x_1, x_2 \in K$ such that $d_X(x_1, x_2) \leqslant \delta$. Similarly (or by the symmetry of G)

$$|G(x_2, y_1) - G(x_2, y_2)| \leqslant A\, d_X(y_1, y_2)^\alpha \tag{7.5}$$

for $x_2 \in K$ and $y_1, y_2 \in L$ such that $d_X(y_1, y_2) \leqslant \delta$. We may of course arrange that the constants A, α, δ are the same in (7.4) and (7.5). For (x_1, y_1) and (x_2, y_2) in $K \times L$ it follows that

$$\begin{aligned} |G(x_1, y_1) - G(x_2, y_2)| &\leqslant A \left(d_X(x_1, x_2)^\alpha + d_X(y_1, y_2)^\alpha \right) \\ &\leqslant 2A\, d_{X\times X}\big((x_1, y_1), (x_2, y_2)\big)^\alpha, \end{aligned}$$

where e.g., $d_{X\times X}\big((x_1,y_1),(x_2,y_2)\big) = \big(d_X(x_1,x_2)^2 + d_X(y_1,y_2)^2\big)^{\frac{1}{2}}$ is the distance between (x_1,y_1) and (x_2,y_2) in $X \times X$. □

The following lemma and proposition, to be used in the proof of Theorem 7.4, are essentially known; they hold in the framework of [FP 1978, Chapter I] (in the symmetric case).

Lemma 7.1. *For any regular domain V in X and any point $a \in V$, the potential of the harmonic measure ω_a^V is majorized by $G(\cdot,a)$, hence ω_a^V has finite energy. Moreover, $G\omega_a^V$ is finite and continuous because*

$$G\omega_a^V(x) = \begin{cases} G(x,a) & \text{for } x \in \complement V \\ H^V_{G(x,\cdot)}(a) & \text{for } x \in V. \end{cases}$$

Proof. For any $x \in X$,

$$G\omega_a^V(x) = \int G(x,\cdot)d\omega_a^V \leqslant G(x,a)$$

because $G(x,\cdot) = G(\cdot,x) = G\varepsilon_x$ is superharmonic in X ($G\varepsilon_x$ being the potential of the Dirac measure ε_x). Since the probability measure ω_a^V is carried by ∂V, it follows that

$$\int G\omega_a^V d\omega_a^V \leqslant \int G(\cdot,a)d\omega_a^V \leqslant \max_{\partial V} G(\cdot,a) < \infty.$$

For $x \in V$ the asserted expression for $G\omega_a^V(x)$ follows from (7.1) applied to the finite continuous function $G(x,\cdot)_{|\partial V}$. For $x \in \complement\overline{V}$ the asserted expression $G(x,a)$ for $G\omega_a^V(x)$ follows from the meanvalue property (7.2) applied to the harmonic function $u = G(x,\cdot)$ in $\complement\{x\} \supset \overline{V}$.

In the remaining, more delicate case where $x \in \partial V$, we use the fact that $\omega_a^V = \varepsilon_a^{\complement V}$, the measure on X obtained by sweeping (balayage) the measure ε_a onto $\complement V$, cf. [CC 1972, Theorem 7.1.2], and we obtain by the symmetry of G

$$\begin{aligned} G\omega_a^V(x) &= \int G\varepsilon_x\, d\varepsilon_a^{\complement V} = \hat{R}^{\complement V}_{G\varepsilon_x}(a) = \hat{R}^{\complement V}_{G\varepsilon_a}(x) \\ &= \int G\varepsilon_a\, d\varepsilon_x^{\complement V} = G(x,a). \end{aligned} \tag{7.6}$$

Here $\widehat{R}_u^A$ denotes the superharmonic function $\geqslant 0$ obtained by sweeping a superharmonic function $u \geqslant 0$ on a set A, cf. [Her 1962, p. 19]. For the second and fourth equations in (7.6), see [Her 1962, Théorème 10.1]. For the third equation, see [Her 1962, Théorème 31.1], and use the symmetry of G. The last equation in (7.6) follows from $\varepsilon_x^{\complement V} = \varepsilon_x$, valid when $x \in \partial V$ because V is regular, see e.g., [CC 1972, Corollary 7.1.4]. □

Proposition 7.5. (a) *The Green kernel G_U on a domain $U \subset X$ satisfies*

$$G(x,y) = G_U(x,y) + H_U(x,y), \qquad x, y \in U,$$

where $H_U(\cdot, y)$ is harmonic and $\geqslant 0$ in U for each $y \in U$. The symmetric kernel H_U is Hölder continuous on $U \times U$.

(b) *If U is regular then, for each $y \in U$, $H(\cdot, y) = G\omega_y^U$ in U, and hence $G_U(x,y) \to 0$ as x approaches ∂U.*

Proof. Ad (a). By the above lemma, for any $a \in U$ and any regular domain V such that $a \in V \subset \overline{V} \subset U$, ω_a^V has finite energy, both relative to X and to U; and so by Theorem 7.3 (3), $G\omega_a^V \in L_0^{1,2}(X)$ and $G_U\omega_a^V \in L_0^{1,2}(U)$ $(\subset L_0^{1,2}(X))$, and furthermore

$$E(G\omega_y^V, \varphi) = \int \varphi d\omega_y^V = E_U(G_U\omega_y^V, \varphi), \qquad \varphi \in \mathrm{Lip}_c(U),$$

(where we have replaced X by U in the latter equation). This shows that

$$G\omega_y^V = G_U\omega_y^V + h_y^V \quad \text{in } U,$$

first a.e. in U with h_y^V $(\in L_0^{1,2}(U))$ weakly harmonic, and then everywhere in U with h_y^V harmonic, $G\omega_y^V$ and $G_U\omega_y^V$ being finite and continuous, by Lemma 7.1. Because $G\omega_y^V \geqslant 0$ and $G_U\omega_y^V$ is a potential on U, it follows from F. Riesz' theorem (cf. e.g., [CC 1972, p. 38]) that $h_y^V \geqslant 0$ in U. In view of Lemma 7.1 (with a replaced by $y \in V$) we have $G(\cdot, y) = G_U(\cdot, y) + h_y^V$, first in $U \setminus V$, then in $U \setminus \{y\}$ by varying V, and finally in all of U. (The last assertion is trivial if $G_U(y,y) = \infty$, and follows otherwise by the continuity principle, stated in (6) above, and applied to the potentials $G\varepsilon_y$ on X and $G_U\varepsilon_y$ on U.) This establishes the decomposition $G = G_U + H_U$ as stated, with $H_U(\cdot, y) = h_y^V$.

Given a compact set $K \subset U$, choose a relatively compact open neighbourhood L of K in U, and write $c = \max\{G(x,y) : x \in \partial L, y \in K\}$ $(< \infty)$. For $y \in K$ we have $H_U(\cdot, y) \leqslant G(\cdot, y) \leqslant c$ on ∂L, hence $H_U(\cdot, y) \leqslant c$ in L, by the boundary maximum principle [Br 1958b] (see Chapter 2 above). In particular, $H_U \leqslant c$ on $K \times K$, and so H is locally bounded in $U \times U$. This leads to the Hölder continuity of H_U by application of Theorem 6.3, much as in Proposition 7.4.

Ad (b). The function $p := G\varepsilon_y - G\omega_y^U$ is superharmonic in U because $G\omega_y^U$ is harmonic there by Lemma 7.1, which also shows that $p \to 0$ at ∂U. It follows that $p > 0$ in U, and that p is an $\mathcal{H}$-potential on U, both by classical minimum principles for superharmonic functions, valid in any Brelot harmonic space, [Br 1959, p. 1.04]. Clearly, p is harmonic in $U \setminus \{y\}$, and hence

$$p = G\varepsilon_y - G\omega_y^U = \alpha G_U \varepsilon_y$$

for some constant $\alpha = \alpha(y) > 0$, as noted in the beginning of the proof of Theorem 7.3. Thus

$$G\varepsilon_y = G_U\varepsilon_y + H(\cdot, y) = \alpha G_U \varepsilon_y + G\omega_y^U.$$

Here $H(\cdot, y)$ and $G\omega_y^U$ are harmonic in all of U, and $G_U\varepsilon_y$ and $\alpha G_U\varepsilon_y$ are $\mathcal{H}$-potentials on U. By the uniqueness in Riesz' theorem we conclude that $\alpha = 1$, $H(\cdot, y) = G\omega_y^U$. Hence $G_U\varepsilon_y = p \to 0$ at ∂U. □

It may be added that $H_U(x, y) > 0$ for all $x, y \in U$ except if $G_U = G$, that is, if $\complement U$ is polar.

Example 7.1. Given an admissible n-polyhedron X and a point $a \in X$ (which we may assume is a vertex), consider the closed star $S = \overline{\mathrm{st}}(a)$, embedded in a Euclidean space V with origin 0 corresponding to a. We may assume that $S^\circ = \mathrm{st}(a)$ (cf. Remark 4.2). Fix $\varrho > 0$ so that $B^e(a, \varrho) \Subset S^\circ$, and write

$$\Omega := B^e(a, \varrho)^\circ = \{x \in S^\circ : |x| < \varrho\},$$

where $|\cdot|$ is the norm on V, and g^e indicates the Euclidean Riemannian structure on X, so that $|x| = d_X^e(a, x)$ when $x \in S$. The corresponding Green function $G_\Omega^e(x, a)$ for Ω with pole at $a = 0$ is given outside the pole, up to a constant factor (depending only on X, a; and > 0 when $n \neq 1$), by

$$G_\Omega^e(x, a) = \begin{cases} |x|^{2-n} - \varrho^{2-n}, & n \neq 2 \\ \log(\varrho/|x|), & n = 2. \end{cases} \tag{7.7}$$

In fact, the function $x \mapsto p(x) > 0$ given by the right hand side of (7.7) for $x \in S^\circ \setminus \{a\}$ is clearly a classical harmonic function in the interior of each n-simplex of S and has normal derivative 0 along the interior of each

$(n-1)$-simplex of S containing a. Since $p \in W^{1,2}_{\text{loc}}(\Omega \setminus \{a\})$, it follows from Remark 5.3 (or directly by application of Green's formula) that p is weakly harmonic in $\Omega \setminus \{a\}$, and indeed harmonic there, being continuous. If $n \geqslant 2$, p clearly becomes superharmonic in Ω when we define $p(a) = \infty$. Because $p(x) \to 0$ as $|x| \to \varrho$, p is a *potential* on Ω, for the only harmonic minorant $\geqslant 0$ of p is 0, by the boundary minimum principle (see Chapter 2). If $n = 1$, $p(x)$ is already defined at a by the right hand side of (7.7), and $-p$ is then a potential on Ω, cf. Example 8.1.

We proceed to estimate the Green kernel G on an admissible Riemannian n-polyhedron (X, g). Inspired by the study by Littman, Stampacchia and Weinberger [LSW 1963, §7] for the particular case where X is a Euclidean domain endowed with a Riemannian metric with bounded measurable coefficients, we begin by comparing the Green kernel G_U on suitable domains $U \subset X$ with the corresponding Green kernel G^e_U with respect to the Euclidean Riemannian metric g^e (Lemma 7.2 below). The proof uses, among other things, that the sets $J_\alpha = \{x \in U : G_U(x, y) > \alpha\}$, for suitable $y \in U$, have capacity $\operatorname{cap} J_\alpha = 1/\alpha$.

Lemma 7.2. *For every point $a \in X$ there are constants $\varrho, c > 0$ depending only on X as a polyhedron, on a, and on the ellipticity constant Λ of (X, g), such that*

$$c^{-1} G^e_U(x, y) \leqslant G_U(x, y) \leqslant c\, G^e_U(x, y) \qquad \text{for } x, y \in B^e(a, \varrho),$$

whereby $U = B^e(a, 5\varrho)^\circ$.

Here the Green kernel G^e_U and the ball $B^e(a, \varrho)$ are with respect to the Euclidean Riemannian structure g^e on X.

Proof. Relative to a given triangulation of X we may assume that a is a vertex. The only requirement as to $\varrho > 0$ will be that $B^e(a, 5\varrho) \Subset S := \operatorname{st}(a)$. Set $U = B^e(a, 5\varrho)^\circ$. Fix for a while $y \in B^e(a, \varrho)$, and write for brevity

$$u(x) = G_U(x, y), \qquad x \in U.$$

We begin by estimating u in terms of the capacity of the open sets

$$J_\alpha = \{x \in U : u(x) > \alpha\}, \qquad 0 < \alpha < u(y).$$

Each J_α is relatively compact in U because $u(x) \to 0$ at ∂U, by Proposition 7.5 (b), U being regular (Remark 7.1). Recall from Chapter 2 that, for the regular Dirichlet space $L^{1,2}_0(U)$,

$$\operatorname{cap} J_\alpha = \min\{E(v) : v \in L^{1,2}_0(U), v \geqslant 1 \text{ a.e. in } J_\alpha\},$$

cf. [De 1970, p. 162], [FP 1978, 3°]. The minimizing function v here is an E-potential $p \in L_0^{1,2}(U)$ called the *equilibrium potential* for J_α. Taking p quasicontinuous and superharmonic (see the end of the proof of Theorem 7.2) we have

$$p = G_U\mu = \min\{1, u/\alpha\},$$

where μ is the *equilibrium measure* on J_α, a positive Radon measure supported by the boundary $\partial J_\alpha \subset U$ in view of Theorem 7.3 (2) because $p = 1$ (harmonic) on J_α. From the quasicontinuity of p it follows that $p = 1$ q.e. on $\overline{J}_\alpha$. To see that $p = \min\{1, u/\alpha\}$, denote this minimum by q, and note that q is likewise a potential on U (being superharmonic and majorized by the potential u/α). Moreover, u, and hence q, is continuous even on ∂J_α (which does not contain the pole y of u because $y \in J_\alpha$). It follows that $q = 1$ on $\overline{J}_\alpha$, while q is harmonic ($= u/\alpha$) in $U \setminus \overline{J}_\alpha$ (not containing y). The bounded potentials p and q on U therefore both equal 1 q.e. on $\overline{J}_\alpha$, and both are harmonic in $U \setminus \overline{J}_\alpha$. Consequently, $p = q$ (everywhere in U), by the axiom of domination (property (5) stated after Theorem 7.3).

We are now prepared to show that $\operatorname{cap} J_\alpha = 1/\alpha$ for $0 < \alpha < u(y)$:

$$\begin{aligned}\operatorname{cap} J_\alpha = E(p) &= \int_{\partial J_\alpha} p d\mu = \frac{1}{\alpha}\int_{\partial J_\alpha} u d\mu \\ &= \frac{1}{\alpha}\int_U G_U(x,y)\, d\mu(x) = \frac{1}{\alpha}\int_U G_U(y,x)\, d\mu(x) \\ &= \frac{1}{\alpha}(G_U\mu)(y) = \frac{1}{\alpha}p(y) = \frac{1}{\alpha}.\end{aligned} \tag{7.8}$$

For $r \leqslant 2\varrho$ consider the ball $B^e(y,r) \subset B^e(a, 3\varrho) \subset B^e(a, 5\varrho)^\circ = U$. Now take

$$\alpha = \min_{B^e(y,r)} u = \min_{\partial B^e(y,r)} u.$$

The latter equation follows from the minimum principles for superharmonic functions, cf. Chapter 2, in this case the Green function u on U. Similarly we show that $0 < \alpha < u(y)$, as required above. Because $B^e(y,r) \subset J_{\alpha-\varepsilon}$ for $0 < \varepsilon < \alpha$ we obtain by (7.8)

$$\operatorname{cap} B^e(y,r)^\circ \leqslant \inf_\varepsilon \operatorname{cap} J_{\alpha-\varepsilon} = 1/\alpha = 1/\min_{\partial B^e(y,r)} u. \tag{7.9}$$

Then write

$$\beta = \max_{\partial B^e(y,r)} u \ (\geqslant \alpha > 0),$$

whereby $\beta < u(y)$ also if $u(y) < \infty$, for then $u \leqslant u(y)$ in U, by Frostman's maximum principle (stated before Proposition 7.4), and indeed $u < u(y)$ by the local maximum principle applied to the harmonic function u in $U \setminus \{y\}$.

Note that the open set J_β is connected; for in any component Z of J_β not containing y we would have $u = \beta$ on ∂Z ($\subset \partial J_\beta$) in U, hence $u = \beta$ in Z (where u is harmonic); but this is impossible since $Z \subset J_\beta$. Moreover, J_β does not meet $\partial B^{\mathrm{e}}(y,r)$ (where $u \leqslant \beta$), and since $y \in J_\beta \cap B^{\mathrm{e}}(y,r)$, we conclude that $J_\beta \subset B^{\mathrm{e}}(y,r)$, and so $J_\beta \subset B^{\mathrm{e}}(y,r)^\circ$. Hence

$$\operatorname{cap} B^{\mathrm{e}}(y,r)^\circ \geqslant \operatorname{cap} J_\beta = 1/\beta = 1/\max_{\partial B^{\mathrm{e}}(y,r)} u,$$

and, invoking (7.9),

$$\min_{\partial B^{\mathrm{e}}(y,r)} u \leqslant \frac{1}{\operatorname{cap} B^{\mathrm{e}}(y,r)^\circ} \leqslant \max_{\partial B^{\mathrm{e}}(y,r)} u, \qquad r \leqslant 2\varrho. \tag{7.10}$$

By Proposition 6.2 (with ϱ replaced by the present 2ϱ, and K by $B^{\mathrm{e}}(a,\varrho)$),

$$\max_{\partial B^{\mathrm{e}}(y,r)} u \leqslant c_1 \min_{\partial B^{\mathrm{e}}(y,r)} u, \qquad r \leqslant 2\varrho,$$

where c_1 depends only on X, Λ, and a. Inserting (7.9) and (7.10) here we obtain for any $x, y \in B^{\mathrm{e}}(a,\varrho)$, writing $d_X^{\mathrm{e}}(x,y) = r$ ($\leqslant 2\varrho$) so that $x \in \partial B^{\mathrm{e}}(y,r)$,

$$\frac{1}{c_1} \frac{1}{\operatorname{cap} B^{\mathrm{e}}(y,r)^\circ} \leqslant u(x) = G_U(x,y) \leqslant c_1 \frac{1}{\operatorname{cap} B^{\mathrm{e}}(y,r)^\circ}. \tag{7.11}$$

Considering instead of G_U the Green kernel G_U^{e} and the capacity $\operatorname{cap}^{\mathrm{e}}$ on U with respect to the Euclidean Riemannian structure g^{e}, we obtain as in (7.11) for the same x, y, r

$$\frac{1}{c_1} \frac{1}{\operatorname{cap}^{\mathrm{e}} B^{\mathrm{e}}(y,r)^\circ} \leqslant G_U^{\mathrm{e}}(x,y) \leqslant c_1 \frac{1}{\operatorname{cap}^{\mathrm{e}} B^{\mathrm{e}}(y,r)^\circ}. \tag{7.12}$$

For any $v \in L_0^{1,2}(U)$ we have $\Lambda^{-n-2} E^{\mathrm{e}}(v) \leqslant E(v) \leqslant \Lambda^{n+2} E^{\mathrm{e}}(v)$, where E^{e} is the energy on (U, g^{e}) and $n = \dim X$. It follows by the definition of capacity in Chapter 2, Dirichlet structures on a space, that

$$\Lambda^{-n-2} \operatorname{cap}^{\mathrm{e}} A \leqslant \operatorname{cap} A \leqslant \Lambda^{n+2} \operatorname{cap}^{\mathrm{e}} A \tag{7.13}$$

for any relatively compact open set A in U. Combining (7.11), (7.12), (7.13) leads to the inequalities of the lemma with $c = c_1^2 \Lambda^{n+2}$. □

In the following lemma let V denote a Euclidean space into which the star $S = \operatorname{st}(a)$ of a point $a \in X$ is affinely embedded (cf. Chapter 4). Taking a as origin in V we consider a homothety $\psi : V \to V$ with centre $a = 0$ and factor $0 < \alpha \leqslant 1$; thus $\psi(x) = \alpha x$, $x \in V$.

Lemma 7.3. *For any open set $U \Subset S$,*

$$G^e_{\alpha U}(\alpha x, \alpha y) = \alpha^{2-n} G^e_U(x,y) \qquad \text{for } x,y \in U,\ 0 < \alpha \leqslant 1.$$

Proof. For fixed $y \in U$ the function $x \mapsto G^e_{\alpha U}(\alpha x, \alpha y)$ is a potential on U and harmonic in $U \setminus \{y\}$, ψ being a Riemannian isometry up to the scaling factor α. (If u is a harmonic minorant $\geqslant 0$ of $x \mapsto G^e_{\alpha U}(\alpha x, \alpha y)$ on U then $u \circ \psi^{-1}$ is a harmonic minorant $\geqslant 0$ of $G^e_{\alpha U}(\cdot, \alpha y)$ on αU, and so $u \circ \psi^{-1} = 0$, $u = 0$.) By the uniqueness property (see the beginning of the proof of Theorem 7.3),

$$G^e_{\alpha U}(\alpha x, \alpha y) = c\, G^e_U(x,y) \tag{7.14}$$

for all $x \in U$, whereby c might depend on y, but actually does not, by the symmetry of G. It remains to show that $c = \alpha^{2-n}$.

Consider a measure $\lambda \geqslant 0$ on U of finite energy $\int_U G^e\lambda\, d\lambda$ (e.g., the equilibrium measure for some open set $\Subset U$), and a function $\varphi \in \mathrm{Lip}_c(U)$. The measure $\psi^*\lambda$ on $\psi(U) = \alpha U$ likewise has finite energy, and $\varphi \circ \psi^{-1} \in \mathrm{Lip}_c(\alpha U)$. By Theorem 7.3 (3)

$$\int_U \langle \nabla_e G^e_U \lambda, \nabla_e \varphi \rangle d\mu_e = \int_U \varphi\, d\lambda, \tag{7.15}$$

where ∇_e and μ_e are the gradient and the volume measure on X with respect to the Euclidean Riemannian structure g^e. Via ψ, (7.15) transforms into

$$\frac{\alpha^{2-n}}{c} \int_{\alpha U} \langle \nabla_e [G^e_{\alpha U}(\psi^*\lambda)], \nabla_e(\varphi \circ \psi^{-1}) \rangle\, d\mu_e = \int_{\alpha U} (\varphi \circ \psi^{-1})\, d(\psi^*\lambda),$$

on account of (7.14), and so indeed $c = \alpha^{2-n}$, by (7.15) applied with αU, $\psi * \lambda$, $\varphi \circ \psi^{-1}$ in place of U, λ, φ. □

The following theorem can also be derived from [BM 1995, Theorem 1.3] by application of Lemma 4.4 and Proposition 7.5 (a) above.

Theorem 7.4. *For every compact subset K of X there is a constant $C = C(X, g, K) > 0$ such that, when $\dim X = n > 2$,*

$$C^{-1} d_X(x,y)^{2-n} \leqslant G(x,y) \leqslant C d_X(x,y)^{2-n}$$

for $x, y \in K$, $x \neq y$.

For $n = 2$ the same holds with $d_X(x,y)^{2-n}$ replaced by $\log(R/d_X(x,y))$, $R > \operatorname{diam} K$. For $n = 1$, G is Lip continuous (in particular locally bounded) in $X \times X$.

Proof. Suppose there is no such constant C. In the case $n > 2$ there are then sequences (x_j), (y_j) in K such that $x_j \neq y_j$ and

$$G(x_j, y_j)/d_X(x_j, y_j)^{2-n} \to 0 \text{ or } \infty. \tag{7.16}$$

We may assume that $x_j \to x_0 \in K$ and $y_j \to y_0 \in K$. Clearly $x_0 = y_0$; or otherwise $G(x_j, y_j) \to G(x_0, y_0)$ with $0 < G(x_0, y_0) < \infty$, by Theorem 7.3. Let s denote the carrier of $a := x_0 = y_0$, and write $S = \overline{\mathrm{st}}(a) = \overline{\mathrm{st}}(s^\circ)$. Choose $\varrho > 0$ so that $B^e(a, 5\varrho) \Subset \mathrm{st}(a)$, and write

$$U := B^e(a, 5\varrho)^\circ. \tag{7.17}$$

We may further assume that all $x_j, y_j \in B^e(a, \varrho)$. By Lemma 7.2,

$$c^{-1} G_U^e(x,y) \leqslant G_U(x,y) \leqslant c G_U^e(x,y) \qquad \text{for } x, y \in B^e(a, \varrho), \tag{7.18}$$

with $c = c(X, g, a)$. By Proposition 7.5, G_U differs on $B^e(a,\varrho) \times B^e(a,\varrho)$ only by a bounded function from G. Hence (7.16) holds with G replaced by G_U, and even by G_U^e in view of (7.18). Furthermore, d_X may be replaced by the locally equivalent distance d_X^e, and so (7.16) takes the form

$$G_U^e(x_j, y_j)/d_X^e(x_j, y_j)^{2-n} \to 0 \text{ or } \infty. \tag{7.19}$$

With S embedded as usual in a Euclidean space V with origin at a and norm $|\cdot|$, consider the cone $Y := [0, \infty[\, S$ generated by S. The Euclidean metric $e(x,y) = |x - y|$ on the admissible n-polyhedron Y embedded in V induces on Y a Riemannian metric g_Y^e with associated distance function d_Y^e (see Chapter 4). Let $p : V \to \mathbb{R}s$ denote orthogonal projection on the Euclidean subspace $\mathbb{R}s$ of V generated by s. Since $x_j, y_j \to a$, and therefore $p(x_j), p(y_j) \to p(a) = a$, we may assume in addition to $x_j, y_j \in B^e(a, \varrho)$ that

$$p(y_j) \in s^\circ \cap B^e(a, \varrho) \ \ (\subset U \subset S \subset Y). \tag{7.20}$$

Proceeding much as in Step 2 of the proof of Theorem 5.1 we consider the homothety $\psi_j : V \to V$ with centre $p(y_j)$ and factor

$$\alpha_j := \varrho^{-1} \max\{|x_j - y_j|, |y_j - p(y_j)|\} > 0 \tag{7.21}$$

since $x_j \neq y_j$. Because $\alpha_j \to 0$ we may assume that all $\alpha_j < 1$, and so by (7.20)

$$U'_j := \psi_j^{-1}(U) \supset \overline{U}, \quad \psi_j^{-1}(Y) = Y.$$

Writing $x'_j = \psi_j^{-1}(x_j)$, $y'_j = \psi_j^{-1}(y_j)$, both in Y, we have $d^e_Y(x'_j, y'_j) = \alpha_j^{-1} d^e_Y(x_j, y_j)$, and so from Lemma 7.3 applied with $X, a\,(= 0), x, y, \alpha, U$ replaced by $Y, p(y_j), x'_j, y'_j, \alpha_j, U'_j$:

$$G^e_U(x_j, y_j)/d^e_Y(x_j, y_j)^{2-n} = G^e_{U'_j}(x'_j, y'_j)/d^e_Y(x'_j, y'_j)^{2-n}. \tag{7.22}$$

Here $d^e_Y(x_j, y_j) = d^e_S(x_j, y_j) = d^e_X(x_j, y_j)$.[3] We therefore conclude from (7.19) that the right hand side of (7.22) also tends to 0 or ∞.

Because $p(y_j)$ is the centre of ψ_j we have

$$y'_j - p(y_j) = \alpha_j^{-1}(y_j - p(y_j)) \perp \mathbb{R}s \tag{7.23}$$

(including the possibility $y'_j - p(y_j) = 0$, i.e., $y_j \in s^\circ$); and from (7.21)

$$\max\{|x'_j - y'_j|, |y'_j - p(y_j)|\} = \varrho. \tag{7.24}$$

It follows from (7.24) that $|y'_j| \leqslant 2\varrho$ (since $p(y_j) \in B^e(a, \varrho)$), and hence that $|x'_j| \leqslant |y'_j| + \varrho \leqslant 3\varrho$. In particular, $x'_j, y'_j \in B^e(a, 3\varrho) \subset S$ ($\subset U$), by the choice of ϱ before (7.17).

We show presently that the Poincaré inequality (7.3) in Proposition 7.3 holds for Y, and since $\overline{U} \subset U'_j \subset Y$ we have then

$$G^e_U \leqslant G^e_{U'_j} \leqslant G^e_Y$$

and $G^e_Y = G^e_U + H^e_U$ in $U \times U$, by Proposition 7.5. Having shown above that $x'_j, y'_j \in U$ we therefore conclude from (7.22) that

$$G^e_U(x'_j, y'_j)/d^e_Y(x'_j, y'_j)^{2-n} \to 0 \text{ or } \infty. \tag{7.25}$$

After passing to subsequences we may assume that $x'_j \to x'$, $y'_j \to y'$, say, in (Y, d^e_Y) and hence in (X, d^e_X). As in the beginning of the proof it follows

---

[3]Indeed, $x_j, y_j \in B^e(a, \varrho) \subset S$, and so

$$d^e_X(x_j, y_j) \leqslant d^e_X(x_j, a) + d^e_X(a, y_j) = |x_j| + |y_j| \leqslant 2\varrho$$

since e.g., x_j and a are in the same n-simplex of $\operatorname{st}(a)$. Similarly, $d^e_Y(x_j, y_j) \leqslant 2\varrho$. It follows that the geodesic segment in (X, d^e_X) or in (Y, d^e_Y) joining x_j with y_j is contained in the ball $B^e(a, 3\varrho) \subset S$.

from (7.25) that $x' = y'$. This common limit point a' satisfies $a' \perp \mathbb{R}s$, by (7.23) because $p(y_j) \to a = 0$; furthermore, $|a'| = \varrho$ by (7.24) for $j \to \infty$. By the choice of ϱ this shows that $a' \in \mathrm{st}(s^\circ) \setminus \mathbb{R}s$, and so the carrier s' of a' contains s properly. Consequently, $\dim s < \dim s' < n$. Iterating the above construction at most $n - \dim s$ times leads to a contradiction.

As to the Poincaré inequality (7.3) for $Y = [0, \infty[\,\mathrm{st}(a)$, it reduces immediately to the same for $[0, \infty[\,\sigma$ for any n-simplex σ of $\mathrm{st}(a)$. As a very special case of Step 1 leading to Lemma 4.3 we may embed $[0, \infty[\,\sigma$ in the orthant $[0, \infty[^n$ in $\mathbb{R}^n$. Any function $u \in \mathrm{Lip}_c([0, \infty[^n)$ extends to $\mathbb{R}^n$ as a function $u^* \in \mathrm{Lip}_c(\mathbb{R}^n)$ which is symmetric in each coordinate of $x \in \mathbb{R}^n$. Being valid for $\mathbb{R}^n$, (7.3) therefore holds in $[0, \infty[^n$, hence in $[0, \infty[\,\sigma$, and finally in Y.

In the case $n = 2$ replace d_X^{2-n} in (7.16) and (7.19), etc., by $\log(R/d_X)$, etc., and $Y = [0, \infty[\,\mathrm{st}(a)$ by the ball $B := \{x \in Y : |x| < R\}$. As above, (7.16) leads to a contradiction, now because the classical Green kernel G_B for that ball is given by

$$G_B(x, y) = \log \frac{R}{|x - y|} + \frac{1}{2} \log\Big(1 - \frac{2\langle x, y\rangle}{R^2} + \frac{|x|^2|y|^2}{R^4}\Big),$$

and so

$$0 < \lim_{(x,y)\to(0,0)} G_B(x, y)/\log \tfrac{R}{|x-y|} < \infty;$$

and because the Poincaré inequality (7.3) holds for B.

In the remaining case $n = 1$, 'Hölder' may be replaced by 'Lipschitz' both in Proposition 7.4 and in Proposition 7.5 (a) in view of Remark 6.6. It therefore suffices to note that the Green kernel G_B for a Riemannian ball $B = B(a, \varrho)^\circ \Subset \mathrm{st}(a)$ is Lipschitz on $B \times B$ (see the explicit expression for G_B in Example 8.1 below). □

As in [Her 1962, §36] and [Her 1964, p. 218] (where X is a domain in $\mathbb{R}^n$ carrying a Riemannian metric with bounded measurable coefficients) we obtain from Theorem 7.4:

Corollary 7.1. *The following items are independent of the Riemannian structure on an admissible polyhedron X:* (a) *the polar sets,* (b) *the notion of thinness,* (c) *the regular domains and* (d) *the fine topology.*

Ad (a). Follows e.g., from (7.13) in the proof of Theorem 7.4 in view of property (4) stated after Theorem 7.3.

Ad (b). Follows from Theorem 7.4 in view of the following characterization of thinness due to Brelot [Br 1962a], see also [Her 1962, §14]:

A subset A of a Brelot harmonic space X with a Green kernel G is thin at a point $x_0 \in X$ (necessarily $\{x_0\}$ polar if $x_0 \in A$) iff *either* $x_0 \notin \overline{A}$ *or* there exists a potential $p = G\mu$ such that $p(x_0) < \infty$ and

$$p(x) \to \infty \qquad \text{as } x \to x_0,\ x \in A \setminus \{x_0\}.$$

One may clearly require that μ is supported by a given neighbourhood of x_0. This leads to (b).

Ad (c). Follows from (b) because a relatively compact domain V in X is regular if and only if $\complement V$ is not thin at any point of ∂V, [Br 1962b].

Ad (d). The fine topology on $\mathbb{R}^n$, or equally well on a harmonic space X, was introduced by H. Cartan as the weakest topology on X for which all subharmonic functions are continuous [Ca 1946]. He observed that a set $V \subset X$ is a neighbourhood of a point $x_0 \in V$ in the fine topology iff the complement $\complement V$ is thin at x_0. Thus (d) follows from (b).

Corollary 7.2. *If* $\dim X = n > 1$, *every point of* X *is polar (i.e.,* $\{x\}$ *is a polar set). If* $n = 1$, *every point of* X *is nonpolar, and the fine topology on* X *equals the usual topology; hence every subharmonic function is continuous.*

It follows, in fact, from Theorem 7.4 that $G\varepsilon_x(x) = G(x,x) = \infty$ for every $x \in X$ when $n > 1$, and so x is polar. Suppose now that $n = 1$. It follows then from Theorem 7.4 that every potential $G\mu$ with $\operatorname{supp}\mu$ compact is locally bounded, and there are therefore no polar points. Similarly, a set $A \subset X$ cannot be thin at any point of $\overline{A}$, cf. the proof of (b) above. The two topologies are therefore the same, and every subharmonic function is continuous, being continuous in the fine topology by the very definition thereof, cf. the proof of (d) above. Alternatively, it is easily shown directly that (when $n = 1$) subharmonic functions are continuous, in fact locally Lipschitz, cf. Example 8.1, and so the two topologies are the same.

Proposition 7.6. *The* $(n-2)$*-skeleton* $X^{(n-2)}$ *of an admissible Riemannian* n*-polyhedron is a polar set.*

Proof. Recall from Chapter 4 that $X^{(n-2)}$ is the union of all simplexes of dimension $\leqslant n-2$ (in a given triangulation of X). We may assume that $n \geqslant 2$ (since $X^{(n-2)} = \emptyset$ if $n = 1$). We shall prove that every point $a \in X^{(n-2)}$ has an open connected neighbourhood U in X such that $U \cap X^{(n-2)}$ is polar in U and hence in X, polarity being a local property. By subdivision of the given triangulation of X (thus enlarging the skeleton $X^{(n-2)}$) we may arrange that a is a *vertex* of X, and that (4.8) and (4.9) hold. Use the Lip homeomorphism θ from Lemma 4.3 to transport the Riemannian structure g on $S = \overline{\mathrm{st}}(a)$ to the cell complex $P \subset [0,1]^k$ ($k \geqslant n$ being the number of 1-simplexes of X having a as a vertex). Taking $U = S^\circ$, we shall prove that the set $P^\circ \cap P^{(n-2)}$ $(= \theta(S^\circ \cap S^{(n-2)}))$ is polar in P° with respect to $(\theta^{-1})^* g$, or equivalently with respect to the Euclidean Riemannian structure h^e on P, in view of Corollary 7.1 above.

For any $0 \leqslant d \leqslant n$ write

$$F_d = \{x \in]0,1[^k \colon x_i > 0 \ \text{ iff } \ 1 \leqslant i \leqslant d\}.$$

Any finite union of polar sets being polar, it suffices to prove that each open d-face F of P such that $0 \in \overline{F}$ and that $d \leqslant n-2$ (e.g., $F =$ the above F_d) is polar in (P°, h^e). Consider on the (open) star Ω of F_d in P the continuous extended real valued function

$$u(x) = \begin{cases} -\frac{1}{2}\log(x_{d+1}^2 + \cdots + x_k^2), & d = n-2 \\ (x_{d+1}^2 + \cdots + x_k^2)^{2-n+d}, & d < n-2, \end{cases}$$

understood as ∞ for $x_{d+1}^2 + \cdots + x_k^2 = 0$, e.g., for $x \in F_d$. The proof will be completed by showing that u is harmonic in $\Omega \setminus F_d$, for then u is clearly superharmonic in Ω (Definition 5.2).

Consider any open $(n-1)$-face of Ω and any adjacent open n-face of Ω, e.g., the faces F_{n-1} and F_n, respectively. Because $d+2 \leqslant n$, $x_{d+1}^2 + \cdots + x_n^2 > 0$ on $F_n \cup F_{n-1}$; and so u is smooth there, and harmonic in F_n by classical potential theory, the number of variables $x_{d+1}, \cdots, x_n$ being $n-d \geqslant 2$. The normal derivative $\partial u/\partial x_n$ of $u_{|F_n \cup F_{n-1}}$ along F_{n-1} equals 0 since $x_n = 0$ on F_{n-1}. By Remark 5.3 (or directly by Green's formula), u is indeed harmonic in Ω. □

Invoking well known results on polar sets as removable singularities in a harmonic space, see Chapter 2, we obtain the following:

Corollary 7.3. *Every bounded harmonic, resp. lower bounded superharmonic, function on $X \setminus X^{(n-2)}$ has a unique harmonic, resp. superharmonic, extension to X.*

In particular, $X \setminus X^{(n-2)}$ is *connected* (see Chapter 2).

Example 7.2. As in Chapter 4, let (X, g) denote a Riemannian normal n-circuit without boundary, whose singular set $S = S(X)$ is a subpolyhedron of the $(n-2)$-skeleton X^{n-2} (relative to some triangulation). Assume that the Riemannian metric g of X is smooth on $X \setminus S$, usually incomplete. Then Corollary 7.3 identifies the potential theory of $(X, \mathcal{H})$ as in Theorem 7.1 with that of those harmonic functions on $(X \setminus S, g_{|X\setminus S})$ which are bounded near any point of S.

That situation arises when X is a projective algebraic variety as in Example 8.9. The Fubini–Study metric of projective space induces the Bergman metric of $X \setminus S$, which is of course Kähler but incomplete. *If $g_{X\setminus S}$ is Kähler, then any bounded holomorphic function $u : U \to \mathbb{C}$ (U a subdomain of X) is harmonic.* Indeed, $U \setminus S$ is a Kähler manifold, so the restriction of u to it is harmonic [ES 1964, §2C]. The assertion now follows upon application of Corollary 7.3.

As another application of Corollary 7.3 (to be used in Chapter 9) we bring the following simplexwise characterization of local Sobolev $(1,2)$-functions, noting that $W_0^{1,2}(X) \subset L_0^{1,2}(X)$, and so every function $u \in W_{\mathrm{loc}}^{1,2}(X)$ (being locally of class $W_0^{1,2}$) has quasicontinuous versions, cf. the proof of Theorem 7.2.

Proposition 7.7. *A quasicontinuous function u on X is of class $W_{\mathrm{loc}}^{1,2}(X)$ if (and only if) $u_{|s} \in W^{1,2}(s)$ for every topdimensional simplex $s \in \mathcal{S}^{(n)}(X)$.*

Proof. For a given triangulation of X consider the $(n-2)$-skeleton $X^{(n-2)}$ and the connected open set $U = X \setminus X^{(n-2)}$, cf. Proposition 7.6.

Step 1. $u_{|U} \in W_{\mathrm{loc}}^{1,2}(U)$. Given a point $a \in U$, in proving that $u_{|V} \in W^{1,2}(V)$ for some subdomain V of U containing a, we may assume that $u = 0$ off some compact subset of the open star $\mathrm{st}_X(a)$ (otherwise replace u by φu for some $\varphi \in \mathrm{Lip}_c(\mathrm{st}_X(a))$ with $\varphi = 1$ in a neighbourhood V of a; then φu has the same properties as u, and equals u in V). The carrier t of a has dimension $n-1$, the case $\dim t = n$ being trivial.

We may assume that $S := \overline{\mathrm{st}}_X(a)$ is embedded affinely in a Euclidean space in such a way that the n-simplexes $s_1, \cdots, s_k$ containing t are all congruent, cf. Step 1 leading to Lemma 4.3. We may further suppose that $k > 1$, for otherwise s_1 is a neighbourhood of a, and $u \in W^{1,2}(s^1)$ by assumption. Let $\widetilde{u}$ denote the continuation of $u_{|s_k}$ to $S = \bigcup_j s_j$ by symmetry across t. By hypothesis, $u_{|s_k} \in W^{1,2}(s_k)$, and hence $\widetilde{u} \in W^{1,2}(S)$, with $\mathrm{supp}\,\widetilde{u}$ compact in $S^\circ = \mathrm{st}_X(a)$. (Indeed, if $u_\nu \in \mathrm{Lip}(s_k)$ and $u_\nu \to u_{s_k}$ in $W^{1,2}(s_k)$, then the continuation $\widetilde{u}_\nu$ of u_ν to S by symmetry across t is of class $\mathrm{Lip}(S)$, and $\widetilde{u}_\nu \to \widetilde{u}$ in $W^{1,2}(S)$.) Furthermore, $\widetilde{u}$ is quasicontinuous on S, and hence so is $u - \widetilde{u}$, as well as its restriction to $M := (s_j \cup s_k)^\circ$ for each $j < k$; M may be identified with a Euclidean domain. (We use here that the restriction of any function of class $L_0^{1,2}(S^\circ)$ to M is of class $L_0^{1,2}(M)$ with no bigger energy. For any open set $e \Subset S^\circ$ it follows that the capacity of $e \cap M$ relative to M is no bigger than that of e relative to S°.)

Because $(u - \widetilde{u})_{|s_j} \in W^{1,2}(s_j)$ $(\subset W^{1,2}(M))$ and $(u - \widetilde{u})_{|s_k} = 0$, we conclude altogether from the spectral synthesis theorem of Beurling and Deny, applied in M, that $(u - \widetilde{u})_{|s_j^\circ} \in W_0^{1,2}(s_j^\circ)$ for each $j < k$ (cf. [De 1970, p. 172f.], [Fuk 1980, p. 81f.], [AH 1996, Theorem 9.1.3 and p. 279]). For each $j \leqslant k$ there is therefore a sequence $(v_{1,j}, v_{2,j}, \cdots) \subset \mathrm{Lip}_c(s_j^\circ)$ converging to $(u - \widetilde{u})_{|s_j}$ in $W^{1,2}(s_j^\circ) = W^{1,2}(s_j)$. For each ν these $v_{\nu,j}$ $(j = 1, \cdots, k)$ fit together so as to form a function $v_\nu \in \mathrm{Lip}(S)$, and $v_\nu \to u_{|S} - \widetilde{u}$ in $W^{1,2}(S)$. It follows (with $\widetilde{u}_\nu \in \mathrm{Lip}(S)$ from the preceding paragraph) that $\widetilde{u}_\nu + v_\nu \in \mathrm{Lip}(S)$, $\widetilde{u}_\nu + v_\nu \to u_{|S}$ in $W^{1,2}(S)$, and so $u_{|S} \in W^{1,2}(S)$, $u \in W^{1,2}_{\mathrm{loc}}(U)$.

Step 2. From $u \in W^{1,2}_{\mathrm{loc}}(X \backslash X^{(n-2)})$ (Step 1) we shall obtain $u \in W^{1,2}_{\mathrm{loc}}(X)$. A proof of this directly in terms of the polyhedral structure of X, as in Step 1, does not seem at hand; instead we use potential theory.

We may assume that $u \geqslant 0$ because u_+ and u_- inherit the properties required for u. This time consider a point $a \in X^{(n-2)}$ and its *open* star $S := \mathrm{st}_X(a)$, and write $U = S \setminus X^{(n-2)}$. By Step 1, $u_{|U} \in W^{1,2}_{\mathrm{loc}}(U)$, and even $u \in W^{1,2}(U)$ by Lemma 5.1 because the Sobolev norm of $u_{|U}$ is finite by hypothesis. It follows therefore from Theorem 5.2 and Remark 5.5 that there is a (unique) weakly harmonic function $h \geqslant 0$ a.e. on U such that $u_{|U} - h \in W_0^{1,2}(U)$. We may assume that h is continuous and hence *harmonic* and $\geqslant 0$ on U. By Corollary 7.3, h extends (uniquely) to a superharmonic function $s \geqslant 0$ on all of S. Because $u_{|S} - s$ is an extension of $u_{|U} - h \in W_0^{1,2}(U)$, and $S \backslash U \subset X^{(n-2)}$ is polar, $u_{|S} - s$ is quasicontinuous and of class $W_0^{1,2}(S)$. We may therefore replace u (or $u_{|S}$) by s (quasicontinuous) in Step 2 (with X replaced by S), and our task is to prove that $s \in W^{1,2}_{\mathrm{loc}}(S)$, hence actually $s \in W^{1,2}(S)$.

We first show, for given $\lambda > 0$, that

$$s_\lambda := \min\{s, \lambda\} \in W^{1,2}(S).$$

For any domain V of compact closure $\overline{V} \subset S$, s_λ coincides in V with the potential $p = \widehat{R}^V_{s_\lambda}$ on S obtained by sweeping the bounded superharmonic function s_λ onto V (see e.g., [CC 1972, Proposition 5.3.5]); p is harmonic in $S \setminus \overline{V}$ (see [Br 1959, p. 1.09]). Let G denote the Green kernel on $S \times S$ (cf. Remark 7.2). Then $p = G\mu$ for a measure $\mu \geqslant 0$ of compact support

($\subset \overline{V}$), and so $\int G\mu \, d\mu < \infty$. By Theorem 7.3 (3), $p \in L_0^{1,2}(S) = W_0^{1,2}(S)$ (invoking the Poincaré inequality (5.12), cf. Remark 5.5). Thus $(s_\lambda)_{|V} = p_{|V} \in W^{1,2}(V)$, whence $s_\lambda \in W^{1,2}_{\mathrm{loc}}(S)$, and so indeed $s_\lambda \in W^{1,2}(S)$ by Lemma 5.1 because s_λ has finite Sobolev norm (no bigger than that of s).

As $\lambda \to \infty$ (through the integers, say), $s_\lambda \to s$ in $L^2(S, \mu_g)$ because s, like $u_{|S}$, has finite Sobolev norm, and in particular $\int_S s^2 d\mu_g < \infty$. We complete the proof by showing that $s \in W^{1,2}(S)$, and for that it suffices that (s_λ) is Cauchy in energy. For $\lambda' > \lambda$, $s_{\lambda'} - s_\lambda$ is constant off the set $\{\lambda < s < \lambda'\}$, and so, by Remark 5.2, $\nabla s = 0$ a.e. off that set. Thus

$$\int_S |\nabla s_{\lambda'} - \nabla s_\lambda|^2 d\mu_g = \int_{\{\lambda < s < \lambda'\}} |\nabla s|^2 d\mu_g \to 0 \qquad \text{as } \lambda \to \infty$$

because $\int |\nabla s|^2 d\mu_g < \infty$. □

Remark 7.6. In Proposition 7.7 the quasicontinuity of u can be replaced by the following equivalent hypothesis: *For any two n-simplexes s and s' with a common $(n-1)$-face t, the traces of $u_{|s}$ and $u_{|s'}$ on t should be equal.* (Same proof, using in Step 1 [KS 1993, Theorem 12.2.3], with $\mathbb{R}$ as target space, to show that $(u - \widetilde{u})_M \in W^{1,2}(M)$, and hence as before that $(u - \widetilde{u})_{|s_j} \in W_0^{1,2}(s_j)$.)

Remark 7.7. Step 2 in the proof of Proposition 7.7 shows more generally that *every closed polar subset F of X is removable for Sobolev* $(1,2)$-*functions*, i.e., $W^{1,2}(X \setminus F) = W^{1,2}(X)$. A larger class of removable sets in this sense is that of all (closed) sets of $(n-1)$-dimensional Hausdorff measure 0. The proof of this for $X = \mathbb{R}^n$ (see e.g., [AH 1996, Lemma 9.1.10]) can be adapted to the present case of an admissible n-polyhedron X. (In view of the above it suffices to prove that $W^{1,2}_{\mathrm{loc}}(U \setminus F) = W^{1,2}_{\mathrm{loc}}(U)$ for $U = X \setminus X^{(n-2)}$ and F closed in U and of $(n-1)$-dimensional Hausdorff measure 0; and that can be established by an argument similar to Step 1 above.)

Quasitopology and fine topology

For an admissible Riemannian polyhedron (X, g) satisfying the Poincaré-style inequality (7.3) we have introduced the Dirichlet space $L_0^{1,2}(X)$ and its Green kernel G. Furthermore, the concept of capacity, in terms of which there is the notion of quasicontinuous functions (cf. Chapter 2) and similarly that of quasiopen sets:

Definition 7.2. A set $U \subset X$ is *quasiopen* if there are open sets ω of arbitrarily small capacity such that $U \setminus \omega$ is open relative to $X \setminus \omega$.

Every *countable* union and finite intersection of quasiopen sets is quasiopen.

Definition 7.3. A map $\varphi : U \to Y$ from a quasiopen set U to a topological space Y with a countable base of open sets is *quasicontinuous* if there are open sets ω of arbitrarily small capacity such that $\varphi_{|U \setminus \omega}$ is continuous.

Clearly, this amounts to $\varphi^{-1}(V)$ being quasiopen for every open subset V of Y.

Such "quasitopological" concepts are equivalent up to polar sets (i.e., sets of capacity 0) with corresponding true topological concepts relative to the Cartan fine topology (cf. Corollary 7.1 (d)), these latter concepts being distinguished below by the qualification "fine(ly)". Recall that the fine topology is the weakest topology on X for which every subharmonic function is continuous.

Proposition 7.8. (a) *A set $e \subset X$ is polar iff e is finely discrete and its points are polar; e is then finely closed.*

(b) *A set $U \subset X$ is quasiopen iff U differs only by a polar set from some finely open set (e.g., the fine interior of U).*

(c) *A map φ from X (or from a quasiopen subset of X) to a topological space with a countable base is quasicontinuous iff φ is finely continuous quasieverywhere (i.e., off some polar set).*

(d) *The empty set is the only finely open null set.*

Parts (a) and (b) together express that *the quasitopology on X is compatible with the fine topology*, in the sense of [Fug 1971, p. 143], where also (c) is derived easily from (b).

The proof of this compatibility in the present setting is derived from the corresponding proof in [Fug 1971, §5], replacing in n° 5.7 there the cone $\check{G}\lambda$ of bounded co-potentials by the cone $\mathcal{U}$ of potentials $p = G\mu$ of measures $\mu \geqslant 0$ of finite energy $E(p) = \frac{1}{2}\int_X p\, d\mu = \frac{1}{2}\int_X G\mu\, d\mu$ (cf. Theorem 7.3 (3) above). Recall that each such potential is quasicontinuous, and that the axiom of domination holds (property (6) stated after Theorem 7.3).

Ad (d). Choose a finite strict potential $p > 0$ on X (see e.g., [CC 1972, §7.2]), and write $u = p - \widehat{R}_p^{\complement U} \geqslant 0$, where $\widehat{R}_p^{\complement U}$ is obtained by sweeping p on $\complement U$. Then $u > 0$ in U, and $u = 0$ q.e. in $\complement U$; hence $u = 0$ a.e. in X. It follows that $u = 0$ even q.e. according to [De 1970, Théorème 5], and so

$u = 0$ everywhere, by fine continuity, polar sets being finely discrete, by (a) above.

Proposition 7.8 clearly extends to any hypoelliptic Dirichlet space in the sense of Feyel and de La Pradelle [FP 1978].

Remark 7.8. The above quasitopological concepts have a local character and can therefore be defined on any admissible Riemannian polyhedron (X, g) without requiring that (7.3) hold. Say that a set $U \subset X$ is *quasiopen* if X can be covered by open sets X' for which (7.3) holds and such that $U \cap X'$ is quasiopen in X'. Recall that (7.3) always holds in the (open) star $S = \mathrm{st}_X(a)$ of any vertex a of X in a triangulation fine enough so that S is the interior of its closure in X (Remark 5.5).

This local concept of quasiopen sets is equivalent with the global concept from Definition 7.2 in the case where X does satisfy (7.3). That follows immediately from Proposition 7.8 above, simply because the fine topology is a local concept (the subharmonic functions forming a sheaf, cf. Chapter 2), and because the union of countably many polar sets is polar.

Similarly for other quasitopological concepts, such as *quasicontinuity* (Definition 7.3 above). And now Proposition 7.8 holds without assuming (7.3).

In the rest of this book we shall accordingly drop the hypothesis (7.3) without further ado when speaking of quasitopological or other local concepts. (The same can be done in the preceding section, beginning with Corollary 7.1.)

Sobolev functions on quasiopen sets

In our subsequent study of (possibly discontinuous) maps of finite energy from an admissible Riemannian polyhedron (X, g) to a smooth Riemannian manifold or a Riemannian polyhedron enters the class $W^{1,2}(U)$ of Sobolev functions on a set $U \subset X$ which need not be open, but only quasiopen.

The following concept was introduced by Kilpeläinen and Malý for the case $X = \mathbb{R}^n$. Both the definition and the properties quoted below from [KilM 1992, § 2] extend immediately (in the absence of other indications) to the present case of an admissible Riemannian polyhedron (X, g), with the Riemannian measure μ_g replacing Lebesgue measure.

Definition 7.4. A function $u : U \to \mathbb{R}$ defined on a quasiopen set $U \subset X$ is of *class* $W^{1,2}(U)$ if:

(a) U is the union of a sequence of quasiopen sets U_j such that each restriction $u_{|U_j}$ can be extended to a function $\widetilde{u}_j \in W^{1,2}(\widetilde{U}_j)$ on some usual open set $\widetilde{U}_j \supset U_j$; and

(b) $\|u\|^2 := \int_U (u^2 + |\nabla u|^2) d\mu_g$ is finite.

In (b), the gradient $\nabla u \in L^2(U)$ is clearly independent of the choice of cover (U_j) and of extension $\widetilde{u}_j : \widetilde{U}_j \to \mathbb{R}$ of each $u_{|U_j}$, in view of Remark 5.2.

Every function $u \in W^{1,2}(U)$ equals a.e. a quasicontinuous function on U, uniquely determined q.e. in U. Futhermore, $(W^{1,2}(U), \|\cdot\|)$ is a Banach space [KilM 1992, Theorem 2.6].

The restriction to U of the functions of class $W^{1,2}(\widetilde{U})$ on usual open sets $\widetilde{U} \supset U$ are norm dense in $(W^{1,2}(U), \|\cdot\|)$. In particular, for a usual open set $U \subset X$, $(W^{1,2}(U), \|\cdot\|)$ as defined in Definition 7.4 is the same as in the usual sense.

The chain rule (Lemma 5.2) extends immediately to Sobolev functions on a quasiopen set $U \subset X$. For example, if $V \subset \mathbb{R}^n$ is open, if $\varphi \in C^1(V)$ has bounded gradient, and if $u = (u_1, \cdots, u_p) : U \to V$ is of class $[W^{1,2}(U)]^p$, then $\varphi \circ u \in W^{1,2}(U)$ and

$$\nabla(\varphi \circ u) = [(D_\alpha \varphi) \circ u] \nabla u^\alpha \qquad \text{a.e. in } U, \tag{7.26}$$

cf. [KilM 1992, Lemma 2.5]. Under the weaker hypothesis that $\varphi \in \mathrm{Lip}(V)$ it similarly follows that $\varphi \circ u \in W^{1,2}(U)$.

When passing from open sets $U \subset X$ to quasiopen sets, the space $\mathrm{Lip}_c(U)$ of "test functions" on U is replaced by the following space $D(U)$, [KilM 1992, p. 378]:

Definition 7.5. For a quasiopen set $U \subset X$, $D(U)$ denotes the space of all bounded quasicontinuous functions $\lambda \in W^{1,2}(X)$ vanishing outside some quasiopen set $V \Subset U$ and for which there exists a quasicontinuous function $v \in W^{1,2}(X)$ such that $0 \leqslant v \leqslant 1$, $v = 0$ on $\complement U$, and $v = 1$ on V.

For any quasiopen set $U \subset X$, define

$$W_0^{1,2}(U) = \bigcap \{ W_0^{1,2}(G) : G \text{ open in } X,\, G \supset U \}.$$

$W_0^{1,2}(U)$ is the closure of $D(U)$ in $W^{1,2}(U)$. A quasicontinuous function $u \in W^{1,2}(U)$ belongs to $W_0^{1,2}(U)$ iff the extension of u by 0 in $\complement U$ is quasicontinuous on X [KilM 1992, Theorem 2.10].[4]

---

[4]The proof given there uses that $W^{1,2}_{\mathrm{loc}}(X) = W^{1,2}(X)$ in the case of a Euclidean space (X, g), in order to reduce to the case where $\overline{U}$ is compact; but that can always be achieved by multiplying with certain Lip functions and applying Lemma 5.1.

The class $D(U)$ also leads to a good substitute for the concept of partition of unity [KilM 1992, Lemma 2.4]; and that enters in the proof of most of the results on $W^{1,2}(U)$ quoted above.

Subharmonicity of convex functions

The following is a partial extension of a result of Greene and Wu, [GW 1973]. (For definitions of the notions continuous and simplexwise smooth Riemannian metric, see above in Chapter 4, Riemannian polyhedron.)

Proposition 7.9. *Let (X, g) be an admissible Riemannian polyhedron with empty boundary, and suppose that g is continuous and simplexwise smooth. Every continuous convex function $u \in W^{1,2}_{\text{loc}}(X)$ is then subharmonic.*

Proof. As shown in [GW 1973], u is subharmonic in the interior s° of each n-simplex s ($n = \dim X$). By Corollary 7.3 above, it remains to verify the inequality $\sum_{j=1}^k D_j u \geqslant 0$ (Remark 5.3) for the inner normal derivatives $D_j u$ of $u_{|s_j}$ along any open $(n-1)$-simplex t° of X, $s_1, \cdots, s_k$ denoting the n-simplexes of X containing t. Here $k \geqslant 2$ because $b(X) = \emptyset$.

Fix $i, j \in 1, \cdots, k$ with $i \neq j$. For any point $p \in t^\circ$ consider the geodesics γ_i, γ_j in s_i, s_j emanating from p and perpendicular to t. Then γ_i (reversed) followed by γ_j is a geodesic segment γ in the Riemannian polyhedron $s_i \cup s_j$. In fact, for any points $p_i \in \gamma_i$, $p_j \in \gamma_j$ sufficiently close to p, the balls B_i, B_j centred at p_i, p_j and of radii $d_X(p, p_i), d_X(p, p_j)$ are entirely contained in s_i, resp. s_j, and hence have disjoint interiors. Any path joining p_i and p_j has therefore a length no less than $d_X(p, p_i) + d_X(p, p_j) \geqslant d_X(p_i, p_j)$.

Because $u \circ \gamma$ is convex, $D_i u(p) + D_j u(p) \geqslant 0$. When (i, j) ranges over all the $k(k-1)$ ordered pairs as stated above, we obtain by addition

$$2(k-1)\sum_{j=1}^{k} D_j u(p) = \sum_{i \neq j} \big(D_i u(p) + D_j u(p)\big) \geqslant 0.$$

□

Remark 7.9. For the case of a smooth Riemannian manifold with empty boundary the above proposition is due to Greene and Wu [GW 1973]. That would break down in the presence of a boundary. Thus id : $[0, 1] \to [0, 1]$ is convex, but fails to be subharmonic at 1, cf. Remark 5.3.

8

Examples of Riemannian polyhedra and related spaces

This chapter offers a variety of examples of Riemannian polyhedra. We illustrate the importance of admissibility – but have not insisted here on the *global* uniform ellipticity of the Riemannian metric (condition (4.1)).

Example 8.1. 1-dimensional Riemannian polyhedra

Every 1-dimensional polyhedron is admissible in the sense of Chapter 4. Let X be a 1-dimensional Riemannian polyhedron. Every function $u \in W^{1,2}(X)$ is locally uniformly Hölder continuous with exponent $\frac{1}{2}$ (after correction on a null set). Indeed, it is sufficient to verify that for compact X: Any two sufficiently nearby points $x_1, x_2 \in X$ are either in the same 1-simplex or in two with a common vertex. In either case, by the Cauchy-Schwarz inequality,

$$|u(x_1) - u(x_2)|^2 \leqslant d_X(x_1, x_2) \sum_{s \in \mathcal{S}^1(X)} \int |u'|^2.$$

More generally, every map $\varphi \in W^{1,2}_{\text{loc}}(X, Y)$ of (X, g) into a locally compact geodesic space (Y, d_Y) has a locally uniformly Hölder continuous version with exponent $\frac{1}{2}$ (see the paragraph immediately preceding Proposition 12.4).

The fine topology on X equals the usual (metric) topology, and $\emptyset$ is the only polar set (Corollary 7.2). It follows that every sub/superharmonic function is continuous and finite valued.

The harmonic space structure $\mathcal{H} = \mathcal{H}_X$ as defined in Chapters 2 and 5–7 can be described and characterized as follows in view of Remark 5.3:

For any open set $U \subset X$, a continuous function $u : U \to \mathbb{R}$ is of class $\mathcal{H}(U)$, resp. $\mathcal{S}(U)$ (the superharmonic functions on U), iff:

(a) For any 1-simplex s^1 of X meeting U, u is an affine, resp. convex, function of the Riemannian path length on s^1 in each component of $U \cap s^1$.

(b) For any 0-simplex $s^0 = p \in U$, let s_j^1 $(1 \leqslant j \leqslant k = k(p))$ denote the collection of all 1-simplexes of X sharing the common vertex p. Then

$$\sum_{j=1}^{k} D_j u(p) = 0, \quad \text{resp.} \sum_{j=1}^{k} D_j u(p) \geqslant 0, \tag{8.1}$$

where $D_j u(p) = du(p)/d\sigma_j$; here $\sigma_j = d_X(p, \cdot)$ denotes Riemannian path length from p in s_j^1.

As an application it is easily verified that the symmetric *Green kernel* G^B for a ball $B = B(p, \varrho)^\circ \Subset \mathrm{st}(p)$ is given (up to a constant factor > 0) in terms of $|x| = d_X(p, x)$, $|y| = d_X(p, y)$ by

$$G^B(x, y) = \begin{cases} (\varrho - |x|)(\varrho - |y|) \\ (\varrho - |x|)(\varrho + (k-1)|y|) \end{cases}$$

for $x, y \in B$, the former expression applying when x and y lie on different 1-simplexes of $\mathrm{st}(p)$, the latter when x and y lie on the same 1-simplex and $|x| > |y|$. It follows that $G^B \in \mathrm{Lip}(B \times B)$.

The fact that the above map $U \mapsto \mathcal{H}(U)$ is a Brelot harmonic sheaf (see Chapters 2 and 7) can alternatively be verified directly; and it can be shown that this remains true if equation (8.1) is replaced by $\sum a_j D_j u(p) = 0$ for given constants $a_j > 0$ $(j = 1, \cdots, k)$, allowed to depend on p. If, however, at some vertex p, $a_j = 0$ is allowed for some, but not all j, then $\mathcal{H}$ becomes a Bauer harmonic sheaf, but not a Brelot sheaf.

It is elementary to verify that every complete 1-dimensional Riemannian polyhedron (X, d_X) has *nonpositive curvature*: $\mathrm{curv}(X, d_X) \leqslant 0$, in the sense of Definition 2.8.

Example 8.2. The need for dimensional homogeneity

Let s^j be a Euclidean j-simplex, and form the polyhedron $X = s^1 \cup s^2 \subset \mathbb{R}^2$, where $s^1 \cap s^2$ is a common vertex $p = s^0$. X is not dimensionally homogeneous, and therefore not admissible. Each of the homogeneous parts s^j, $j = 1, 2$, is an admissible Riemannian polyhedron with the induced Euclidean metric. Let $\mathcal{H}_{s^2}$ denote the Brelot harmonic sheaf on s^2 as defined in Chapter 7. We show below that *there does not exist any Brelot harmonic sheaf $\mathcal{H}$ on X compatible with $\mathcal{H}_{s^2}$* in the sense that $\mathcal{H}(U) = \mathcal{H}_{s^2}(U)$ for any set $U \subset s^2$ which is open in X (and hence in s^2). Briefly, this is because p is a polar point for $\mathcal{H}_{s^2}$, according to Corollary 7.2, hence a removable singularity for bounded $\mathcal{H}_{s^2}$-functions.

Suppose there does exist a harmonic sheaf $\mathcal{H}$ on X which is compatible with $\mathcal{H}_{s^2}$. Consider a regular domain $U \subset X$ (with respect to $\mathcal{H}$) containing p and such that $\overline{U} \subset \mathrm{st}_X(p)$. Let $u \in \mathcal{H}(U)$ extend continuously to $\overline{U}$ so that $u = 1$ on $s^1 \cap \partial U$ and $u = 0$ on the rest of ∂U. By the local maximum principle for harmonic functions on a Brelot harmonic space [Br 1959] (cf. Chapter 2) it follows that $0 < u(p) < 1$; but on the other hand $u(p) = 0$, as we shall now see. Being $\mathcal{H}_{s^2}$-harmonic in $U \cap s^2 \setminus \{p\}$ (an open set in X contained in s^2) by the required compatibility, and continuous in $U \cap s^2$, u is in $\mathcal{H}_{s^2}(U \cap s^2)$ by a well known result on removable singularities [Br 1959] (cf. Chapter 2) valid in any Brelot harmonic space (such as the Riemannian polyhedron s^2). Because $u = 0$ on $\partial_{s^2}(U \cap s^2) = (\partial U) \cap s^2$, we conclude from the boundary minimum principle (cf. Chapter 2) that $u \equiv 0$ in $U \cap s^2$, in particular at p, a contradiction.

Example 8.3. The need for local chainability

(Cf. [GMacP 1980, Fig. 5].) Let X be a Riemannian 2-torus with one meridian circle pinched to a point p.

As a polyhedron (indeed a circuit), X is not locally 1-chainable at p, and therefore not admissible. Again, *there is no harmonic structure on X extending the usual harmonic structure on $X \setminus \{p\}$*. To see this, take two meridian circles C_0, C_1, one on either side of p, forming the boundary of a double cone X_0 in X which we may assume is regular in X. Assign constant values $f = 0$ to C_0 and $f = 1$ to C_1. Then f is continuous, but has no harmonic extension to X_0, for such an extension would have to be both 0 and 1 at p, this point being polar (Corollary 7.2) in either half-cone (an admissible Riemannian polyhedron, cf. Examples 4.3 and 8.8), so p would be a removable singularity for bounded harmonic functions there according to [Br 1959] (as in Example 8.2 above).

Example 8.4. Manifolds as polyhedra

Relative to a smooth triangulation, every connected smooth *Riemannian manifold* (possibly with boundary) is a Riemannian polyhedron, indeed a Riemannian normal circuit, and so is every connected *triangulable* Riemannian Lipschitz manifold M (cf. Example 4.2 and Subexample 4.2). That includes the case where M has a triangulable Riemannian Lipschitz boundary (e.g., a smooth Riemannian M whose boundary has corners).

Example 8.5. A kind of connected sum of polyhedra

Let M_0, M_1 be admissible n-dimensional polyhedra, and $f_i : U \to M_i$ proper embeddings ($i = 0, 1$) of the open unit ball U in $\mathbb{R}^n$ onto the interiors of two n-dimensional simplexes of M_0 and M_1, respectively. Use those to attach the cylinder $(\partial U) \times I$ to the disjoint union $M_0 \cup M_1$, via the identification

$$f_1 \circ f_0^{-1} : f_0(\partial U) \to f_1(\partial U).$$

The resulting space $M_0 \widetilde{\#} M_1$ is an admissible n-polyhedron. (That is like the connected sum $M_0 \# M_1$, except that *its* construction starts with the disjoint union $(M_0 \setminus \operatorname{int} f_0(U)) \cup (M_1 \setminus \operatorname{int} f_1(U))$. But even when M_0 and M_1 are manifolds, $M_0 \widetilde{\#} M_1$ is not a circuit.) See [Mil 1959].

If M_0, M_1 are Riemannian with metric structures g_0, g_1, then $M_0 \widetilde{\#} M_1$ *is an admissible Riemannian polyhedron* with metric g given by g_i on M_i and

$$dt^2 + (1-t) f_0^* g_{0|f_0(\partial U)} + t f_1^* g_{1|f_1(\partial U)}$$

on $(\partial U) \times \{t\}$ for each $t \in I$.

Example 8.6. Riemannian joins of Riemannian manifolds

If X and Y are subsets of Euclidean spaces V and W, respectively, with V and W embedded as orthogonal subspaces of a larger Euclidean space, the *join* $X * Y$ is the set of all points of all segments xy joining $x \in X$ to $y \in Y$; i.e.,

$$X * Y = \{z = (x, y, t) \in V \times W \times I : z = (1-t)x + ty\}.$$

If (X, d_X) and (Y, d_Y) are *metric subspaces* of the above spaces V and W, respectively, a metric d_{X*Y} on $X * Y$ is defined by

$$d_{X*Y}^2(z, z') = d_X^2(x, x') + d_Y^2(y, y') + |t - t'|^2.$$

If (X, d_X) and (Y, d_Y) are *length spaces,* then so is $(X * Y, d_{X*Y})$: Given z, z', let α be a path of X joining x, x' and of length $L(\alpha)$ close to $d_X(x, x')$, and let β be a path of Y joining y, y' with $L(\beta)$ close to $d_Y(y, y')$. Then the following path γ in $X * Y$ joins z to z':

$$\gamma(s) = \big(1 - ((1-s)t + st')\big)\alpha(s) + \big((1-s)t + st'\big)\beta(s),$$

and $L(\gamma)$ is close to $d_{X*Y}\ (z, z')$.

The *join* of two *simplicial complexes* P, Q is the simplicial complex $P * Q$ whose set of vertices is the disjoint union of those of P and Q. A simplex of $P * Q$ is the span of a simplex of P and one of Q. Thus, if s is an i-simplex of P and t a j-simplex of Q, then their span $s * t$ is an $(i+j+1)$-simplex of $P * Q$; its boundary is $(b(s) * t) \cup (s * b(t))$.

Let R denote the set of all midpoints of all segments from points of P to those of Q. Then $R \cong P \times Q$, and $P * Q - P - Q \cong R *]0,1[$. For the above, cf. [EiSt 1952, p. 74].

Now suppose M and N are smooth *Riemannian submanifolds* of V and W as above. *Their join* $M * N$ *is a length space* with d_M, d_N the indicated intrinsic distance of the Riemannian structure (and $M * N$ is a geodesic space if M and N are complete).

Relative to smooth triangulations of M and N, $M * N$ *is a Riemannian normal* $(m+n+1)$*-circuit*, in particular an admissible Riemannian polyhedron. (Presumably, this is still true if M, N are themselves smooth Riemannian normal circuits.)

In [CD 1993, Appendix] a detailed description is given of the Riemannian join of two finite Riemannian polyhedra whose simplexes are spherical (i.e., of constant positive curvature).

Here is a simple model to serve for most of the following examples:

Example 8.7. Riemannian orbit spaces

In [Sc 1983, §§1–2] Scott discusses in detail the Riemannian orbit spaces $X = M/\Gamma$ of discrete groups Γ acting isometrically (hence $\pm$ holomorphically) on the 2-dimensional space forms M. In that special situation, X is a topological manifold, perhaps with boundary. It has a naturally induced Riemannian metric, possibly with singularities, of three types (cone points, reflector lines and corner reflections). Thus X *is an admissible Riemannian polyhedron*, and even a *Riemannian normal circuit.*

The following is a closely related situation (see also Subexample 8.13 (i)): With all structures smooth, let $(M, \mathcal{F}, g)$ be a compact oriented 3-manifold, and:

(i) $\mathcal{F}$ a foliation by oriented circles;

(ii) g a Riemannian metric on M relative to which $\mathcal{F}$ is a horizontally conformal foliation by geodesics.

Then the leaf space $X = M/\mathcal{F}$ is a Riemannian 2-orbifold with a cone point corresponding to each critical leaf of $\mathcal{F}$; X has a unique Riemann surface structure, and the leaf map $\pi : M \to X$ is a harmonic morphism [BW 1992]. (For harmonic morphisms, see Definition 13.1 below.)

Example 8.8. Conical singular Riemannian spaces

(Cf. [Chee 1980; 1983].) Let X be a compact n-circuit with triangulation

T, and singular set $S(X)$, a subcomplex of the $(n-2)$-skeleton $X^{(n-2)}$ of X. Cheeger studied the (generally incomplete) Riemannian metrics g on the complement $X \setminus S(X)$ which are conical around $S(X)$, or more universally: around $X^{(n-2)}$.

For any j-simplex σ of T the dual linking $(n-j-1)$-cell Q of the first barycentric subdivision of T determines a local product structure on small neighbourhoods of interior points of σ, homeomorphic to

$$P = C^a_{0,1}(Q) \times]0,1[^j,$$

where $C^a_{0,1}(Q)$ is a linear truncated cone over Q with vertex a. We endow P with the metric

$$h_p = r^2 h_Q + dr^2 + \sum_{i=1}^{j} dr_i^2 \quad (0 < r, r_i < 1).$$

A Riemannian metric g on $X \setminus S(X)$ is said to be *conical* if locally it has such a form near each simplex of $S(X)$, in the strict sense – not just locally quasi-isometric to a conical metric. X *is then a Riemannian normal n-circuit,* and so an admissible Riemannian polyhedron.

The geometry of Riemannian spaces X has been studied in a more general framework by Liu and Shen [LS 1998], with special attention to convergence of the Riemannian metric of $X \setminus S(X)$ near $S(X)$ in metric conical charts.

Subexample 8.8. Let $\pi : V \to M$ be an oriented vector bundle of Euclidean spaces over a compact manifold M. Its *Thom space* is the one-point compactification $X = V \cup \{\infty\}$, expressed as the space $X = D \cup \operatorname{cone} \partial D$ formed as the union of the closed ball bundle D of V with the cone over its boundary. X is a normal circuit whose singular set $S(X) = \infty$, the vertex of the cone. For instance, the quadric cone X given in homogeneous coordinates $[z_0, z_1, z_2, z_3]$ of $\mathbb{C}P^3$ (the complex projective 3-space) by $z_0^2 + z_1^2 + z_2^2 = 0$ is a normal 4-circuit with isolated singular point $[0,0,0,1]$. X is homeomorphic to the Thom space of the tangent bundle of the 2-sphere. [McC 1977].

Example 8.9. Normal analytic spaces with singularities

(Cf. [Ło 1991].) Let X be a complex analytic space in the sense that X is a topological space with a maximal open cover (θ_i, U_i) such that:

(a) $\theta_i : U_i \to V_i \subset \mathbb{C}^{n_i}$ is a homeomorphism onto an analytic subset V_i (i.e., given locally by the zero locus of a finite number of holomorphic functions), and

(b) every $\theta_j \circ \theta_i^{-1}$ is holomorphic.

Let X^0 denote the set of regular points of X (i.e., points having neighbourhoods which are manifolds). Then X^0 is a disjoint union of open closed submanifolds, and is open and dense in X. Its complement $S = X \setminus X^0$ of singular points is an analytic nowhere dense subspace of X. The space X is *irreducible* iff X^0 is a connected manifold; thus $\dim X = \dim X^0$. Say X is *normal* if, for every open $U \subset X$ and every nowhere dense analytic subspace $A \subset U$, every holomorphic function on $U \setminus A$ which is locally bounded near A has a holomorphic extension to U.

Every analytic space X has a *normalization* $\pi : \hat{X} \to X$, unique up to isomorphism; thus π is a finite (i.e., proper and with finite fibres) holomorphic map of a normal analytic space $\hat{X}$ onto X whose restriction to the regular set of $\hat{X}$ is biholomorphic, with dense image [Ło 1991, Chapter VI.4].

Now let X be a complex analytic space whose irreducible components have complex dimension n. Then there is a semi-analytic triangulation of X into a locally finite simplicial complex with its singular set triangulated as a subcomplex. X is an orientable $2n$-circuit without boundary (Giesecke [Gie 1964] and Łojasiewicz [Ło 1964]). And normality as a complex analytic space is topologically equivalent to normality as a circuit [Hae 1984, §1.6].

As an explicit example of a non-normal complex analytic space, take the pinched torus X in $\mathbb{C}P^2$ given (in homogeneous coordinates) by $x^3 + y^3 = xyz$; its singular set S is the point $[0, 0, 1]$. As a circuit, X is not normal (Example 8.3).

Finally, suppose that X is a normal complex analytic subset of a Kähler manifold M. There is a triangulation of X with the interior of each top dimensional simplex s holomorphically embedded in M [Gie 1964, (6.1)]. That induces a Riemannian metric on each simplex, and so X *becomes a Riemannian normal circuit.* Furthermore, g restricted to the interior of s is a Kähler metric.

Subexample 8.9 (i). If X is a normal complex analytic space of complex dimension 1, then X is a Riemann surface (by Example 4.1).

Subexample 8.9 (ii). Suppose that G denotes a group of holomorphic automorphisms of a compact analytic space X; and that G acts properly discontinuously, i.e.,

(a) each pair of points x', x'' in X in different G-orbits have neighbourhoods U', U'' such that $sU' \cap sU'' = \emptyset$ for all $s \in G$ (that insures that the quotient space $G \backslash X$ is Hausdorff), and

(b) for any $a \in X$, its isotropy group G_a is finite; and there is an open neighbourhood U of a which is G_a-stable and for which the following implication holds: $s \in G$, $x \in U$, $sx \in U$ implies $s \in G_a$. (That ensures that the image of U in $G\backslash X$ is isomorphic to $G_a|U$.)

A theorem of H. Cartan [Ca 1957] asserts that $G\backslash X$ *is a complex analytic space, and normal if* X *is. The orbit map* $\pi : X \to G\backslash X$ *is holomorphic.*

Now assume that X is a complex manifold of $\dim_{\mathbb{C}} X = n$. Then $Y = G\backslash X$ *is also a complex orbifold;* i.e., each point $y \in Y$ has a neighbourhood biholomorphic to a neighbourhood of the origin in $G_y\backslash\mathbb{C}^n$, the isotropy group G_y of y being a finite group of linear automorphisms of $\mathbb{C}^n$.

Subexample 8.9 (iii). Fix a set of positive integers $Q = (q_0, \cdots, q_r)$, and form the finite group G_Q whose elements are $(g_0, \cdots, g_r)$, where each $g_j = \exp(2\pi i b_j/q_j)$ (for integers b_j with $0 \leqslant b_j < q_j$). Define the isometric action of G_Q on $\mathbb{C}^{r+1} \setminus \{0\}$ by

$$((g_0, \cdots, g_r), (z_0, \cdots, z_r)) \longmapsto (g_0 z_0, \cdots, g_r z_r).$$

Its orbit space $\mathbb{P}(Q) = G_Q \setminus (\mathbb{C}^{r+1} \setminus \{0\})$ *is a weighted projective space* [Dol 1982]. *It is a normal, irreducible, projective algebraic variety.* All of its singularities are cyclic quotients, so in particular, $\mathbb{P}(Q)$ is a Riemannian orbifold, as in Example 8.13 below.

Subexample 8.9 (iv). The Grassmannian $G^k(\mathbb{C}^n)$ is the space of k-dimensional subspaces of $\mathbb{C}^n$. It is a compact homogeneous projective (hence Kählerian) manifold. ($G^1(\mathbb{C}^n)$ is the complex projective space $\mathbb{C}P^{n-1}$.) $G^k(\mathbb{C}^n)$ has a natural decomposition into open complex cells, called Schubert varieties, whose closures are complex analytic spaces with explicit and simply described singularities [GrHa 1978, Chapter 1, §5].

There are analogous complex cellular decompositions of compact homogeneous Kählerian manifolds; the closures of these cells are geometrically significant (usually singular) complex analytic subspaces [Bor 1954].

Subexample 8.9 (v). Suppose that X is a compact projective irreducible complex analytic space. (Thus X is a connected algebraic variety.) Then an application of Hironaka's theorem [Hir 1964] on the resolution of singularities gives a compact connected Kähler manifold (M, g) with $\dim M = \dim X$, and a surjective holomorphic map $\pi : M \to X$. Such resolutions can often be used to recognize harmonic maps in a holomorphic framework.

Subexample 8.9 (vi). Assume that X has a Riemannian metric g as in Example 7.2, and that $g_{|X\backslash S}$ is Kähler. Then any bounded holomorphic function $u : U \to \mathbb{C}$ is harmonic. Indeed, $U \setminus S$ is a Kähler manifold, so the restriction of u to U is harmonic (see e.g., [ES 1964, §2C]). The assertion now follows upon application of Corollary 7.3.

Remark 8.1. Similarly, a real (semi-)analytic or (semi-)algebraic set is triangulable [Har 1975], [Hir 1975]. The key ingredients are

- that such sets admit suitable stratifications (in the sense of Thom and Whitney),
- that these are triangulable; see [Ver 1984, Theorem 7.8].

That monograph provides a thorough treatment and fine perspective on stratifications and triangulations.

Example 8.10. The Kobayashi distance

(Cf. [Kob 1970; 1976; 1998].) Let U denote the open unit disc in $\mathbb{C}$, and ϱ_U its Poincaré metric. Given two points p, q of a complex analytic space X, join them by a chain of holomorphic discs, i.e., points $a_1, b_1, \cdots, a_k, b_k$ in U, $p = p_0, \cdots, p_k = q$ in X, and holomorphic maps $f_1, \cdots, f_k : U \to X$ with $f_i(a_i) = p_{i-1}$, $f_i(b_i) = p_i$ $(1 \leqslant i \leqslant k)$.

Define the *Kobayashi distance*

$$d_X(p, q) = \inf \sum_{i=1}^{k} \varrho_U(a_i, b_i),$$

the infimum taken over all chains of holomorphic discs joining p to q. Say that the complex analytic space X is *hyperbolic* if d_X is a metric. In that case ([Bar 1972], [Roy 1971a]), its topology coincides with that originally given on X, and (X, d_X) *is a length space.* A normalization of a hyperbolic space is hyperbolic [Kob 1970, Chapter 7, Corollary 4.2].

That notion of hyperbolicity of Kobayashi metrics is also applicable to certain almost complex analytic spaces, as well (in a real context) as to simplicial polyhedra [Gro 1999, §1.8 bis$_+$].

In the case where M is a hyperbolic complex manifold, the metric d_M is associated to the Riemann–Finsler structure (as in Example 3.3):

$$f_M(x, v) = \inf\{1/r : \exists h : \{|z| < r\} \to M, \text{ holomorphic with } h(0) = x, h'(0) = v\}$$

for $v \in T_x(M)$ [Roy 1971a;b]. If M is complete hyperbolic, then f_M is continuous on $T(M)$. Further regularity of f_M is rather delicate. For instance, if M is a polydisc in $\mathbb{C}^n$ $(n > 1)$, then f_M is not C^1 on $T(M) \setminus \{0\}$ [Far 1990].

Subexample 8.10 (i). A bounded domain $X \subset \mathbb{C}^n$ is hyperbolic [Kob 1970, Chapter 4, Corollary 4.5]. If the Kobayashi metric of X is complete, then X is a domain of holomorphy.

Subexample 8.10 (ii). If X is a hyperbolic complex manifold and Γ a discrete group of holomorphic automorphisms of X acting properly discontinuously, then the orbit space X/Γ is again a complex orbifold. Γ is a group of d_X-isometries; its quotient distance is therefore a metric, and coincides with the Kobayashi metric of X/Γ. Thus X/Γ is hyperbolic [Kob 1970, Chapter VII, §§5,6].

Subexample 8.10 (iii). Let U be a domain in $\mathbb{C}^n$, and $(f_\alpha)_{\alpha \in A}$ a family of holomorphic functions on U such that the set

$$U_1 = \{z \in U : |f_\alpha(z)| < 1 \text{ for all } \alpha \in A\}$$

is open. A *generalized analytic polyhedron* is a component of U_1. *Every generalized analytic polyhedron is a complete hyperbolic space.*

Subexample 8.10 (iv). General reference: [Nag 1988]. Let $\mathcal{T}_p$ denote the Teichmüller space of a closed oriented surface of genus $p \geqslant 2$. $\mathcal{T}_p$ is a $(3p-3)$-complex dimensional cell. Ahlfors and Bers have shown that $\mathcal{T}_p$ has a natural complex structure. Royden [Roy 1971a;b] identified Kobayashi's distance $d_{\mathcal{T}_p}$ with Teichmüller's metric, thereby proving that $\mathcal{T}_p$ is a complete hyperbolic manifold. (It is not nonpositively curved, however [Mas 1975].) Furthermore he showed that the group of holomorphic automorphisms of $\mathcal{T}_p$ coincides with the Teichmüller modular group $\mathcal{M}_p$. Thus the Riemann space $\mathcal{R}_p = \mathcal{T}_p/\mathcal{M}_p$ *is a geodesic normal hyperbolic space*, the latter by Subexample 8.10 (ii) because $\mathcal{M}_p$ acts properly discontinuously on $\mathcal{T}_p$.

Royden [Roy 1971b] has represented Teichmüller's metric $d_{\mathcal{T}}$ through a complete Riemann–Finsler structure. See also O'Byrne [O'B 1971].

Example 8.11. Riemannian branched coverings

Suppose that $\pi : M \to (N, h)$ is a smooth branched cover of a Riemannian manifold (N, h) by a smooth manifold M. Thus π is a smooth surjective map, and there are closed 2-codimensional submanifolds A in M and B in N:

$$\begin{array}{ccc} A & \xrightarrow{i} & M \\ {\scriptstyle \pi_{|A}}\downarrow & & \downarrow{\scriptstyle \pi} \\ B & \xrightarrow{j} & N \end{array}$$

such that $A = \pi^{-1}(B)$, with $\pi_{|M\setminus A} \to N \setminus B$ and $\pi_{|A} \to B$, both covering maps. B is called the *branching locus of* π. (See [Vä 1966] for the special role played by the codimension 2.)

Remark 8.2. Depending on the context, sometimes greater precision is required in the definition of smooth branched covering; e.g.,

(a) that B be a locally flat submanifold of N, i.e., each point $y \in B$ has a neighbourhood V of the form $\mathbb{R}^{n-2} \times \mathbb{R}^2$, where $V \cap B = \mathbb{R}^{n-2} \times 0$;

(b) that A and B have tubular neighbourhoods between which π takes a canonical rotational form [Bra 1979].

The induced tensor field $g = \pi^* h$ is positive semidefinite on M, and definite on $M \setminus A$. The restriction $\pi : (M \setminus A, g_{|M\setminus A}) \to (N \setminus B, h_{|N\setminus B})$ is a Riemannian cover. Similarly for $\pi : (A, i^* g) \to (B, j^* h)$. We shall call $\pi : M \to (N, h)$ *a Riemannian branched covering* (following [CD 1993, §5]).

Together, g and $i^* g$ induce a metric d_M on M, relative to which (M, d_M) is a length space. Completing earlier work by Gromov [Gro 1987], Charney and Davis [CD 1993, Theorem 5.3] give geometric conditions on (N, h) to insure that $\operatorname{curv}(M, d_M) \leqslant c$ holds iff $\operatorname{curv}(N, h) \leqslant c$.

There are locally finite triangulations of M and N, with A and B triangulated as subcomplexes, and with respect to which π is a simplicial map. Of course, M is a normal circuit, and thus an admissible Riemannian polyhedron, because for any top dimensional simplex s, the restriction of π to s is an isometry onto $\pi(s)$ of N. See also Example 13.2.

Subexample 8.11 (i). The $(n-1)$-st *suspension* $\pi_r : S^n \to S^n$ of the map $\pi_r : S^1 \to S^1$ defined by $\pi_r(z) = z^r$, is a Riemannian branched cover of the Euclidean n-sphere, with branching locus a great $(n-2)$-sphere ($n \geqslant 2$).

Subexample 8.11 (ii). Complex conjugation in $\mathbb{C}^3$ induces an involution $\sigma : \mathbb{C}P^2 \to \mathbb{C}P^2$ with real projective space $\mathbb{R}P^2 \hookrightarrow \mathbb{C}P^2$ as fixed point set. The quotient map $\pi : \mathbb{C}P^2 \to \mathbb{C}P^2/\sigma$ endows the quotient with a smooth manifold structure; in fact $\mathbb{C}P^2/\sigma$ is explicitly diffeomorphic to S^4 (Hitchin, unpublished). Then $\pi : \mathbb{C}P^2 \to S^4$ is a Riemannian branched covering, with $A = \mathbb{R}P^2$.

Subexample 8.11 (iii). Take a configuration L of k lines ($k \geqslant 3$) in $\mathbb{C}P^2$ not all passing through the same point, and any branching order $r \geqslant 2$. This determines *Kummer's branched* (over L) *cover* $\pi : X_r(L) \to \mathbb{C}P^2$ [BHH 1987, §§1.5, 3], with $X_r(L)$ represented as the complete intersection of $k-3$ hypersurfaces of degree r, and thus a normal complex analytic space. The space $X_r(L)$ is simply connected.

Subexample 8.11 (iv). Suppose $\varphi : X \to Y$ is a finite, surjective, continuous open closed map between polyhedra. Its local degree (if finite) at a point $x \in X$ is

$$\deg(\varphi; x) = \inf_U \sup\{\operatorname{card}\big(\varphi^{-1}(\varphi(z)) \cap U\big) : z \in U\},$$

where U ranges over the neighbourhoods of x. Define the degree (if finite) of φ:

$$\deg(\varphi) = \sup\{\operatorname{card} \varphi^{-1}(y) : y \in Y\}.$$

Say that $\varphi : X \to Y$ is a *finite branched cover* if $\deg(\varphi) < \infty$ and for each $y \in Y$,

$$\deg(\varphi) = \sum \{\deg(\varphi; x) : x \in \varphi^{-1}(y)\}.$$

Let A_φ denote the branch set of φ (i.e., the set of points in X at which φ fails to be a local homeomorphism). If $\varphi : X \to Y$ is a finite, surjective, continuous open closed map and X a normal circuit, then $\deg(\varphi) < \infty$, and $\varphi(A_\varphi)$ does not locally separate Y. Furthermore, Y is also a normal circuit, and φ a finite branched cover [Ed 1976, Theorem 3.5].

A surjective proper holomorphic map $\varphi : X \to Y$ with finite fibres between connected normal analytic spaces is an *analytic branched covering.* There is a proper analytic subspace $B \subset Y$ of pure complex codimension one such that $\varphi_{|X \setminus \varphi^{-1}(B)} \to Y \setminus B$ is unbranched, [BPV 1984, I.16]. X and Y can be triangulated as normal circuits (Example 8.9), relative to which the map becomes simplicial [Gie 1964]. If $Y = (N, h)$ is a Kähler manifold, then h induces a metric d_X on X relative to which (X, d_X) becomes an admissible Riemannian polyhedron, by Example 8.9.

Let X be a connected normal analytic subspace of $\dim X = n$, in $\mathbb{C}P^r$. Then there is a finite holomorphic branched covering map $X \to \mathbb{C}P^n$ [Sh 1974, p. 52].

If $\varphi : \mathbb{C}P^n \to Y$ *is a nonconstant holomorphic map onto a normal analytic space, then* φ *is an analytic branched cover,* [RV 1961] and [RS 1960].

Example 8.12. The quotient M/K

Let K be a compact group of isometries of a complete smooth Riemannian manifold (M, g), and let $\pi : M \to M/K$ be the projection onto the orbit space M/K.

There is a smooth triangulation of M for which π induces an admissible Riemannian polyhedral structure on M/K such that the associated intrinsic distance $d_{M/K}(y_1, y_2)$ between elements y_1 and y_2 of M/K equals the (constant) intrinsic distance in M between the corresponding compact orbits $\pi^{-1}(y_1)$ and $\pi^{-1}(y_2)$, cf. Subexample 8.12 (ii); in particular, M/K admits a triangulation such that every open simplex consists of points of the same orbit type [Ver 1984, Corollary 7.10]. (As in Example 8.9, M/K has a suitable stratified structure.)

Furthermore, with the distance $d_{M/K}$ the polyhedron M/K is a geodesic space whose curvature is bounded below by the infimum of the sectional curvatures of M. The polyhedral structure determines a Brelot harmonic sheaf $\mathcal{H}_{M/K}$ on M/K as described in Theorem 7.1.

On the other hand, π carries the Brelot harmonic sheaf $\mathcal{H}_M$ (the local solutions to the Laplace–Beltrami equation on M) *onto a Brelot harmonic sheaf* $\mathcal{H}'_{M/K} = \pi_* \mathcal{H}_M$ *on* M/K. It is obtained by assigning to each open set $V \subset M/K$ the linear space $\mathcal{H}'_{M/K}(V)$ of all functions $u : V \to \mathbb{R}$ such that $u \circ \pi \in \mathcal{H}_M(\pi^{-1}(V))$. Accordingly, $\pi : (M, \mathcal{H}_M) \to (M/K, \mathcal{H}'_{M/K})$ is a *harmonic morphism*, by the very definition of such morphisms in [CC 1965, §3], cf. Definition 13.1 below; and this characterizes the Brelot harmonic sheaf $\mathcal{H}'_{M/K}$. The proof of these assertions will be given after Subexample 8.12 (iii).

The two harmonic structures $\mathcal{H}_{M/K}$ and $\mathcal{H}'_{M/K}$ on M/K are generally distinct, cf. Subexample 8.12 (iii).

Subexample 8.12 (i). [Ed 1976], [Jä 1966]. In the case where K is a *finite* group of isometries acting on (M, g), conditions are known to ensure that the orbit map $\pi : M \to N = M/K$ is a smooth branched cover of a smooth manifold; and of a geodesic space (N, d_N).

Subexample 8.12 (ii). Here is a simple example of a fibration with singular fibres (a general picture of which can be found in [CT 1968]). If $K = S^1$ is the circle group acting on M as above, then the orbit space M/K is a Riemannian normal $(n-1)$-circuit ($n = \dim M$) [EdFi 1976, Lemma 2.6], and thus an admissible Riemannian polyhedron. Therefore, for $K = S^1$ we have the Brelot harmonic structure $\mathcal{H}_{M/K}$ from Theorem 7.1. A concrete example follows:

Subexample 8.12 (iii). Take $M = S^2$ with its standard metric, $K = S^1$ = the group of orientation preserving isometries of M leaving a given point $p \in M$ (and $-p$) fixed. If θ denotes the geodesic distance from p, M/K can be identified as a metric space with the interval $0 \leqslant \theta \leqslant \pi$, which we shall view as a Riemannian 1-circuit with just one 1-simplex, embedded in $\mathbb{R}$ and equipped with the Riemannian metric

$$ds = d\theta.$$

The corresponding Brelot harmonic sheaf $\mathcal{H}_{M/K}$ from Theorem 7.1 consists of the linear combinations of 1 and $\theta \mapsto \theta$, i.e., the affine functions of θ, as in Example 8.1. However, the constants are the only functions harmonic on an interval containing 0 or π (cf. Remark 5.3); but the function $\theta \mapsto \theta$ is subharmonic at 0 and superharmonic at π. Every point of M/K is non-polar (for the sheaf $\mathcal{H}_{M/K}$) by Corollary 7.2.

In addition to the above harmonic sheaf $\mathcal{H}_{M/K}$ we have, as mentioned above, the harmonic sheaf $\mathcal{H}'_{M/K} = \pi_*\mathcal{H}_M$ on $M/K = [0, \pi]$ induced by $\mathcal{H}_M$ so that the projection $M \to M/K$ becomes a harmonic morphism, i.e., pulls harmonic functions back to harmonic functions. The harmonic functions on $M/K = [0, \pi]$ are here the solutions to

$$\frac{1}{\sin\theta}\frac{d}{d\theta}\left(\sin\theta\frac{du}{d\theta}\right) = 0,$$

and these are the linear combinations of the constant 1 and the function $\log\cot\theta/2$, the latter being (finite and) harmonic only for $0 < \theta < \pi$. Thus the constants are the only functions harmonic on an interval containing 0 or π; but the function $\log\cot(\theta/2)$ is superharmonic at 0 and subharmonic at π (when given the value $+\infty$ at 0 and $-\infty$ at π), so 0 and π are *polar* points for the sheaf $\mathcal{H}'_{M/K}$, while the remaining points of $M/K = [0, \pi]$ are nonpolar.

The above two harmonic sheaves on M/K are obviously different, and neither contains the other, in agreement with Corollary 13.1 (ii).

Example 8.12 (continued). We now prove the assertions about the sheaf $\mathcal{H}'_{M/K} = \pi_*\mathcal{H}_M$ on M/K. For simplicity we now write $\mathcal{H}_{M/K}$ in place of $\mathcal{H}'_{M/K}$. First a more general setup:

Let $(X, \mathcal{H}_X)$ be a Brelot harmonic space for which the constant functions are harmonic. Let Y be a Hausdorff space, and $\pi : X \to Y$ a *surjective, continuous, open map.* Then Y is connected, locally connected and locally compact, because X is so. For open $V \subset Y$ define:

$$v \in \mathcal{H}_Y(V) \quad \Longleftrightarrow \quad v \circ \pi \in \mathcal{H}_X(\pi^{-1}(V)). \tag{8.2}$$

Then $\mathcal{H}_Y(V)$ is a linear subspace of $C(V)$, and $\mathcal{H}_Y$ $(= \pi_* \mathcal{H}_X)$ is a sheaf containing the constants on Y. We show that this sheaf has the *Harnack monotone convergence property.* Consider a domain $V \Subset Y$ and an increasing sequence of nonnegative functions $v_n \in \mathcal{H}_Y(V)$, and write $v = \sup_n v_n$. Then

$$v \circ \pi = \sup_n (v_n \circ \pi), \quad 0 \leqslant v_n \circ \pi \in \mathcal{H}_X(\pi^{-1}(V)).$$

For a component U of $\pi^{-1}(V)$ such that $v \circ \pi \not\equiv \infty$ in U, we have $v \circ \pi_{|U} \in \mathcal{H}_X(U)$, hence $v < \infty$ on $\pi(U)$, by Proposition 7.2 or Theorem 7.1. For a component U' of $\pi^{-1}(V)$ such that $v \circ \pi \equiv \infty$ in U', we have of course $v \equiv \infty$ in $\pi(U')$. Thus $\pi(U) \cap \pi(U') = \emptyset$. Denote by $\widetilde{U}$, resp. $\widetilde{U}'$, the union of all components U, resp. U', of $\pi^{-1}(V)$ as stated. Then $\widetilde{U}, \widetilde{U}'$ are open in X, hence $\pi(\widetilde{U}), \pi(\widetilde{U}')$ are open in Y. Moreover, $v \circ \pi_{|\widetilde{U}} \in \mathcal{H}_X(\widetilde{U})$, and

$$\pi(\widetilde{U}) = \{y \in V : v(y) < \infty\}, \quad \pi(\widetilde{U}') = \{y \in V : v(y) = \infty\}.$$

Because V is connected, we conclude that *either* $\pi(\widetilde{U}) = \emptyset$ and so $v \equiv \infty$ in V, *or else* $\pi(\widetilde{U}') = \emptyset$, $\widetilde{U}' = \emptyset$, $\widetilde{U} = \pi^{-1}(V)$, $v \circ \pi \in \mathcal{H}_X(\pi^{-1}(V))$, $v \in \mathcal{H}_Y(V)$, by (8.2). This proves the Harnack convergence property.

We do not know whether, in this generality, $(Y, \mathcal{H}_Y)$ has a base of regular domains (not even if π is also proper). In the affirmative case, $(Y, \mathcal{H}_Y)$ is a Brelot harmonic space, and the implication $\Longrightarrow$ in (8.2) expresses that $\pi : (X, \mathcal{H}_X) \to (Y, \mathcal{H}_Y)$ is a harmonic morphism (Definition 13.1).

As to the stated uniqueness, if $\mathcal{H}$ denotes any Brelot harmonic sheaf on Y such that $\pi : (X, \mathcal{H}_X) \to (Y, \mathcal{H})$ is a harmonic morphism, then clearly $\mathcal{H} \subset \mathcal{H}_Y$, and consequently $\mathcal{H} = \mathcal{H}_Y$, by the maximality (under inclusion) of any Brelot harmonic sheaf on a given space, [CC 1965, p. 23] and Corollary 13.1 (vi) below.

Consider now the original situation where $X = (M, g)$ is a *complete smooth Riemannian manifold* with a given *compact group K of isometries*, and where $Y = M/K$ is the *orbit space* (with the quotient topology), and finally $\pi : M \to M/K$ is the orbit map, which is indeed surjective, continuous and open. Every orbit $\pi^{-1}(\pi(x)) = Kx$ $(x \in M)$ being compact, π is moreover *proper*, by general topology.

In that situation every orbit Kx is a smooth embedded *submanifold* of M, see e.g., [Di 1970, III, Théorème 16.10.7]. (We do not assume that Kx be connected, nor that its components have equal dimension; thus the orbits need not foliate M.)

The intrinsic distance d_X on $X = (M, g)$ induces via π a metric d_Y on $Y = M/K$, compatible with the quotient topology on Y, as indicated above:

$$d_Y(y_1, y_2) = \min\{d_X(x_1, x_2) : x_i \in \pi^{-1}(y_i),\ i = 1, 2\}.$$

We show that an open metric ball $V = B_Y(y_0, r)^\circ$ in Y is *regular* for $\mathcal{H}_Y = \pi_*\mathcal{H}_X$ ($= \mathcal{H}'_{M/K}$ in our original notation) when r is sufficiently small. Its preimage $U = \pi^{-1}(V)$ is a relatively compact, open tubular neighbourhood of the orbit $\pi^{-1}(y_0) \subset M$, and $\partial U = \pi^{-1}(\partial V)$ is a smooth embedded submanifold of M of codimension 1 when r is small enough. It follows (see references in the text preceding Corollary 7.1 (b)) that $M \setminus U$ is *not thin* at any point of ∂U, and so U is regular for $\mathcal{H}_X$, see [Br 1962a, p. 28], [Bau 1966, Satz 4.3.1]. From that we shall see that $\pi(U) = V$ is regular for $\mathcal{H}_Y = \pi_*\mathcal{H}_M$.

Given a function $f \in C(\partial V)$, we have $f \circ \pi \in C(\pi^{-1}(\partial V)) = C(\partial U)$. By regularity of U there is a unique function $u \in C(\overline{U})$ such that

$$u_{|U} \in \mathcal{H}_X(U) \ \text{ and } \ u_{|\partial U} = f \circ \pi. \tag{8.3}$$

For any $\kappa \in K$, $\kappa U = U$ because $U = \pi^{-1}(V)$ is the union of the orbits $\pi^{-1}(y)$, $y \in V$. Furthermore, $u \circ \kappa_{|U} \in \mathcal{H}_X(U)$ because the isometry κ is a harmonic morphism of $(X, \mathcal{H}_X)$. Moreover,

$$u \circ \kappa \in C(\kappa^{-1}(\overline{U})) = C(\overline{U}),$$

and

$$u \circ \kappa_{|\partial U} = f \circ \pi \circ \kappa_{|\partial U} = f \circ \pi$$

since $\pi \circ \kappa = \pi$ and $\kappa(\partial U) = \partial U$. By the uniqueness of u in (8.3) we infer that

$$u \circ \kappa = u \ \text{ for every } \ \kappa \in K,$$

and so there is a function v on $\pi(\overline{U}) = \overline{V}$ such that $v \circ \pi = u$, hence $v_{|V} \circ \pi = u_{|U}$, and by (8.2)

$$v_{|V} \in \mathcal{H}_Y(V).$$

Moreover $v \in C(\overline{V})$; in fact, for $\alpha \in \mathbb{R}$, $\{v \geqslant \alpha\} = \pi(\{u \geqslant \alpha\})$ is compact because $\{u \geqslant \alpha\}$ is a closed subset of the compact set $\overline{U}$, and similarly $\{v \leqslant \alpha\}$ is compact. Furthermore,

$$v_{|\partial V} = f$$

because $v_{|\partial V} \circ \pi = (v \circ \pi)_{|\partial U} = u_{|\partial U} = f \circ \pi$, by (8.3). Finally, $f \geqslant 0$ implies $v \geqslant 0$ because $f \circ \pi \geqslant 0$ implies $u \geqslant 0$, by regularity of U. This shows that $V = B_Y(y_0, r)^\circ$ is indeed regular in $(Y, \mathcal{H}_Y)$ for small enough r, thus completing the proof that $Y = M/K$, endowed with the sheaf $\mathcal{H}_Y = \pi_* \mathcal{H}_M$ $(= \pi_* \mathcal{H}_X)$ from (8.2) above (originally denoted $\mathcal{H}'_{M/K}$), is a Brelot harmonic space.

Assume now that the group K is also *connected.* Let M^0 be the union of the K-orbits in M of maximal rank. As discussed in [ER 1993, Chapter IV, §3], M^0 is an open dense submanifold of M. The harmonic morphism $\pi : M \to M/K$ contains an open dense harmonic Riemannian submersion $\pi^0 : M^0 \to M^0/K$ induced from that on M.

Example 8.13. Riemannian orbifolds

The construction in Example 8.12 can be adjusted to apply to Riemannian orbifolds (roughly speaking, spaces modelled on quotients of domains in $\mathbb{R}^n$ by finite groups of isometries) [Mol 1988, §3.6]: *A Riemannian orbifold has a natural geodesic structure, if complete, and natural harmonic space structures.* Especially important are those Riemannian orbifolds with curvature bounded above [CD 1993, §5].

Subexample 8.13 (i). If $\mathcal{F}$ is a *Riemannian foliation* (i.e., any two leaves are locally equidistant) of a smooth Riemannian manifold (M, g) with closed leaves, then *the leaf space $M/\mathcal{F}$ is a Riemannian orbifold* [Rei 1961]. For instance, let K be a closed subgroup of isometries of a Riemannian manifold (M, g), all of whose orbits have the same dimension. Then their components form such a Riemannian foliation [Rei 1983, IV, Example 4.10].

More generally, any closed, connected subgroup of isometries of (M, g) determines a model singular Riemannian foliation, in the sense of [Mol 1988, Chapter 6].

Query. Let (M, g) be a Riemannian manifold, and $\mathcal{F}$ a Riemannian foliation with closed leaves, all of which are minimal submanifolds. Is the leaf-quotient map $\pi : M \to M/\mathcal{F}$ harmonic [a harmonic morphism] in the sense of Definition 12.1 [Definition 13.1]? That is so over the regular points of $M/\mathcal{F}$ (i.e., where the leaves have no $\mathcal{F}$-holonomy), by [EV 1998, §5].

Subexample 8.13 (ii). Let $\mathfrak{S}^r M$ denote the r-fold *symmetric power* of the Riemannian manifold (M, g): $\mathfrak{S}^r M = (M \times \cdots \times M)/\mathfrak{S}_r$, where $\mathfrak{S}_r$ denotes the symmetry group on r factors. It acts isometrically, so $\mathfrak{S}^r M$ *is a Riemannian orbifold* (nonsingular if $\dim M = 2$; see [DuL 1971] and [MacD 1962]). In particular, with M the Euclidean n-sphere S^n $(n > 0)$ and r large, $\mathfrak{S}^r S^n$ serves as a classifying space for n-dimensional integral cohomology, by a theorem of Dold and Thom [DT 1958]. Thus for any compact polyhedron X, the homotopy classes of maps $X \to \mathfrak{S}^r S^n$ are in natural bijective correspondence with the cohomology classes in $H^n(X, Z)$ [Sp 1966, Chapter 8, §1].

Example 8.14. Buildings of Bruhat–Tits

Here we describe a construction which associates Riemannian polyhedra X of nonpositive curvature to certain algebraic groups. In fact, X are Euclidean buildings associated to Tits systems. For motivation, background, and relevant literature, see [Ti 1974], [Bro 1989], [BruTi 1972] – especially, the formal analogy between Euclidean buildings and Riemannian symmetric spaces of noncompact type.

Step 1. [Bou 1974, Chapter IX, §3], [Bou 1981, Chapter IV–VIII]. The valuation $v_p : \mathbb{Q} \to \mathbb{R}$ associated to a prime p is defined by letting $v_p(x)$ be the exponent of p in the prime factorization of x; associated to v_p is the ultrametric

$$d_p(x, y) = \exp(-v_p(x - y)).$$

Thus

$$d_p(x, z) \leqslant \sup(d_p(x, y), d_p(y, z)) \leqslant d_p(x, y) + d_p(y, z).$$

The completion of $\mathbb{Q}$ with respect to d_p is the field $\mathbb{Q}_p$ of *p-adic numbers.* $\mathbb{Q}_p$ is locally compact and totally disconnected.

A subgroup $G \subset GL(n, \mathbb{Q})$ is *algebraic* over $\mathbb{Q}_p$ if it is the set of invertible matrices whose entries annihilate some collection of polynomials in n^2 indeterminates with entries in $\mathbb{Q}_p$. Let G_p be the group of invertible $n \times n$ matrices with coefficients in $\mathbb{Q}_p$ and which satisfy the polynomial equations defining G. We refer to G_p as a *p-adic Lie group.*

Step 2. As a special case, $G_p = SL(n, \mathbb{Q}_p)$, the group of invertible $n \times n$ matrices of determinant 1, with coefficients in $\mathbb{Q}_p$. Let B denote the subgroup of upper triangular matrices in G_p; and N the normalizer of the subgroup D of diagonal matrices. Set $W = N/D$, which is the permutation group on n elements; and $S = \{s_1, \cdots, s_{n-1}\}$, where s_j denotes the permutation of the lines in $(\mathbb{Q}_p)^n$ spanned by e_j and e_{j+1}, leaving the other lines fixed. Then (W, S) is a *Coxeter system,* in the sense of Step 3 below.

Associating the complex (W, S) to the group G_p is a model example of a general method of associating a Euclidean building to a suitable algebraic group over a field.

Step 3. A *Coxeter matrix* on a finite set S of n elements is a generating set S and the relations

$$m_{ss'} = 1 \text{ if } s = s', \quad m_{ss'} \geqslant 2 \text{ if } s \neq s'.$$

See [Bro 1989]. Associated to $(m_{ss'})$ is the group W (the *Coxeter group*) defined by the presentation

$$W = \{S : (ss')^{m_{ss'}} = 1\}.$$

Call (W, S) a *Coxeter system of rank* n. The natural map $S \to W$ is injective. The *Coxeter complex* of (W, S) is defined as follows: Set $m = n - 1$; let Δ_S be an open m-simplex whose vertices are the elements of S; and for any subset T of S let σ_T be the open face of Δ_S spanned by the vertices $s \in S \setminus T$. Define $X = (W \times \Delta_S)/\sim$, where $(w_1, x_1) \sim (w_2, x_2)$ iff (i) $x_1 = x_2$ and (ii) if σ_T is the open simplex containing x_1, then $w_1^{-1} w_2 \in W_T$, the subgroup generated by T. Then X is a simplicial complex on which W acts simplicially.

Step 4. Consider now a simplicial complex X all of whose maximal (open) simplexes (called *chambers* here) have the same dimension m. Suppose X is expressible as a union of subcomplexes A (called *apartments*) such that,

(a) every apartment is a Coxeter complex of dimension m;
(b) any two simplexes of X are contained in an apartment;
(c) given any two apartments A, A' with a common chamber, there is an isomorphism $A \to A'$ fixing $A \cap A'$ pointwise.

Such a complex is called a (Bruhat–Tits) *building.* The group W acts on X simplicially and preserves the rank (i.e., the isomorphism type of the apartments). Furthermore,

(i) every simplex of a building is contained in one of dimension m;
(ii) any two chambers are (globally) $(m - 1)$-chainable by chambers.

Step 5. Each apartment A carries a canonical Euclidean metric g_A such that the isomorphisms in (c) above are isometries. Such an X will be called a *Euclidean building.* Here it is appropriate to relax the hypothesis of local finiteness on X, requiring instead that X have only finitely many isometry types of simplexes. The required properties of X as a complete geodesic space are retained [Bri 1991]. *These g_A together extend to a complete geodesic structure on X. Furthermore, X is contractible and has nonpositive curvature.* (See [Bro 1989, §VI.3], [BruTi 1972].) Verifying nonpositivity of curvature is equivalent to this: Let d denote the metric of the geodesic structure of X. For any points $w, x, y \in X$ (with x and y sufficiently close) and $t \in [0,1]$ let $z = (1-t)x + ty$ be the unique geodesic segment joining x to y. It is required (compare the reference to Reshetnyak towards the end of Chapter 2) that

$$d^2(w,z) \leqslant (1-t)d^2(w,x) + td^2(w,y) - t(1-t)d^2(x,y).$$

If X is associated to an algebraic group G as at the end of Step 2, then G acts on X simplicially and isometrically.

Thus in may respects Euclidean buildings are like Riemannian symmetric spaces of noncompact type; with W behaving like a Weyl group.

Subexample 8.14. The 1-dimensional Euclidean buildings are just the trees all of whose edges have the same length and all of whose vertices have the same number of adjacent edges [Serre 1980].

Part III

Maps between polyhedra

9

Energy of maps

For maps φ of a Riemannian domain into an arbitrary metric space Y a concept of energy $E(\varphi)$ was defined and investigated by Korevaar and Schoen [KS 1993]. We propose an extension of this concept to maps φ from an admissible Riemannian polyhedron (X, g) with simplexwise smooth Riemannian metric. Again, $E(\varphi)$ is defined as the limit of a suitable approximate energy expressed in terms of the distance function d_Y on Y (Definition 9.1).

We show that the maps $\varphi : X \to Y$ of finite energy are precisely those quasicontinuous maps whose restriction to each top dimensional simplex of X has finite energy in the sense of Korevaar-Schoen, and $E(\varphi)$ is then the sum of the energies of these restrictions (Theorem 9.1).

In the particular case that the target Y is a Riemannian C^1-manifold the energy of a map $\varphi : X \to Y$ is given (under an additional hypothesis) by the expected explicit expression, which generally involves the concept of Sobolev functions on quasiopen sets (Theorem 9.2). Similarly when Y is a Riemannian polyhedron with continuous Riemannian metric (Theorem 9.3).

When (Y, h) is an embedded Riemannian polyhedron in a Euclidean space, the elementary integral expression for energy, used by Gromov and Schoen [GS 1992] and by Chen [Chen 1995], is identical with that of Definition 9.3 below (see Remark 9.9).

Energy density and energy

In this section let (X, g) be an admissible m-dimensional Riemannian polyhedron with *simplexwise smooth Riemannian metric.* This means that the restriction of g to each m-simplex s (relative to a given triangulation of X) shall have C^∞-smooth components g_{ij} up to the boundary $b(s)$. We do not require that g be continuous across lower dimensional simplexes. Cf. Chapter 4, Riemannian metric.

All properties discussed in the sequel remain unaltered after a smooth subdivision of T, cf. [KS 1993, Theorem 1.12.3], or Chapter 14 below.

The target (Y, d_Y) is an arbitrary metric space. Denote $L^2_{\text{loc}}(X, Y)$ the space of all μ_g-measurable maps $\varphi : X \to Y$ having separable essential range and for which $d_Y(\varphi(\cdot), q) \in L^2_{\text{loc}}(X, \mu_g)$ for some point q (and therefore by the triangle inequality: for any $q \in Y$). For $\varphi, \psi \in L^2_{\text{loc}}(X, Y)$ define their distance $D(\varphi, \psi)$ by

$$D^2(\varphi, \psi) = \int_X d_Y^2(\varphi(x), \psi(x))\, d\mu_g(x). \tag{9.1}$$

Two maps $\varphi, \psi \in L^2_{\text{loc}}(X, Y)$ are said to be *equivalent* if $D(\varphi, \psi) = 0$, i.e., if $\varphi(x) = \psi(x)$ μ_g-a.e. If X is compact then $D(\varphi, \psi) < \infty$ and D is a metric on $L^2_{\text{loc}}(X, Y) = L^2(X, Y)$, complete provided that Y is complete, cf. [KS 1993, §1.1].

Throughout this section we understand energy density and energy in the sense described below, extending that of Korevaar-Schoen [KS 1993]; see also Chapter 14.

The *approximate energy density* of a map $\varphi \in L^2_{\text{loc}}(X, Y)$ is defined for $\varepsilon > 0$ by

$$e_\varepsilon(\varphi)(x) = \int_{B_X(x,\varepsilon)} \frac{d_Y^2(\varphi(x), \varphi(x'))}{\varepsilon^{m+2}} d\mu_g(x'). \tag{9.2}$$

The function $e_\varepsilon(\varphi) \geqslant 0$ is of class $L^1_{\text{loc}}(X, \mu_g)$, cf. [KS 1993, §1.2], which applies because φ equals μ_g-a.e. a Borel measurable map $X \to Y$.

Definition 9.1. The *energy* $E(\varphi)$ of a map φ of class $L^2_{\text{loc}}(X, Y)$ is

$$E(\varphi) = \sup_{f \in C_c(X,[0,1])} \left(\limsup_{\varepsilon \to 0} \int_X f e_\varepsilon(\varphi)\, d\mu_g \right) \quad (\leqslant \infty). \tag{9.3}$$

Note that the supremum in (9.3) equals the limit ($\leqslant \infty$) of the corresponding increasing net with the upper directed index set $C_c(X, [0,1])$ (ordered pointwise).

We say that φ is *locally of finite energy*, and write $\varphi \in W^{1,2}_{\text{loc}}(X, Y)$, if $E(\varphi_{|U}) < \infty$ for every relatively compact domain $U \subset X$, or equivalently if X can be covered by domains $U \subset X$ such that $E(\varphi_{|U}) < \infty$. (The equivalence is established by use of a continuous partition of unity of X.)

As a consequence of Lemma 4.4 on ball volumes, *every Lip continuous map* $\varphi : X \to Y$ *is of class* $W^{1,2}_{\text{loc}}(X, Y)$.

If X is compact we may write $W^{1,2}(X, Y)$ in place of $W^{1,2}_{\text{loc}}(X, Y)$, and this is then the space of all maps $X \to Y$ of finite energy.

For each m-simplex s of (X,T), $(s^\circ, g_{|s^\circ})$ may be regarded as a *Riemannian domain* in the sense of [KS 1993, §1.1] (see Chapter 1, Maps between polyhedra), because the smooth Riemannian metric $g_{|s}$ can be extended to such a metric on an open neighbourhood of s in the Euclidean m-space containing s (embedded in a Euclidean space, cf. Remark 4.1). A map $\varphi : s \to Y$ clearly has the same energy in the above sense as in the sense of [KS 1993]. Thus $W^{1,2}(s,Y) = W^{1,2}(s^\circ, Y)$, the latter in the sense of [KS 1993].

Theorem 9.1. *Let (X,g) be an admissible m-dimensional Riemannian polyhedron with simplexwise smooth Riemannian metric, and (Y,d_Y) a metric space.*

A map $\varphi \in L^2_{\mathrm{loc}}(X,Y)$ is locally of finite energy iff there is a function $e(\varphi) \in L^1_{\mathrm{loc}}(X)$ such that $e_\varepsilon(\varphi) \to e(\varphi)$ as $\varepsilon \to 0$, in the sense of weak convergence of measures:

$$\lim_{\varepsilon \to 0} \int_X f e_\varepsilon(\varphi)\, d\mu_g = \int_X f e(\varphi)\, d\mu_g, \qquad f \in C_c(X). \tag{9.4}$$

A map $\varphi \in L^2_{\mathrm{loc}}(X,Y)$ is locally of finite energy iff (a) *and* (b) *below both hold, or equivalently iff* (a) *and* (b') *both hold:*

(a) *For each m-simplex s of X, $\varphi_{|s^\circ}$ has finite energy.*
(b) *The map $\varphi : (X,g) \to (Y,d_Y)$ has a quasicontinuous version.*
(b') *For each $(m-1)$-simplex t of X the trace $\mathrm{tr}_t\, \varphi_{|s^\circ}$ is the same for all m-simplexes s having t as a face.*

Recall that a map $\varphi : (X,g) \to (Y,d_Y)$ is said to be *quasicontinuous* if φ has a continuous restriction to closed sets $F \subset X$ whose complements have arbitrarily small capacity, cf. [De 1970, Chapter 4], [Fuk 1980, §3.1]; see also Chapter 2, Dirichlet structures on a space; and Chapter 7, Quasitopology and fine topology. For the *trace operator* $\mathrm{tr}_t : W^{1,2}(s^\circ, Y) \to L^2(t,Y)$, see [KS 1993, §1.12].

The limit function $e(\varphi)$ from (9.4) is called the *energy density* of the map $\varphi \in W^{1,2}_{\mathrm{loc}}(X,Y)$. The *energy* $E(\varphi)$ equals $\int_X e(\varphi)\, d\mu_g$, as seen from (9.3).

Remark 9.1. When $\varphi \in W^{1,2}_{\mathrm{loc}}(X,Y)$, the restriction of $e(\varphi)$ to the interior s° of an m-simplex s of X equals a.e. the energy density of $\varphi_{|s^\circ}$:

$$e(\varphi)_{|s^\circ} = e(\varphi_{|s^\circ}) \qquad \text{for } s \in \mathcal{S}^{(m)}(X). \tag{9.5}$$

The *energy* of φ (Definition 9.1) equals

$$E(\varphi) = \int_X e(\varphi)\, d\mu_g = \sum_{s \in \mathcal{S}^{(m)}(X)} E(\varphi_{|s^\circ}) \quad (\leqslant \infty).$$

Proof of Theorem 9.1 and Remark 9.1. Each of the properties (a) and (b) is clearly preserved under postcomposition of φ with a Lipschitz map ψ of Y into another metric space Z; and so is (a) $\wedge$ (b') because $\mathrm{tr}_t(\varphi_{|s^\circ})$ is the pointwise limit of $\varphi_{|s^\circ}$ along almost every Euclidean normal to t in s, [KS 1993, p. 606], hence $\mathrm{tr}_t(\psi \circ \varphi)_{|s^\circ} = \psi \circ \mathrm{tr}_t\, \varphi_{|s^\circ}$.

For any point $q \in Y$ this preservation applies to the Lip function $\psi = d_Y(\cdot, q) : Y \to \mathbb{R}$; we show that, consequently, the function

$$u_q := d_Y(\varphi(\cdot), q) \quad \text{is of class } W^{1,2}_{\mathrm{loc}}(X)$$

in the sense of Chapter 5.

Step 1. (a) $\wedge$ (b) $\iff$ (a) $\wedge$ (b'). From (a) $\wedge$ (b) follows $u_q \in W^{1,2}_{\mathrm{loc}}(X)$ according to [KS 1993, Theorem 1.6.2] and Proposition 7.7 above because $u_q = \psi \circ \varphi$ has the properties corresponding to (a) and (b). By Remark 7.5, $\mathrm{tr}_t(u_q)_{|s^\circ}$ is the same for all m-simplexes $s \supset t$, and so therefore is $\mathrm{tr}_t(\varphi_{|s^\circ})$ because $d_Y(\mathrm{tr}_t(\varphi_{|s^\circ}), q) = \mathrm{tr}_t(u_q)_{|s^\circ}$ (see the first paragraph of the present proof). This shows that (a) $\wedge$ (b) $\implies$ (a) $\wedge$ (b').

From (a) $\wedge$ (b') it follows similarly by Remark 7.5 that $u_q \in W^{1,2}_{\mathrm{loc}}(X)$, and in particular u_q has a quasicontinuous version u_q^*. From that, a simple compactness argument will lead to (b), noting that a point $y \in \overline{\varphi(X)}$ is uniquely determined by its distances from given points q_j forming a dense sequence in the separable range $\varphi(X)$ of φ. There is a null set N such that, for every j, $u_{q_j} = u^*_{q_j}$ off N, and a polar set $e \subset X$ such that each $u^*_{q_j}$ is finite and finely continuous off e, by Proposition 7.8 (c). For any $x \in X$ denote $\mathcal{U}_x$ the class of all *essential fine neighbourhoods* U of x (i.e., sets U which differ only by a null set from some fine neighbourhood of x). $\mathcal{U}_x$ is a filter because every set $U \in \mathcal{U}_X$ is nonvoid, by Proposition 7.8 (d).

We first show that, for every $x \notin e$, the image filter base $\varphi(\mathcal{U}_x)$ converges in $\overline{Y}$, the 1-point compactification of the locally compact geodesic space Y:

$$\exists\, \varphi^*(x) = \operatorname*{ess\,fine\,lim}_{\xi \to x} \varphi(\xi) := \lim \varphi(\mathcal{U}_x) \in \overline{Y}.$$

For each $q \in Y$ the function $u_q = d_Y(\cdot, q)$ extends by continuity to $\overline{Y}$ and becomes infinite at ∞_Y (the point at infinity of Y) if Y is noncompact. If

$y \in \overline{Y}$ is a cluster point of $\varphi(\mathcal{U}_x)$ then, for each j, $d_Y(y, q_j)$ ($\leqslant \infty$) is a cluster value of the filter base $d_Y(\varphi(\mathcal{U}_x), q_j) = u_{q_j}(\mathcal{U}_x)$, hence equal to the limit $u^*_{q_j}(x) < \infty$ of that filter base. It follows that $d_Y(y, q_j) < \infty$, i.e., $y \in Y$, and in fact $y \in \overline{\varphi(X)}$. If $z \in \overline{Y}$ is another cluster value of $d_Y(\varphi(\mathcal{U}_x), q_j)$ then likewise $z \in \overline{\varphi(X)}$ and $d_Y(z, q_j) = u^*_{q_j}(x)$; and so $d_Y(y, q_j) = d_Y(z, q_j)$ for all j, i.e., $y = z$.

Next we show that $\varphi^*(x) = \varphi(x)$ for $x \in X \setminus (N \cup e)$. Taking $y = \varphi^*(x)$ ($\in \overline{\varphi(X)}$) above, we obtain for every j, since $u^*_{q_j} = u_{q_j}$ off N,

$$d_Y(\varphi^*(x), q_j) = \lim_{\mathcal{U}_x} d_Y(\varphi(\mathcal{U}_x), q_j) = \lim_{\mathcal{U}_x} u_{q_j}(\mathcal{U}_x) = \lim_{\mathcal{U}_x} u^*_{q_j}(\mathcal{U}_x) = u^*_{q_j}(x);$$

the last equation holds because $u^*_{q_j}$ is finely continuous at $x \notin e$. And since $d_Y(\varphi(x), q_j) = u_{q_j}(x) = u^*_{q_j}(x)$ off N, we get $d_Y(\varphi^*(x), q_j) = d_Y(\varphi(x), q_j)$ outside the null set $N \cup e$, hence $\varphi^*(x) = \varphi(x)$ outside $N \cup e$.

Replacing, for every $x \notin e$, $\mathcal{U}_x$ by the filter $\mathcal{U}^*_x$ of actual fine neighbourhoods of x, we similarly find from the fine continuity of $u^*_{q_j}$ that $\lim \varphi^*(\mathcal{U}^*_x)$ exists in Y when $x \notin e$. This means that φ^* is finely continuous in $X \setminus e$, and therefore also quasicontinuous in X, by Proposition 7.8 (c). Thus (a) $\wedge$ (b') $\Rightarrow$ (b), and altogether (a) $\wedge$ (b) $\iff$ (a) $\wedge$ (b').

Step 2. If (9.4) holds for some function $e(\varphi) \in L^1_{\text{loc}}(X)$ then $E(\varphi) = \int_X e(\varphi)\, d\mu_g$. Indeed, $E(\varphi) \leqslant \int e(\varphi)\, d\mu_g$, by (9.3). The opposite inequality follows, likewise by (9.3), from $\int f e(\varphi)\, d\mu_g \leqslant E(\varphi)$ for $f \in C_c(X, [0,1])$.

Step 3. Again, if (9.4) holds for some function $e(\varphi) \in L^1_{\text{loc}}(X)$, we show that $\varphi \in W^{1,2}_{\text{loc}}(X, Y)$, which in turn implies (a)$\wedge$(b'). For each open star S relative to the given triangulation of X it follows from Step 2 that $E(\varphi_{|S}) = \int_S e(\varphi)\, d\mu_g < \infty$. This shows that $\varphi \in W^{1,2}_{\text{loc}}(X, Y)$. If $\varphi \in W^{1,2}_{\text{loc}}(X, Y)$ then $\varphi_{|s^\circ} \in W^{1,2}_{\text{loc}}(s^\circ, Y)$ for each $s \in \mathcal{S}^{(m)}(X)$. This establishes (a).

Furthermore, for distinct $s_1, s_2 \in \mathcal{S}^{(m)}(X)$ with $s_1 \cap s_2 = t \in \mathcal{S}^{(m-1)}(X)$, $(M, g) := (s_1 \cup s_2)^\circ$ may be regarded as a Riemannian domain in $\mathbb{R}^m$. It follows as above that $\varphi_{|M} \in W^{1,2}(M, Y)$, and so $\operatorname{tr}_t \varphi_{|s_1^\circ} = \operatorname{tr}_t \varphi_{|s_2^\circ}$ by a straightforward adaptation of the construction of the trace in [KS 1993, §1.12]. This establishes (b').

Step 4. Finally suppose that (a) $\wedge$ (b') holds, and let us verify (9.4) with $e(\varphi)$ determined by (9.5). We may assume that $\operatorname{supp} f$ is contained in the (open) star $\operatorname{st}_X(a)$ of some vertex a. By Lemma 4.3 the closed star $S = \overline{\operatorname{st}}_X(a)$ may be identified with an m-dimensional cell complex $P \subset [0,1]^k$ (k being the number of vertices of S other than a). Here $a \in S$ corresponds to $0 \in [0,1]^k$, and each m-cell of P is a cube containing 0 and of the form

$$Q_J = \left\{x \in [0,1]^k : x_j = 0 \text{ for } j \notin J\right\},$$

where $J \subset \{1, \cdots, k\}$ and card $J = m$. Only m-tuples J such that Q_J is a cell of P will be considered.

Fix a function $f \in C_c(P^\circ, [0,1])$, the interior P° being understood relatively to $\mathbb{R}_+ P$ as in (4.8). Let Z denote the union of all $(m-1)$-faces of P containing 0 (as a vertex), and let V be a prescribed neighbourhood of $Z \cap P^\circ$ in P°. Consider a continuous function $\eta : P^\circ \to [0,1]$ such that

$$\operatorname{supp}(1-\eta) \subset V, \quad \operatorname{supp}\eta \subset P^\circ \setminus Z.$$

Then $\eta f \in C_c(P^\circ, [0,1])$ has compact support in $P^\circ \setminus Z = \bigcup_J Q_J^\circ$, and so, by (a),

$$\lim_{\varepsilon \to 0} \int_{P^\circ} \eta f e_\varepsilon(\varphi)\, d\mu_g = \int_{P^\circ} \eta f e(\varphi)\, d\mu_g$$

according to [KS 1993, Theorem 1.5.1] applied to each of the Riemannian domains Q_J°. Here $e(\varphi)$ is defined a.e. in P by

$$e(\varphi)_{|Q_J^\circ} = e(\varphi_{|Q_J^\circ}),$$

the energy density of $\varphi_{|Q_J^\circ} : Q_J^\circ \to Y$. To establish (9.4) (now for P in place of X) it remains to show for given $\delta > 0$ that

$$\limsup_{\varepsilon \to 0} \int_{P^\circ} (1-\eta) f e_\varepsilon(\varphi)\, d\mu_g < \delta \tag{9.6}$$

for a suitable choice of the above neighbourhood $V = V(\delta)$ of $Z \cap P^\circ$ in P°.

In verifying this we may replace the given Riemannian metric g on P (i.e., on S) by the Euclidean Riemannian metric g' (induced by the Euclidean metric on $\mathbb{R}^k \supset P$), and hence $d\mu_g$ by $d\mu'$ $(= \prod_{j \in J} dx_j$ on $Q_J)$, often suppressed in integrals. Indeed, if Λ denotes the ellipticity constant for $g_{|P}$ (cf. (4.1) and (4.2)) we have $d\mu_g \leqslant \Lambda^m d\mu'$ on P, and $B_P(x,\varepsilon) \subset B'_P(x, \Lambda\varepsilon)$ (ball with respect to g') for every $x \in P$ and small $\varepsilon > 0$; it follows that $e_\varepsilon(\varphi_{|P}) \leqslant \Lambda^{2m+2} e'_{\Lambda\varepsilon}(\varphi_{|P})$.

In the rest of the proof we therefore refer tacitly to the Euclidean Riemannian structure on P (omitting the dashes), and the associated distance d_P, relative to which P is a geodesic space (viz. a compact length space, see Chapter 4, The intrinsic distance).

By a *chain* we shall understand the union of a finite sequence $Q_0, \cdots, Q_p$ of distinct m-cells Q_i of P (i.e., cubes of the above form Q_J) such that, for each $i = 1, \cdots, p$, $\dim(Q_i \cap Q_{i-1}) = m - 1$ and $(Q_i \setminus Q_{i-1}) \cap Q_j = \emptyset$ for $j < i - 1$. There is only a finite number N of such chains.

Any two m-cells Q, Q' of P can be joined by a chain $A = Q_0 \cup \cdots \cup Q_p$, $Q_0 = Q$, $Q_p = Q'$. Since (P, d_P) is admissible this is easily shown by joining a point x of Q to a point x' of Q' by a (minimizing) geodesic segment xx' in P, and noting that the intersection of xx' with each m-cell of P is either a single straight line segment or a single point, if not empty.

Every chain $A = Q_0 \cup \cdots \cup Q_p$ is Lip homeomorphic to a cube in $\mathbb{R}^m$. For the construction of a Lip homeomorphism $\alpha : A \to [-1,1]^m$ it suffices, by recursion with respect to p, to construct (if $p \geqslant 1$) a Lip homeomorphism of $Q_p \cup Q_{p-1}$ onto Q_{p-1} which (if $p \geqslant 2$) leaves the $(m-1)$-dimensional face $Q_{p-1} \cap Q_{p-2}$ of Q_{p-1} pointwise fixed. Here we may replace Q_{p-1} by $[0,1]^m = [0,1]^2 \times [0,1]^{m-2}$, $Q_p \cup Q_{p-1}$ by $[-1,1] \times [0,1] \times [0,1]^{m-2}$, and the face $Q_{p-1} \cap Q_{p-2}$ by $[0,1] \times \{0\} \times [0,1]^{m-2}$. We only need to consider the case $m = 2$ (for if $\kappa : E_1 \to E_0$ is a bi-Lip bijection between sets $E_1, E_0 \subset \mathbb{R}^2$, then $\kappa \times \mathrm{id} : E_1 \times [0,1]^{m-2} \to E_0 \times [0,1]^{m-2}$ is likewise a bi-Lip bijection). When $m = 2$, we thus consider the rectangle $E_1 = [-1,1] \times [0,1]$ and its right half $E_0 = [0,1]^2$. Now, E_1 is composed of three triangles t_1, t_2, t_3, all with 0 as a vertex, the opposite sides being $\{1\} \times [0,1]$, $[-1,1] \times \{1\}$ and $\{-1\} \times [0,1]$. We therefore divide E_0 into 3 triangles s_1, s_2, s_3, again with 0 as a common vertex, now with opposite sides $\{1\} \times [0,1]$ (as before), $[\frac{1}{2}, 1] \times \{1\}$ and $[0, \frac{1}{2}] \times \{1\}$. The obvious affine maps $t_i \to s_i$ $(i = 1,2,3)$ are bi-Lip bijections which combine to produce the desired bi-Lip bijection $\kappa : E_1 \to E_0$ extending the identity map of $t_1 = s_1$, in particular that of the side $[0,1] \times \{0\}$ thereof.

From (a) $\wedge$ (b') follows by p applications of [KS 1993, Theorem 1.12.3] that $\varphi \circ \alpha^{-1}$ is a finite energy map $[0,1]^m \to Y$.

Every cube is a Lip ball. (For example the map $x \mapsto (\|x\|_2/\|x\|_\infty)x$ is a Lip homeomorphism of the closed unit ball U in $\mathbb{R}^m$ onto $[-1,1]^m$, as seen by replacing $\|\cdot\|_1$ by $\|\cdot\|_2$ in the proof of Sublemma 4.3.) Therefore *we also have a Lip homeomorphism $\beta : A \to U$, and $\varphi \circ \beta^{-1}$ has finite energy.* We may assume that $\overline{V}$ does not contain the point $\beta^{-1}(0)$, and so $\beta(A \cap \overline{V})$ does not meet ϱU for some radius $\varrho > 0$. By reflection in the unit sphere ∂U the finite energy map $\varphi \circ \beta^{-1} : U \to Y$ extends to a finite energy map φ^* of $X^* := (2 - \varrho)U^\circ$ to Y, again by [KS 1993, Theorem 1.12.3].

By the same reflection, $\beta(A \cap V)$ extends to an open set $V^* \Subset X^*$; and the functions $f \circ \beta^{-1}$ and $\eta \circ \beta^{-1}$ extend to functions f^* and η^*, both of class $C(X^*, [0,1])$. The above extension of the unit ball U to the larger, open ball X^* (a Riemannian domain) was designed to achieve that $\mathrm{supp}(1 - \eta^*) \subset V^*$ $(\Subset X^*)$. (The quantities $\beta, X^*, V^*, f^*, \eta^*, \varphi^*$ all depend on the chain A.)

Fix temporarily an m-cell Q of P. For any point $x \in Q$, any chain A starting with Q, and any $\varepsilon > 0$ such that $|\beta|\varepsilon < d_{X^*}(\beta(A), \mathbb{R}^m \setminus X^*)$, whereby $|\beta|$ denotes a bi-Lip constant of β, we obtain from (9.2), writing $\beta(x) = x^*$ and $\varepsilon^* = |\beta|\varepsilon$:

$$\int_{A\cap B_P(x,\varepsilon)} \frac{d_Y^2(\varphi(x),\varphi(x'))}{\varepsilon^{m+2}} dx' \leqslant |\beta|^{2m+2} e_{\varepsilon^*}(\varphi^*)(x^*),$$

where $e_{\varepsilon^*}(\varphi^*)$ denotes the ε-approximate energy density of $\varphi^* : X^* \to Y$.

Integrate with respect to $x \in Q$ and make $\varepsilon \to 0$, hence $\varepsilon^* \to 0$, while writing $\lambda = \beta^{3m+2}$:

$$\begin{aligned}
&\limsup_{\varepsilon\to 0} \int_Q (1-\eta(x))f(x)dx \int_{A\cap B_P(x,\varepsilon)} \frac{d_Y^2(\varphi(x),\varphi(x'))}{\varepsilon^{m+2}} dx' \\
&\leqslant \lambda \lim_{\varepsilon\to 0} \int_{X^*} (1-\eta^*)f^* e_{\varepsilon^*}(\varphi^*) = \lambda \int_{X^*} (1-\eta^*)f^* e(\varphi^*) \\
&\leqslant \lambda \int_{V^*} e(\varphi^*) \leqslant \delta/N,
\end{aligned}$$

since $f^*(1-\eta^*) \in C_c(X^*)$, $\mathrm{supp}(1-\eta^*) \subset V^*$, $e(\varphi^*) \in L^1(X^*)$, and since V can be chosen so that V and hence V^* have arbitrarily small measures.

Now add over all chains A that start with Q, and recall that these chains cover P. Adding finally over all m-cells Q of P leads to (9.6). □

Remark 9.2. Every map $\varphi \in W^{1,2}_{\mathrm{loc}}(X,Y)$ of a 1-*dimensional* Riemannian polyhedron (X,g) into a locally compact geodesic space (Y,d_Y) has a *continuous* version (no curvature requirement for Y). Indeed, the locally compact separable metric space $\overline{\varphi(X)}$ has a countable base. A quasicontinuous version φ^* of φ is therefore finely continuous q.e., by Proposition 7.8 (c), and hence continuous in the usual sense everywhere in X, by Corollary 7.2. Actually, $\varphi \in W^{1,2}_{\mathrm{loc}}(X,Y)$ has a locally uniformly *Hölder* continuous version (see the text preceding Proposition 12.4).

Corollary 9.1. *If $\varphi \in W^{1,2}_{\mathrm{loc}}(X,Y)$, and if ψ denotes a Lip map of Y into another metric space Z with Lip constant $\lambda(\psi)$, then*

$$e(\psi\circ\varphi) \leqslant \lambda(\psi)^2 e(\varphi), \quad E(\psi\circ\varphi) \leqslant \lambda(\psi)^2 E(\varphi).$$

Indeed, $\psi\circ\varphi$ is quasicontinuous, and $e_\varepsilon(\psi\circ\varphi) \leqslant \lambda(\psi)^2 e_\varepsilon(\varphi)$, cf. (9.2).

Corollary 9.2. $W^{1,2}_{\mathrm{loc}}(X,\mathbb{R}) = W^{1,2}_{\mathrm{loc}}(X)$. *Explicitly, a function $u : X \to \mathbb{R}$ is locally of finite energy in the sense of Definition 9.1 (with $Y = \mathbb{R}$) iff $u \in W^{1,2}_{\mathrm{loc}}(X)$ in the sense of Chapter 5. If that is the case, the energy density of u is*

$$e(u) = c_m|\nabla u|^2 = c_m g^{ij}\partial_i u\, \partial_j u, \tag{9.7}$$

where $c_m = \omega_m/(m+1)$, ω_m being the volume of the unit ball in $\mathbb{R}^m$. If $u, v \in W^{1,2}_{\mathrm{loc}}(X)$ then, as $\varepsilon \to 0$,

$$\int_{B_X(x,\varepsilon)} \frac{(u(x') - u(x))(v(x') - v(x))}{\varepsilon^{m+2}}\, d\mu_g(x') \to c_m\langle \nabla u(x), \nabla v(x)\rangle \tag{9.8}$$

in the sense of weak convergence as measures.

The former part, including (9.7), follows by combining Theorem 9.1 and Proposition 7.7 whilst applying [KS 1993, Theorem 1.6.2]. Next, (9.8) is derived from (9.7) by polarization.

Corollary 9.3. $W^{1,2}_{\mathrm{loc}}(X,\mathbb{R}^q) = [W^{1,2}_{\mathrm{loc}}(X)]^q$. *Explicitly, a map $\varphi : X \to \mathbb{R}^q$, $\varphi = (\varphi^1.\cdots,\varphi^q)$, is locally of finite energy iff $\varphi^1,\cdots,\varphi^q \in W^{1,2}_{\mathrm{loc}}(X)$. In that case the energy density of φ is*

$$e(\varphi) = c_m|\nabla\varphi|^2 := c_m \sum_{k=1}^{q} |\nabla\varphi^k|^2. \tag{9.9}$$

Indeed, if $\varphi \in W^{1,2}_{\mathrm{loc}}(X,\mathbb{R}^q)$ then $\varphi^1,\cdots,\varphi^q \in W^{1,2}_{\mathrm{loc}}(X,\mathbb{R}) = W^{1,2}_{\mathrm{loc}}(X)$ by the two preceding corollaries. The converse follows from

$$e_\varepsilon(\varphi) = \sum_{k=1}^{q} e_\varepsilon(\varphi^k)$$

by letting $\varepsilon \to 0$, and that leads to (9.9) as well.

Lemma 9.1. (Lower semicontinuity of energy.) *Let (X,g) and (Y,d_Y) be as in Theorem 9.1. For any E-bounded sequence $(\varphi_i) \subset W^{1,2}_{\mathrm{loc}}(X,Y)$ converging in $L^2_{\mathrm{loc}}(X,Y)$ to a map φ, this map has finite energy $E(\varphi) \leqslant \liminf_i E(\varphi_i)$, and*

$$\int_X f e(\varphi)\, d\mu_g \leqslant \liminf_i \int_X f e(\varphi_i)\, d\mu_g, \qquad f \in C^+(X). \tag{9.10}$$

Proof. For each $s \in \mathcal{S}^{(m)}(X)$ it follows from [KS 1993, Theorem 1.6.1] that $\varphi_{|s^\circ} \in W^{1,2}(s^\circ, Y)$ and

$$\int_{s^\circ} f e(\varphi_{|s^\circ})\, d\mu_g \leqslant \liminf_i \int_{s^\circ} f e((\varphi_i)_{|s^\circ})\, d\mu_g \tag{9.11}$$

for any $f \in C^+(s^\circ)$. (Indeed, such f is the pointwise supremum of the family of all $f' \leqslant f$, $f' \in C_c^+(s^\circ)$.)

For any $t \in \mathcal{S}^{(m-1)}(X)$ and any $s \in \mathcal{S}^{(m)}(X)$ containing t it follows from [KS 1993, Theorem 1.12.2] that $\operatorname{tr}_t(\varphi_i)_{|s} \to \operatorname{tr}_t \varphi_{|s}$ in $L^2(t)$, and so $\operatorname{tr}_t \varphi_{|s}$ is independent of the particular s, by Theorem 9.1. Consequently, $\varphi \in W^{1,2}_{\text{loc}}(X, Y)$. After summation over all $s \in \mathcal{S}^{(m)}(X)$, (9.11) leads easily to (9.10) (even with $f \geqslant 0$ lower semicontinuous). Finally, take $f = 1$ in (9.10). □

Lemma 9.2. (Precompactness.) *Suppose in addition that X is compact and that (Y, d_Y) is a complete metric space in which closed balls are compact, e.g., a complete geodesic space. Let a sequence $(\varphi_i) \subset W^{1,2}(X, Y)$ satisfy*

$$E(\varphi_i) + \int_X d_Y^2(\varphi_i(x), q)\, d\mu_g(x) \leqslant c$$

for some constant c and some fixed point $q \in Y$. Then (φ_i) has a subsequence which converges in $L^2(X, Y)$ to a map $\varphi \in W^{1,2}(X, Y)$.

The lemma follows easily from [KS 1993, Theorem 1.13] applied to the restrictions of φ to the (finitely many) m-simplexes of (X, T), using Theorem 9.1 as in the proof of Lemma 9.1 above. If (Y, d_Y) is a complete geodesic space, the Hopf-Rinow theorem (cf. Chapter 2) applies.

Proposition 9.1. (Weak Poincaré inequality for maps.) *Let X be an admissible Riemannian polyhedron with simplexwise smooth Riemannian metric g, and suppose that the metric space (Y, d_Y) can be (globally) bi-Lip embedded in some $\mathbb{R}^q$. For any relatively compact subset X_0 of X there is then a constant $C = C(X, X_0)$ such that, for any map $\varphi \in W^{1,2}_{\text{loc}}(X, Y)$ and any ball $B_X(x, r) \subset X_0$,*

$$\min_{y \in Y} \int_{B(x, \frac{1}{2}\Lambda_{X_0}^{-2} r)} d_Y^2(y, \varphi(x'))\, d\mu_g(x') \leqslant C \Lambda_{X_0}^{6m} \tilde{\Lambda}_Y^4 r^2 \int_{B(x,r)} e(\varphi)\, d\mu_g,$$

where $\tilde{\Lambda}_Y$ denotes a bi-Lip constant for the embedding $Y \hookrightarrow \mathbb{R}^N$:

$$\tilde{\Lambda}_Y^{-1}|y-y'| \leqslant d_Y(y,y') \leqslant \tilde{\Lambda}_Y|y-y'|, \qquad y,y' \in Y. \tag{9.12}$$

As to the ellipticity constant $\Lambda = \Lambda_{X_0}$ for the restriction of g to the compact closure of X_0 in X, see (4.2). See also the definition of the meanvalue $\overline{\varphi}$ of φ after Lemma 10.4. The stated Poincaré inequality is a consequence of the following version thereof:

$$\begin{aligned} \frac{1}{2}\int_b\int_b d_Y^2(\varphi(x_1),\varphi(x_2))\, d\mu_g(x_1)\, d\mu_g(x_2) \\ \leqslant C|b|\Lambda_{X_0}^{6m}\tilde{\Lambda}_Y^4 r^2 \int_B e(\varphi)\, d\mu_g, \end{aligned} \tag{9.13}$$

where we write $B = B_X(x,r)$, $b = B_X(x, \frac{1}{2}\Lambda_{X_0}^{-2}r)$ and $|b| = \mu_g(b)$.

As noted in the text preceding Theorem 5.1 it follows from [St 1996, Theorem 2.4] that there is a corresponding *strong Poincaré inequality*, in which b is replaced by B in (9.13) and hence in Proposition 9.1, at the expense of replacing $C(X,X_0)\Lambda_{X_0}^{6m}$ by some constant $C(X,X_0,\Lambda_{X_0})$. (This will not be used in our exposition.)

Remark 9.3. Suppose the metric space (Y,d_Y) is a Riemannian polyhedron with its intrinsic distance. The hypothesis of bi-Lip embeddability of Y in Proposition 9.1 can be replaced by one of the following assumptions on (Y,d_Y) and/or the map $\varphi \in W^{1,2}_{\text{loc}}(X,Y)$:

(i) Y is compact; or

(ii) φ is locally bounded, and closed balls in Y are compact (e.g., Y is a complete geodesic space); or

(iii) φ is continuous.

This follows from Remark 4.1 because (9.12) is only used for $y, y' \in \varphi(X)$.

Proof of Proposition 9.1. It follows from (9.2) and (9.12) that

$$\tilde{\Lambda}_Y^{-2} e_\varepsilon(\widetilde{\varphi}) \leqslant e_\varepsilon(\varphi) \leqslant \tilde{\Lambda}_Y^2 e_\varepsilon(\widetilde{\varphi}), \tag{9.14}$$

where $\widetilde{\varphi}$ stands for φ regarded as a map $X \to \mathbb{R}^q$. In view of Definition 9.1 this shows that $\widetilde{\varphi} \in W^{1,2}_{\text{loc}}(X,Y)$. By Corollary 9.3, $\varphi^1, \cdots, \varphi^q \in W^{1,2}_{\text{loc}}(X)$, and also invoking Theorem 9.1 we obtain in the sense of weak convergence of measures, as $\varepsilon \to 0$,

$$e_\varepsilon(\widetilde{\varphi}) \to e(\widetilde{\varphi}) = c_m \sum_{k=1}^q |\nabla\varphi^k|^2 = c_m|\nabla\widetilde{\varphi}|^2,$$

writing $|\nabla\widetilde{\varphi}|^2 = \sum_k |\nabla\varphi^k|^2$. From (9.14) we therefore have

$$\tilde{\Lambda}_Y^{-2} c_m |\nabla\widetilde{\varphi}|^2 \leqslant e(\varphi) \leqslant \tilde{\Lambda}_Y^2 c_m |\nabla\widetilde{\varphi}|^2. \tag{9.15}$$

Now apply Theorem 5.1 (in the equivalent form that appears in the first paragraph of its proof) to each component φ^k, and add over $k = 1, \cdots, q$ to obtain

$$\begin{aligned}\frac{1}{2}\int_b\int_b |\widetilde{\varphi}(x_1) - \widetilde{\varphi}(x_2)|^2 d\mu_g(x_1)\, d\mu_g(x_2)& \\ = |b|\int_b |\widetilde{\varphi} - \overline{\widetilde{\varphi}}|^2 d\mu_g \leqslant c|b|\Lambda_{X_0}^{6m} r^2 \int_B |\nabla\widetilde{\varphi}|^2 d\mu_g.&\end{aligned}$$

This implies (9.13) with $C = c/c_m$ on account of the former inequality in (9.15) together with the latter inequality in (9.12), applied to the points $y = \varphi(x_1)$ and $y' = \varphi(x_2)$. □

Remark 9.4. The above Poincaré inequality is a companion to that of [KS 1997, Theorem 1.4.1] for maps $M \to Y$ of a *compact smooth domain* (M, g) to a complete metric space (Y, d_Y).

Energy of maps into Riemannian manifolds

In this section the domain is an arbitrary admissible Riemannian polyhedron (X, g) (g just measurable with local elliptic bounds, in the absence of any other indication), while the target is a *Riemannian C^1-manifold* (N, h) without boundary; $\dim X = m$, $\dim Y = n$. Riemannian manifolds are always understood to be *separable*, i.e., to have a countable base of open sets.

Via Nash's isometric embedding theorem [Nash 1956], [Gro 1986], [GRok 1970], there is an immediate concept of energy density and energy of a map $\varphi : X \to N$ (cf. Corollary 9.3). We also give an equivalent covariant definition in terms of Sobolev (1,2)-functions on quasiopen sets (Definition 7.4 and Definition 9.2 below). Finally we relate this energy concept to that of Definition 9.1 when both are defined.

Here and in the remaining chapters we omit the customary factor $\frac{1}{2}$ in the definition of energy.

A chart $\eta : V \to \mathbb{R}^n$ is of course bi-Lip iff the components $h_{\alpha\beta}$ of $h_{|V}$ have elliptic bounds:

$$\Lambda_V^{-2} \sum_{\alpha=1}^{n} (\xi^\alpha)^2 \leqslant h_{\alpha\beta}\xi^\alpha\xi^\beta \leqslant \Lambda_V^2 \sum_{\alpha=1}^{n} (\xi^\alpha)^2.$$

The restriction of any chart $\zeta : W \to \mathbb{R}^n$ to a domain $V \Subset W$ is bi-Lip. In the sequel *every chart is tacitly assumed to be bi-Lip.*

Definition 9.2. Relative to a given countable atlas on a Riemannian C^1-manifold (N, h), a map $\varphi : (X, g) \to (N, h)$ is of class $W^{1,2}_{\mathrm{loc}}(X, N)$, or *locally of finite energy*, if:

(i) φ has a quasicontinuous version (to be used below); and
(ii) its components $\varphi^1, \cdots, \varphi^n$ in charts $\eta : V \to \mathbb{R}^n$ are of class $W^{1,2}(U)$ for every quasiopen set $U \subset \varphi^{-1}(V)$ of compact closure in X; and
(iii) the *energy density* $e(\varphi)$ of φ, defined a.e. in each of the quasiopen sets $\varphi^{-1}(V)$ covering X by

$$e(\varphi) = \mathrm{trace}_g\, \varphi^* h = (h_{\alpha\beta} \circ \varphi)\langle \nabla\varphi^\alpha, \nabla\varphi^\beta \rangle,$$

is locally integrable over (X, μ_g).

The *energy* of $\varphi \in W^{1,2}_{\mathrm{loc}}(X, N)$ is defined by $E(\varphi) = \int_X e(\varphi)\, d\mu_g$ $(\leqslant \infty)$.

See the text following Definiton 7.4 regarding the space $W^{1,2}(U)$, and the gradient operator ∇ on it, for a quasiopen set $U \subset X$.

If X is compact we may write $W^{1,2}(X, N)$ in place of $W^{1,2}_{\mathrm{loc}}(X, N)$, and it suffices to take $U = \varphi^{-1}(V)$ in (ii).

If $\varphi(K)$ is compact for every compact $K \subset Y$, then (iii) is a consequence of (i) and (ii) because only finitely many patches V meet $\varphi(K)$. If φ is even continuous, $W^{1,2}(U)$ becomes the usual Sobolev (1,2)-space on the set U, which may now be taken to be open, like $\varphi^{-1}(V)$.

In any event, $\varphi : X \to N$ *is of class* $W^{1,2}_{\mathrm{loc}}(X, N)$ *iff the restriction of* φ *to every domain* $\Subset X$ *has finite energy.*

The above definition is independent of the choice of (countable) atlas on N. In proving that, we may assume X is compact. Suppose that $\varphi \in W^{1,2}(X, N)$, relative to a given atlas $\mathcal{V}$ on N, and let us first show that the same holds relative to any other atlas $\mathcal{W}$. Given a chart $\eta : V \to \mathbb{R}^n$ from $\mathcal{V}$ and a chart $\zeta : W \to \mathbb{R}^n$ from $\mathcal{W}$, consider any open set Z of compact closure in $V \cap W$, and note that N can be covered by countably many such sets Z. We shall prove first that $\zeta \circ \varphi_{|\varphi^{-1}(Z)} \in [W^{1,2}(\varphi^{-1}(Z))]^n$. By hypothesis, the quasiopen set $\varphi^{-1}(V)$ is the union of a sequence of quasiopen sets U such that each $(\eta \circ \varphi)_{|U}$ extends to a function $u \in [W^{1,2}(\widetilde{U})]^n$, $\widetilde{U}$ open in X, $\widetilde{U} \supset U$. The diffeomorphism

$$(\zeta \circ \eta^{-1})_{|\eta(V \cap W)} : \eta(V \cap W) \to \zeta(V \cap W)$$

between open subsets of $\mathbb{R}^n$ coincides on the compact set $\overline{\eta(Z)} \subset \eta(V \cap W)$ with a Lip C^1-map $\theta : \mathbb{R}^n \to \mathbb{R}^n$, and the map $\zeta \circ \varphi = (\zeta \circ \eta^{-1}) \circ (\eta \circ \varphi)$, considered on $U \cap \varphi^{-1}(Z)$, therefore extends to $\theta \circ u \in [W^{1,2}(\widetilde{U})]^n$. The quasiopen sets $U \cap \varphi^{-1}(Z)$ having the union $\varphi^{-1}(Z)$ $(\subset \varphi^{-1}(V))$, this shows that the restriction of $\zeta \circ \varphi$ to $\varphi^{-1}(Z)$ is indeed of class $[W^{1,2}(\varphi^{-1}(Z))]^n$.

Finally, as to the identity a.e. in $\varphi^{-1}(V \cap W)$ of $e(\varphi)$ from Definition 9.2 and the function $e'(\varphi)$ defined analogously relative to the chart $\zeta : W \to \mathbb{R}^n$, it suffices to prove that identity in each set $U \cap \varphi^{-1}(Z)$ for the above sets Z and U. By the definition of θ above, the components h'_{ij} $(i, j = 1, \cdots, n)$ of h relative to $\zeta : W \to \mathbb{R}^n$ are related in Z to the $h_{\alpha\beta}$ by

$$h_{\alpha\beta} = h'_{ij}[(\partial_\alpha \theta^i) \circ \eta]\,[(\partial_\beta \theta^j) \circ \eta].$$

Because $\eta \circ \varphi = u$ and $\zeta \circ \varphi = \theta \circ u$ in $U \cap \varphi^{-1}(Z)$, it follows that indeed

$$\begin{aligned} e(\varphi) &= (h'_{ij} \circ \varphi)[(\partial_\alpha \theta^i) \circ u]\,[(\partial_\beta \theta^j) \circ u]\langle \nabla u^\alpha, \nabla u^\beta \rangle \\ &= (h'_{ij} \circ \varphi)\langle \nabla(\theta^i \circ u), \nabla(\theta^j \circ u) \rangle = e'(\varphi) \end{aligned}$$

a.e. in $U \cap \varphi^{-1}(Z)$. Here the second equation follows from $[(\partial_\alpha \theta^i) \circ u]\nabla u^\alpha = \nabla(\theta^i \circ u)$, valid a.e. in $\widetilde{U}$, by Lemma 5.2.

Remark 9.5. If, for example, N is simply connected with nonpositive curvature then there is a single chart $\eta : N \to \mathbb{R}^n$ (see e.g., [KN 1969, p. 102]), and therefore (if η is bi-Lip) condition (ii) does not involve quasiopen sets. A similar situation is described in the following lemma.

Lemma 9.3. *Suppose that the Riemannian C^1-manifold (N, h) without boundary is isometrically C^1-embedded as a Riemannian submanifold of some $\mathbb{R}^q$. A map $\varphi : X \to N$ is of class $W^{1,2}_{\text{loc}}(X, N)$ in the sense of Definition 9.2 iff the components $\varphi^1, \cdots, \varphi^q$ of $\varphi : X \to N \hookrightarrow \mathbb{R}^q$ are of class $W^{1,2}_{\text{loc}}(X)$. In the affirmative case, the energy density of φ is*

$$e(\varphi) = |\nabla \varphi|^2 := \sum_{i=1}^{q} |\nabla \varphi^i|^2. \tag{9.16}$$

Proof. We may assume X compact. Denote by $\iota : N \hookrightarrow \mathbb{R}^q$ the Riemannian C^1-embedding in question, and write $\iota(y) = (y^1, \cdots, y^q)$ for $y \in N$. As a map between metric spaces, ι is contractive:

$$|\iota(y) - \iota(y')| \leqslant d_N(y, y') \qquad \text{for } y, y' \in N. \tag{9.17}$$

For each n-tuple $J = (j_1, \cdots, j_n)$ of distinct indices from $\{1, \cdots, q\}$ write

$$\eta_J(y) = (y^{j_1}, \cdots, y^{j_n}), \qquad y \in N.$$

By the implicit function theorem there is a countable cover of N by open sets $V \subset N$ such that, for some $J = J(V)$, $(\eta_J)_{|V}$ is a bi-Lip C^1-diffeomorphism of V into $\mathbb{R}^J$ ($\simeq \mathbb{R}^n$). The pairs (V, η_J) then form an atlas $\mathcal{V}$ on N.

Suppose first that φ has finite energy in the sense of Definition 9.2. By (i), φ and hence $\widetilde{\varphi} = \iota \circ \varphi : X \to \mathbb{R}^q$ have quasicontinuous versions. For each chart $\eta_J : V \to \mathbb{R}^J$ from $\mathcal{V}$ we have, by (ii),

$$\eta_J \circ \varphi_{|\varphi^{-1}(V)} \in [W^{1,2}(\varphi^{-1}(V))]^n$$

in the sense of Definition 7.4 (extended to maps into $\mathbb{R}^n$). For any open set $Z \Subset V$ the map $\iota \circ \eta^{-1} \in C^1(\eta_J(V))$ is Lip on $\eta_J(Z)$. By the extended chain rule (Chapter 7, Sobolev functions on quasiopen sets) it follows that

$$\widetilde{\varphi}_{|\varphi^{-1}(Z)} = (\iota \circ \eta_J^{-1}) \circ (\eta_J \circ \varphi_{|\varphi^{-1}(Z)}) \in [W^{1,2}(\varphi^{-1}(Z))]^q.$$

This means that each component φ^i, $i = 1, \cdots, q$, of $\widetilde{\varphi} = \iota \circ \varphi$ is of class $W^{1,2}(X)$ in the sense of Definition 7.4 (hence also in the usual sense), because the quasiopen sets $\varphi^{-1}(Z)$ form a countable cover of X.

Conversely, suppose that $\widetilde{\varphi} = \iota \circ \varphi = (\varphi^1, \cdots, \varphi^q)$ is of class $[W^{1,2}(X)]^q$ in the usual sense. By Proposition 7.7, $\widetilde{\varphi}$ and hence $\varphi = \iota^{-1} \circ \widetilde{\varphi}$ have quasicontinuous versions, and for each pair $(V, \eta_J) \in \mathcal{V}$

$$\eta_J \circ \varphi_{|\varphi^{-1}(V)} = (\varphi^{j_1}, \cdots, \varphi^{j_n}) \in [W^{1,2}(\varphi^{-1}(V))]^n.$$

This shows that φ satifies (i) and (ii) from Definition 9.2 relative to the above atlas $\mathcal{V}$. Because ι is a Riemannian isometry, the covariant components $h_{\alpha\beta}$ of h relative to the chart $\eta_J : V \to \mathbb{R}^n$ are (in the case $j_\alpha = \alpha$, say)

$$h_{\alpha\beta} = \delta_{\alpha\beta} + \sum_{n<i\leqslant q} \frac{\partial y^i}{\partial y^\alpha} \frac{\partial y^i}{\partial y^\beta} \qquad \text{in } V \text{ for } \alpha, \beta \in \{1, \cdots, n\}.$$

This leads to (9.16) a.e. in $\varphi^{-1}(V)$, by the extended chain rule; and (iii) in Definition 9.2 follows from (9.16). □

Remark 9.6. Lemma 9.1 (lower semicontinuity of energy), Lemma 9.2 (precompactness), and Proposition 9.1 (Poincaré inequality) all carry over to the present setting of maps $\varphi : X \to N$, using Definition 9.2. For Lemma 9.2 and Proposition 9.1, this follows from Theorem 9.2 below because (4.1) and (4.2) allow us to compare with the same assertions relative to the Euclidean Riemannian metric g^e on X, which is indeed simplexwise smooth, so that Lemma 9.2 and Proposition 9.1 apply in their previous setting.

Concerning Lemma 9.1 in the present setting, Nash's embedding theorem allows us to apply Lemma 9.3. When $\varphi_i \to \varphi$ in $L^2(X, N)$ in the sense that $\int_X d_N^2(\varphi_i(\cdot), \varphi(\cdot))\, d\mu_g \to 0$ as $i \to \infty$, it follows from (9.17) in view of expression (9.16) for $e(\varphi)$ that $\varphi_i^k \to \varphi^k$ in $L^2(X)$ for each $k = 1, \cdots, q$. In the Hilbert space $[W^{1,2}(X)]^q$ the Sobolev norms $\|\varphi_i\|$, where $\|\varphi_i\|^2 = \int(|\varphi_i|^2 + |\nabla\varphi_i|^2) d\mu_g$, remain bounded, by hypothesis. We may therefore assume that φ_i converges weakly in $[W^{1,2}(X)]^q$, necessarily to φ,[5] which is thus of class $[W^{1,2}(X)]^q$ as well. The norm on a Hilbert space being weakly lower semicontinuous, it follows that indeed

$$E(\varphi) + \int |\varphi|^2 \leqslant \liminf_i \Big(E(\varphi_i) + \int |\varphi_i|^2\Big) = \liminf_i E(\varphi_i) + \int |\varphi|^2.$$

The more general inequality (9.10) reduces easily to the case where $f \in C^+(X)$ is bounded. If f is also bounded away from 0, the proof is similar, noting that $\|\varphi\|_f^2 := \int(|\varphi|^2 + f|\nabla\varphi|^2)$ then defines a Hilbert norm $\|\cdot\|_f$ on $W^{1,2}(X)$ equivalent to the above Sobolev norm $\|\cdot\|$. Finally, for any bounded $f \in C^+(X)$, apply this to $f + \varepsilon$, $\varepsilon > 0$:

$$\begin{aligned}\int f|\nabla\varphi|^2 &\leqslant \int (f+\varepsilon)|\nabla\varphi|^2 \leqslant \liminf_i \int (f+\varepsilon)|\nabla\varphi_i|^2 \\ &\leqslant \liminf_i \int f|\nabla\varphi_i|^2 + \varepsilon \limsup_i \int |\nabla\varphi_i|^2.\end{aligned}$$

For $\varepsilon \to 0$ this leads to the present version of (9.10) because $E(\varphi_i) = \int |\nabla\varphi_i|^2$ remains bounded, by hypothesis. □

[5]This is a particular case of the following observation: Let $\mathcal{H}_j$ $(j = 0, 1)$ be a Hilbert space with inner product $\langle\cdot,\cdot\rangle_j$ and norm $|\cdot|_j$; and suppose that $\mathcal{H}_1 \subset \mathcal{H}_0$ and $|\varphi|_0 \leqslant |\varphi|_1$ for $\varphi \in \mathcal{H}_1$. (In our setting, $\mathcal{H}_0 = [L^2(X)]^q$, $\mathcal{H}_1 = [W^{1,2}(X)]^q$.) Then

$$|\langle\varphi,\psi\rangle_0| \leqslant |\varphi|_0|\psi|_0 \leqslant |\varphi|_1|\psi|_0$$

for $\varphi \in \mathcal{H}_1$, $\psi \in \mathcal{H}_0$. The linear form $\varphi \mapsto \langle\varphi,\psi\rangle_0$ on $\mathcal{H}_1$ therefore has the form

$$\langle\varphi,\psi\rangle_0 = \langle\varphi,\psi^*\rangle_1, \qquad \varphi \in \mathcal{H}_1,$$

for a fixed $\psi^* \in \mathcal{H}_1$ depending only on ψ. If a sequence $(\varphi_i) \subset \mathcal{H}_1$ converges weakly in $\mathcal{H}_1$ to some $\varphi \in \mathcal{H}_1$, then it follows that $\varphi_i \to \varphi$ likewise holds weakly in $\mathcal{H}_0$.

Remark 9.7. For later use we adapt the chain rule to maps $\varphi \in W^{1,2}_{\text{loc}}(X,N)$ in the sense of Definition 9.2 (instead of $\varphi \in [W^{1,2}(X)]^n$).

If $v \in C^1(V)$ (V open in N) then $v \circ \varphi \in W^{1,2}(U)$ for every quasiopen domain $U \subset \varphi^{-1}(V)$ of compact closure in X; and, in charts $\eta : W \to \mathbb{R}^n$,

$$\nabla(v \circ \varphi) = [(\partial_k v) \circ \varphi]\nabla\varphi^k \qquad \text{a.e. in } U,$$

where $\partial_k v = \partial(v \circ \eta^{-1})/\partial y^k$ $(k = 1, \cdots, n = \dim N)$ are the partials of v as a function of the local coordinates $y^1, \cdots, y^n$ on W.

It suffices to establish the above within $U \cap \varphi^{-1}(Z)$ for any open set $Z \Subset V \cap W$ because U is covered by countably many sets of the form $U \cap \varphi^{-1}(Z)$, and these sets are quasiopen when we use a quasicontinuous version of φ. Writing $v \circ \varphi = (v \circ \eta^{-1}) \circ (\eta \circ \varphi)$ on $U \cap \varphi^{-1}(Z)$, the assertion therefore follows from the chain rule in Lemma 5.2 and its extension in (7.26) because $v \circ \eta^{-1} \in C^1(\eta(V \cap W))$ has bounded partials $\partial_k v$ on $\eta(Z) \Subset \eta(V \cap W) \subset \mathbb{R}^n$. □

By Lemma 9.3, $W^{1,2}_{\text{loc}}(X,\mathbb{R})$ (in the sense of Definition 9.2) equals $W^{1,2}_{\text{loc}}(X)$ from Chapter 5, and $e(u) = |\nabla u|^2$ for every function $u \in W^{1,2}_{\text{loc}}(X)$.

In [KS 1993, p. 562] it is mentioned that the two concepts of energy (Definitions 9.1 and 9.2 above) are consistent when (X, g) is a Riemannian domain and (N, h) a compact Riemannian manifold. We propose to generalize this in an amplified form to the present case of an admissible Riemannian polyhedron (X, g) which we may clearly assume to be compact.

In the following theorem we denote by $E'(\varphi)$ and $e'(\varphi)$ the energy and the energy density of a map $\varphi : X \to N$ in the sense of Definition 9.1, while the corresponding symbols without a dash refer to Definition 9.2.

Theorem 9.2. *Let φ be a map of a compact admissible Riemannian polyhedron X with simplexwise smooth Riemannian metric g into a Riemannian C^1-manifold (N, h) without boundary. Then:*

(a) *$E'(\varphi) \geqslant c_m E(\varphi)$; and if $E'(\varphi) < \infty$ then $e'(\varphi) \geqslant c_m e(\varphi)$ a.e. in X.*

(b) *Suppose either that $\varphi(X)$ has compact closure in N, or that N admits a C^1-embedding into some $\mathbb{R}^q$, whose inverse is globally Lipschitz:*

$$d_N(y, y') \leqslant \tilde{\Lambda}_N |y - y'| \qquad \text{for } y, y' \in N \tag{9.18}$$

and some constant $\tilde{\Lambda}_N \geqslant 1$. Then, if $E(\varphi) < \infty$,

$$e_\varepsilon(\varphi) \to c_m e(\varphi) \qquad \text{pointwise a.e. and in } L^1(X) \text{ as } \varepsilon \to 0.$$

In particular, $E'(\varphi) = c_m E(\varphi)$, *and* $e'(\varphi) = c_m e(\varphi)$ *a.e. in* X.

As in Corollary 9.2, $c_m = \omega_m/(m+2)$, where ω_m is the volume of the unit ball in $\mathbb{R}^m$. The approximate energy density $e_\varepsilon(\varphi)$ was defined in (9.2). For the convergence properties in (b) we merely need that g be simplexwise *continuous.*

In the setting of [KS 1993] the convergence of $e_\varepsilon(\varphi)$ in $L^1(X)$ is stated in [Gre 1998] (without indication of proof). In the same paper, part of that theory is adapted to maps from domains in a Lipschitz submanifold of a Euclidean space with the induced Riemannian structure.

Proof of Theorem 9.2. (a) follows from Corollary 9.3 and Lemma 9.3 in view of Nash's isometric embedding theorem because the components $\varphi^1, \cdots, \varphi^q$ of $\varphi : X \to N \hookrightarrow \mathbb{R}^q$ are of class $W^{1,2}(X)$ by Corollary 9.1, cf. (9.17).

As for (b), it is known that N can be embedded as a closed subset of some $\mathbb{R}^q$. When $\overline{\varphi(X)}$ is compact and therefore contained in a relatively compact subdomain N_0 of N, the restriction of such an embedding $\iota : N \to \mathbb{R}^q$ to the submanifold N_0 is bi-Lip. Thus (9.18) holds for $y, y' \in N_0$. Below we show that this implies (b).

We are thus left with the case where N admits a C^1-embedding $N \hookrightarrow \mathbb{R}^q$ satisfying (9.18), and where $E(\varphi) < \infty$. The second paragraph of the proof of Lemma 9.3 remains in force, and so the components φ^j $(j = 1, \cdots, q)$ of φ in terms of that embedding are of class $W^{1,2}(X)$.

Step 1. Pointwise convergence of $e_\varepsilon(\varphi)$. We may suppose that X is an m-simplex (with the given Riemannian metric g). Consider a compact set $K \subset X$ $(\subset \mathbb{R}^m)$ on which $\varphi : X \to \mathbb{R}^q$ coincides with some C^1-map $K \to \mathbb{R}^q$. Such K exists with m-dimensional Lebesgue measure $|X \setminus K|$ as small as we please [BZ 1974, Theorem 3.6]. Let $\lambda(K)$ denote the Lip constant of $\varphi_{|K} : K \to \mathbb{R}^q$, K being endowed with the intrinsic distance d_X associated with g; it is equivalent to the Euclidean distance on $K \subset \mathbb{R}^m$ (Lemma 4.2).

We may assume that $\varphi : X \to \mathbb{R}^q$ is Borel measurable. For each chart $V \to \mathbb{R}^n$ from a countable atlas $\mathcal{V}$ on N with all V relatively compact, let Z_V denote the null set on X consisting of all points of the set

$$K_V = K \cap \varphi^{-1}(V)$$

which are *not* points of density of that Borel set in $\mathbb{R}^m$, and let $Z = \bigcup_{V \in \mathcal{V}} Z_V$.

According to [Haj 1996], applied to each of the components $\varphi^1, \cdots, \varphi^q \in W^{1,2}(X^\circ)$, there is a function $F \geqslant 0$ of class $L^1(X)$ such that

$$|\varphi(x) - \varphi(x')|^2 \leqslant |x - x'|^2 (F(x) + F(x')), \qquad x, x' \in X^\circ. \tag{9.19}$$

We proceed to show that $e_\varepsilon(\varphi)(x) \to c_m e(\varphi)(x)$ as $\varepsilon \to 0$ for any chart $V \to \mathbb{R}^n$ from $\mathcal{V}$ and *any point $x \in K_V \setminus Z$ which is a Lebesgue point for F*. Write $\varphi(x) = y$. For any number $\eta > 0$ choose $\delta = \delta(y, \eta) > 0$ so that $B_N(y, \delta) \subset V$ and

$$|h_{\alpha\beta}(y') - h_{\alpha\beta}(y)| < \eta \qquad \text{for } y' \in B_N(y, \delta),$$

$\alpha, \beta \in \{1, \cdots, n\}$. For such y' it follows that,

$$d_N^2(y, y') = (1 + O(\eta)) h_{\alpha\beta}(y) \Delta y^\alpha \Delta y^\beta \tag{9.20}$$

as $\eta \to 0$, whereby $\Delta y^\alpha = (y')^\alpha - y^\alpha$, etc.

For any $x' \in B_X(x, \varepsilon) \cap K_V$ the image $y' = \varphi(x') \in V$ satisfies $y' \in B_N(y, \delta)$ provided $\varepsilon < \delta/\lambda(K)$. And then (9.20) applies, leading, after integration with respect to x' over $B_X(x, \varepsilon) \cap K_V$, to

$$\begin{aligned} &\int_{B_X(x,\varepsilon)\cap K_V} d_N^2(\varphi(x), \varphi(x'))\, d\mu_g(x') \\ &\qquad = (1 + O(\eta)) h_{\alpha\beta}(\varphi(x)) \int_{B_X(x,\varepsilon)\cap K_V} \Delta\varphi^\alpha \Delta\varphi^\beta d\mu_g(x') \end{aligned} \tag{9.21}$$

as $\eta \to 0$, $\delta \to 0$, and hence $\varepsilon \to 0$. (We write $\Delta\varphi^\alpha = \varphi^\alpha(x') - \varphi^\alpha(x)$.) On the other hand it follows from (9.18) and (9.19) that

$$\begin{aligned} &\int_{B_X(x,\varepsilon)\setminus K_V} d_N^2(\varphi(x), \varphi(x'))\, d\mu_g(x') \\ &\qquad \leqslant (1 + O(\eta)) \Lambda_X^m \tilde{\Lambda}_N^2 \int_{B_X^e(x, \Lambda_X \varepsilon)\setminus K_V} \Lambda_X^2 \varepsilon^2 (F(x) + F(x'))\, dx' \\ &\qquad = o(\varepsilon^{2+m}) \end{aligned} \tag{9.22}$$

as $\varepsilon \to 0$. Indeed, since x is a point of density for K_V and a Lebesgue point for F,

$$\begin{aligned} &\int_{B_X^e(x,\Lambda_X\varepsilon)\setminus K_V} (F(x) + F(x'))\, dx' \\ &\leqslant \int_{B_X^e(x,\Lambda_X\varepsilon)} |F(x') - F(x)|\, dx' + \int_{B_X^e(x,\Lambda_X\varepsilon)\setminus K_V} 2|F(x)|\, dx' = o(\varepsilon^m). \end{aligned}$$

Adding (9.22) to (9.21), and dividing by ε^{2+m} leads to

$$\begin{aligned} &e_\varepsilon(\varphi)(x) = \\ &\qquad (1 + O(\eta)) h_{\alpha\beta}(\varphi(x)) \int_{B_X(x,\varepsilon)\cap K_V} \frac{\Delta\varphi^\alpha \Delta\varphi^\beta}{\varepsilon^{2+m}} d\mu_g(x') + o(1) \end{aligned} \tag{9.23}$$

as $\eta \to 0$, $\delta \to 0$, and hence $\varepsilon \to 0$.

By the differentiability of $\varphi_{|K}$ we have for $x' \in B_X(x,\varepsilon) \cap K_V$

$$\Delta\varphi^\alpha = d\varphi^\alpha + o(\varepsilon), \quad d\varphi^\alpha = \partial_i \varphi^\alpha(x)\Delta x^i,$$

where $\Delta x^i = (x')^i - x^i$, $i = 1, \cdots, m$. It follows that

$$\int_{B_X(x,\varepsilon)\cap K_V} \frac{\Delta\varphi^\alpha \Delta\varphi^\beta}{\varepsilon^{2+m}}\, d\mu_g(x') \\ = \partial_i\varphi^\alpha(x)\partial_j\varphi^\beta(x)\int_{B_X(x,\varepsilon)} \frac{\Delta x^i \Delta x^j}{\varepsilon^{2+m}}\, d\mu_g(x') + o(1), \tag{9.24}$$

where we have replaced $B_X(x,\varepsilon) \cap K_V$ by $B_X(x,\varepsilon)$ because x is a point of density for K_V with respect to Lebesgue measure, hence also with respect to $d\mu_g$; and furthermore $\Delta x^i \Delta x^j = O(\varepsilon^2)$, and $\mu_g(B_X(x,\varepsilon)) = O(\varepsilon^m)$. Inserting (9.24) in (9.23) leads to

$$e_\varepsilon(\varphi)(x) = o(1) + \\ (1 + O(\eta))h_{\alpha\beta}(\varphi(x))\, \partial_i\varphi^\alpha(x)\, \partial_j\varphi^\beta(x) \int_{B_X(x,\varepsilon)} \frac{\Delta x^i \Delta x^j}{\varepsilon^{2+m}}\, d\mu_g(x'),$$

as $\eta \to 0$ and $\varepsilon \to 0$, and it remains to see that

$$\partial_i\varphi^\alpha(x)\partial_j\varphi^\beta(x)\int_{B_X(x,\varepsilon)} \frac{\Delta x^i \Delta x^j}{\varepsilon^{2+m}} d\mu_g(x') \to c_m\langle \nabla\varphi^\alpha(x), \nabla\varphi^\beta(x)\rangle \tag{9.25}$$

as $\varepsilon \to 0$. For this we may suppose that g is Euclidean at the given point $x \in K_V \setminus Z$, i.e., $g^{ij}(x) = \delta_{ij}$. By the hypothesis that g be *continuous* on the simplex X we thus have for given $0 < \tau < 1$ and for small enough ε

$$B^e(x,(1-\tau)\varepsilon) \subset B_X(x,\varepsilon) \subset B^e_X(x,(1+\tau)\varepsilon), \\ (1-\tau)\, dx' \leqslant d\mu_g(x') \leqslant (1+\tau)\, dx'$$

for $x' \in B_X(x,\varepsilon)$. Writing ξ^i in place of Δx^i we conclude that the left hand member of (9.25) may be written as

$$\partial_i\varphi^\alpha(x)\partial_j\varphi^\beta(x)\int_{|\xi|\leqslant\varepsilon} \frac{\xi^i\xi^j}{\varepsilon^{2+m}}\, d\xi + O(\tau). \tag{9.26}$$

The integral here equals 0 for $i \neq j$ (by antisymmetry in ξ^i), and for $i = j$ it equals

$$\frac{1}{m}\int_{|\xi|\leqslant\varepsilon} \frac{|\xi|^2}{\varepsilon^{2+m}}\,d\xi = \omega_m \int_0^\varepsilon \frac{r^{1+m}}{\varepsilon^{2+m}}\,dr = \frac{\omega_m}{m+2} = c_m,$$

the surface area of the unit sphere in $\mathbb{R}^m$ being $m\omega_m$. For $\tau \to 0$, hence $\varepsilon \to 0$, the expression (9.26) therefore converges to

$$c_m g^{ij}(x)\,\partial_i\varphi^\alpha(x)\,\partial_j\varphi^\beta(x) = c_m\langle\nabla\varphi^\alpha(x), \nabla\varphi^\beta(x)\rangle,$$

as claimed in (9.25), thus establishing $e_\varepsilon(\varphi) \to c_m e(\varphi)$ pointwise a.e. in X.

Step 2. Convergence in $L^1(X)$ when X is a cube in $\mathbb{R}^m$, and $N = \mathbb{R}$ (both with standard metric). Then $\varphi \in W^{1,2}(X)$ extends to $W^{1,2}(\mathbb{R}^m)$ with gradient $\nabla\varphi = (\partial_1\varphi, \cdots, \partial_m\varphi) \in [L^2(\mathbb{R}^m)]^m$. For $x \in X$ and $x' \in B_X(x,\varepsilon)$ write

$$x' = x + r\xi, \quad \xi = (\xi^1, \cdots, \xi^m), \quad |\xi| = 1, \quad r \leqslant \varepsilon.$$

Then

$$(\varphi(x) - \varphi(x'))^2 \leqslant \int_0^r\int_0^r \langle\xi, \nabla\varphi(x+\varrho\xi)\rangle\,\langle\xi, \nabla\varphi(x+\sigma\xi)\rangle\,d\varrho\,d\sigma,$$

$$\begin{aligned}
&e_\varepsilon(\varphi)(x) - c_m e(\varphi)(x)\\
&\quad= \frac{1}{\varepsilon^{2+m}}\int_{B_X(x,\varepsilon)} \Big((\varphi(x)-\varphi(x'))^2 - \frac{\varepsilon^2}{m+2}|\nabla\varphi(x)|^2\Big)\,dx'\\
&\quad\leqslant \frac{1}{\varepsilon^{2+m}}\int_0^\varepsilon r^{m-1}dr\int_{|\xi|=1} d\xi\ \times\\
&\qquad\int_0^r\int_0^r (\langle\xi, \nabla\varphi(x+\varrho\xi)\rangle\,\langle\xi, \nabla\varphi(x+\sigma\xi)\rangle - \langle\xi, \nabla\varphi(x)\rangle^2)\,d\varrho\,d\sigma,
\end{aligned}$$

$$\begin{aligned}
\int_{\mathbb{R}^m} |e_\varepsilon(\varphi) - c_m e(\varphi)|\,dx \leqslant& \int_0^\varepsilon \frac{r^{m-1}dr}{\varepsilon^{m+2}}\int_{|\xi|=1} |\xi^i|\,|\xi^j|\,d\xi\ \times\\
&\int_0^r\int_0^r\int_{\mathbb{R}^m} |\partial_i\varphi(x+\varrho\xi)\,\partial_j\varphi(x+\sigma\xi) - \partial_i\varphi(x)\partial_j\varphi(x)|\,d\varrho\,d\sigma\,dx.
\end{aligned} \tag{9.27}$$

For each $i, j \in \{1, \cdots, m\}$ the integral over $x \in \mathbb{R}^m$ on the right is no bigger than

$$\int_{\mathbb{R}^m} \left(|\partial_i\varphi(x + \varrho\xi) - \partial_i\varphi(x)|\, |\partial_j\varphi(x + \sigma\xi)|\right) dx + \int_{\mathbb{R}^m} |\partial_i\varphi(x)|\, |\partial_j\varphi(x + \sigma\xi) - \partial_j\varphi(x)|) \, dx \tag{9.28}$$
$$\leqslant \|T_{\varrho\xi}(\partial_i\varphi) - \partial_i\varphi\|\, \|\partial_j\varphi\| + \|\partial_i\varphi\|\, \|T_{\sigma\xi}(\partial_j\varphi) - \partial_j\varphi\|,$$

where $\|\cdot\|$ is the L^2-norm on $\mathbb{R}^m$, and $(T_z f)(x) = f(x + z)$ for $f \in L^2(\mathbb{R}^m)$, $z \in \mathbb{R}^m$. For given $\eta > 0$ choose $\varepsilon_0 > 0$ so that $\|T_z(\partial_i\varphi) - \partial_i\varphi\| < \eta$ for $|z| < \varepsilon_0$ and $i = 1, \cdots, m$. For $\varepsilon < \varepsilon_0$ we then have from (9.27) and (9.28), since $\int_{|\xi|=1} d\xi = m\omega_m$,

$$\int_{\mathbb{R}^m} |e_\varepsilon(\varphi) - c_m e(\varphi)|\, dx \leqslant 2m^2 c_m \eta \|\nabla\varphi\|.$$

showing that indeed $e_\varepsilon(\varphi) \to c_m e(\varphi)$ in L^1-norm in the present case.

Step 3. Convergence in $L^1(X)$ in the general case. We shall prove that every point $a \in X$ has a neighbourhood U such that

$$\int_U |e_\varepsilon(\varphi) - c_m e(\varphi)|\, d\mu_g \to 0 \qquad \text{as } \varepsilon \to 0. \tag{9.29}$$

Let S denote the closed star of a in X relative to a triangulation for which a is a vertex. As in the proof of Theorem 9.1 we identify S with its image P under the Lip homeomorphism $\theta : S \to P$ from Lemma 4.3, P being a subcomplex of $[0,1]^k$, where k is the number of vertices of S other than a. Take

$$U = \tfrac{1}{2}P, \quad \varepsilon < \tfrac{1}{2}\Lambda_X^{-1} \tag{9.30}$$

with the ellipticity constant $\Lambda_X = \Lambda$ from (4.1) and (4.2).

Fix an m-cell Q of P. It suffices to establish (9.29) with U replaced by $\frac{1}{2}Q$. In the proof of Theorem 9.1 we showed that any other m-cell Q' in P can be joined to Q by a finite chain $A = Q_0 \cup \cdots \cup Q_p$, $Q_0 = Q$, $Q_p = Q'$. We further constructed a Lip homeomorphism α, now denoted χ, of $A \subset (P, d_P^e)$ onto a cube $X^* \subset \mathbb{R}^m$. We shall use this map χ to reduce (9.29) to Step 2 with $X = X^*$.

In terms of the components $\varphi^j \in W^{1,2}(P)$, $j = 1, \cdots, q$, of the map $\varphi : P \to N \hookrightarrow \mathbb{R}^q$ consider the functions

$$\varphi_A^j = \varphi^j \circ \chi^{-1} \in W^{1,2}(X^*), \qquad j = 1, \cdots, q.$$

Arguing as late in the proof of Theorem 9.1 (where similarly an m-cell Q was fixed), we find for $x \in \frac{1}{2}Q$

$$e_\varepsilon(\varphi)(x) \leqslant \lambda \sum_A \sum_j e_{\varepsilon^*}(\varphi_A^j)(x^*). \qquad x^* = \chi(x), \tag{9.31}$$

where λ is a constant and A ranges over the finite family of all chains in P starting with Q; and for each A we set $\varepsilon^* = \Lambda_X|\chi|\varepsilon$, where $|\chi|$ is a bi-Lip constant of χ.

For brevity write $e_\varepsilon(\varphi) = e_\varepsilon$, $c_m e(\varphi) = e$, and

$$\lambda \sum_A \sum_j e_{\varepsilon^*}(\varphi_A^j) \circ \chi = f_\varepsilon, \quad c_m \sum_A \sum_j e(\varphi_A^j) \circ \chi = f.$$

For $\varepsilon \to 0$ we have shown in Step 1 that $e_\varepsilon \to e$ and $f_\varepsilon \to f$ hold pointwise a.e. in $\frac{1}{2}Q$ (the latter because χ^{-1} preserves null sets); and in Step 2 that $f_\varepsilon \to f$ in norm in $L^1(\frac{1}{2}Q)$ (because postcomposition with χ preserves L^1-convergence). Since $e_\varepsilon \leqslant f_\varepsilon$ in $\frac{1}{2}Q$, by (9.31), we have in that set

$$(e_\varepsilon - e) - (f_\varepsilon - f) \leqslant f - e \leqslant f, \quad |e_\varepsilon - e| - |f_\varepsilon - f| \leqslant f,$$

and so, by dominated convergence together with $f \in L^1$,

$$\limsup_{\varepsilon \to 0} \int_{\frac{1}{2}Q} |e_\varepsilon - e| \, d\mu_g = \limsup_{\varepsilon \to 0} \int_{\frac{1}{2}Q} \big(|e_\varepsilon - e| - |f_\varepsilon - f|\big) \, d\mu_g \leqslant 0.$$

This holds for each m-cell Q of P, and so adds up to (9.29) with $U = \frac{1}{2}P$, cf. (9.30). □

Energy of maps into Riemannian polyhedra

In this section both the domain (X, g) and the target (Y, h) are Riemannian polyhedra. As in the preceeding section (X, g) is admissible, $\dim X = m$, and g is just measurable with elliptic bounds on each m-simplex of X (in the absence of any other indication). The polyhedron Y is not required to be admissible (not even dimensionally homogeneous), but its Riemannian metric h is assumed to be *continuous*, even across nonmaximal simplexes (see Chapter 4, Riemannian polyhedron).

Now consider the (open) star $S = \mathrm{st}_Y(a_0)$ of a vertex a_0 of Y. Let $a_1, \cdots, a_k$ be the other vertices of the closure $\overline{S}$ of S in Y. Each point $y \in \overline{S}$

has barycentric coordinates $y^0, \cdots, y^k \in [0,1]$ with the sum 1, leaving out the barycentric coordinates (equal to 0) with respect to those vertices of Y which are not in $\overline{S}$ (cf. Chapter 4, Simplicial complex).

For any (not necessarily maximal) simplex t of Y containing a_0 (necessarily as a vertex) write

$$J(t) = \{\alpha \in \{1, \cdots, k\} : a_\alpha \in t\}.$$

For $\alpha, \beta \in J(t)$ denote by $h_{\alpha\beta} = h^t_{\alpha\beta}$ the covariant components of $h = h^t$ relative to the barycentric coordinates y^α on t, $\alpha \in J(t)$. Define $h^t_{\alpha\beta} = 0$ for $\alpha, \beta \in \{1, \cdots, k\}$, not both in $J(t)$. In this way $h_{\alpha\beta}$ becomes well defined and continuous on $\overline{S}$ for any $\alpha, \beta \in \{1, \cdots, k\}$.

We now define an energy concept for maps into the polyhedron Y, quite similar to that of Definition 9.2.

Definition 9.3. Relative to a given triangulation of Y, a map $\varphi : (X, g) \to (Y, h)$ is of class $W^{1,2}_{\text{loc}}(X, Y)$, or *locally of finite energy*, if

(i) φ has a quasicontinuous version (to be used below); and
(ii) for every star $S = \text{st}_Y(a_0)$ as above, the components $\varphi^\alpha = y^\alpha \circ \varphi$, $\alpha \in \{1, \cdots, k\}$, are of class $W^{1,2}(U)$ for every quasiopen subset U of $\varphi^{-1}(S)$ having compact closure in X; and
(iii) the *energy density* $e(\varphi)$ of φ, defined a.e. in each of the quasiopen sets $\varphi^{-1}(S)$ covering X by

$$e(\varphi) = (h_{\alpha\beta} \circ \varphi) \langle \nabla\varphi^\alpha, \nabla\varphi^\beta \rangle,$$

is locally integrable over (X, μ_g).

The *energy* of $\varphi \in W^{1,2}_{\text{loc}}(X, Y)$ is defined by $E(\varphi) = \int_X e(\varphi)\, d\mu_g$ ($\leqslant \infty$).

Remark 9.8. If Y is compact, or more generally if $\overline{\varphi(K)}$ is compact in Y for every compact $K \subset X$, then property (iii) in the above definition is a consequence of (i) and (ii) because $\overline{\varphi(K)}$ is covered by finitely many open stars in X.

The various other comments relating to Definition 9.2 (up to Remark 9.5) carry over, mutatis mutandis. In particular, Definition 9.3 is independent of the triangulation of Y as above. Furthermore, if (Y, h) is a Riemannian C^1-manifold, Definitions 9.2 and 9.3 agree (because we may triangulate the manifold Y in such a way that a small neighbourhood of a given point of Y is contained in a maximal simplex of Y as a polyhedron).

Remark 9.9. Lemma 9.3 has a companion for maps φ of X into an embedded Riemannian polyhedron (Y, h) in $\mathbb{R}^q$, whereby the Riemannian isometric embedding $(Y, h) \hookrightarrow \mathbb{R}^q$ is understood to be locally bi-Lip and simplexwise C^1. As in (9.16) one obtains by covariance $e(\varphi) = |\nabla\varphi|^2$ a.e. in X, viz., a.e. in $\varphi^{-1}(t)$ for each (not necessarily maximal) simplex t of Y. This expression for the energy density $e(\varphi)$ from Definition 9.3 above was used in [GS 1992] and [Chen 1995]. We do not know whether every Riemannian polyhedron (Y, h) (h continuous) admits such a Riemannian embedding.

Definition 9.3 is consistent with Definition 9.1 under suitable hypotheses, as will be shown in the following theorem. For this we may assume that X is compact. As in Theorem 9.2 we use a dash to indicate notions of energy in the sense of Definition 9.1, while the symbols without a dash now refer to Definition 9.3 (extending Definition 9.2).

Theorem 9.3. *Let φ be a map of a compact admissible Riemannian polyhedron X with simplexwise smooth Riemannian metric g into a Riemannian polyhedron Y with continuous Riemannian metric h, and suppose that $\varphi(X)$ has compact closure in Y. Then:*

(a) *$E'(\varphi) < \infty$ iff $E(\varphi) < \infty$.*
(b) *If $E(\varphi) < \infty$ then $e_\varepsilon(\varphi) \to c_m e(\varphi)$ pointwise a.e. and in $L^1(X)$ as $\varepsilon \to 0$. In particular, $E'(\varphi) = c_m E(\varphi)$, and $e'(\varphi) = c_m e(\varphi)$ a.e.*

As in Corollary 9.2, $c_m = \omega_m/(m+2)$, where ω_m is the volume of the unit ball in $\mathbb{R}^m$.

Proof. As in Remark 9.8 we may assume that Y is also compact. Relative to a given triangulation $T = (K, \theta)$ of Y let $a_1, \cdots, a_q$ be the vertices of Y and $y^1, \cdots, y^q$ the corresponding barycentric coordinates on Y. The simplexwise affine injection

$$\iota : Y \hookrightarrow \mathbb{R}^q$$

given by $\iota(y) = (y^1, \cdots, y^q)$ is bi-Lip, being an isometry of Y with the barycentric distance d (induced via θ by that on $|K|$), and that distance on Y is equivalent to the given distance on Y, hence also to the intrinsic distance d_Y, cf. Chapter 4.

Ad (a). If $E'(\varphi) < \infty$ then $\iota \circ \varphi \in [W^{1,2}(X)]^q$, by Corollaries 9.1 and 9.3. In the notation in text preceding Definition 9.3 this implies that each y^α is of class $W^{1,2}(\varphi^{-1}(S))$. Since φ has a quasicontinuous version, by Theorem 9.1, this means that φ has the properties (i) and (ii) in Definition 9.3, hence also (iii) according to Remark 9.8, and so $E(\varphi) < \infty$. The converse implication is contained in (b).

Ad (b). Suppose $E(\varphi) < \infty$. Consider any vertex a of Y and the corresponding barycentric coordinate $y : Y \to \mathbb{R}$. According to Definition 9.3, $y \circ \varphi \in W^{1,2}(\varphi^{-1}(S))$ for any open star S in Y (also if $a \notin \overline{S}$, for then $y = 0$ on S). Since the finitely many stars S cover Y, it follows that $y \circ \varphi \in W^{1,2}(X)$, cf. Chapter 7, Sobolev functions on quasiopen sets.

Summing up, we have found a bi-Lip injection $Y \hookrightarrow \mathbb{R}^q$ in terms of which the components $\varphi^1, \cdots, \varphi^q$ are of class $W^{1,2}(X)$. The proof of Theorem 9.2 (b) (Steps 1, 2 and 3) therefore carries over, mutatis mutandis, so as to establish (b) of the present theorem.[6] □

The volume of a map

We conclude this chapter by describing briefly the notion of the volume functional of maps from an admissible Riemannian polyhedron to a complete geodesic space of nonpositive curvature.

First, for a Lipschitz map $\varphi : (M, g) \to (N, h)$ between Riemannian manifolds, where furthermore φ is a Riemannian embedding (i.e., $g = \varphi^* h$). The case where φ is a Riemannian submersion is handled similarly. (Thus at each point x in M the differential $d\varphi(x)$ maps the orthogonal complement of the kernel of $d\varphi(x)$ in $T_x(M)$ isometrically onto $T_{\varphi(x)}(N)$. See [Fe 1969, §3.2].) The *volume density* of φ is the measure

$$dv(\varphi) = \sqrt{\frac{\det \varphi^* h}{\det g}}\, d\mu_g.$$

In local charts, $dv(\varphi) = \sqrt{\det \varphi^* h}\, dx$. See [ES 1964, Chapter I, §1] for various equivalent expressions.

In the presence of a boundary bM, consider deformations (φ_t) of $\varphi = \varphi_0$ through Riemannian embeddings with $(\varphi_t)_{|bM} = \varphi_{|bM}$. The first variation

[6] In Step 1 (pointwise convergence) replace K_V by $K_t = K \cap \varphi^{-1}(t)$, t denoting any (not necessarily maximal) simplex of Y containing a vertex a_0. Likewise replace Z_V by Z_t, similarly defined, and Z by the union of all Z_t. For given $y \in t$ and $\eta > 0$ choose $\delta > 0$ so that $B_Y(y, \delta) \subset S := \mathrm{st}_Y(a_0)$ and

$$|h_{\alpha\beta}(y') - h_{\alpha\beta}(y)| < \eta \quad \text{for} \quad y' \in B_Y(y, \delta),$$

$\alpha, \beta \in \{1, \cdots, k\}$ (cf. the second paragraph of the present section). It follows that (9.20) carries over (with N replaced by Y) for $y, y' \in t$ and summation over $\alpha, \beta \in \{1, \cdots, k\}$. From here on, the rest of the proof in Step 1, and likewise in Steps 2 and 3, carries over.

of the volume functional $V(\varphi) = \int_M dv(\varphi)$ is

$$\frac{dV(\varphi_t)}{dt}\bigg|_{t=0} = -\int_M \left\langle \tau(\varphi), \frac{d\varphi_t}{dt}\bigg|_{t=0}\right\rangle dv(\varphi_0).$$

Then amongst smooth Riemannian embeddings, the extremals of V are those which are harmonic (i.e., minimal embeddings).

Now for maps $\varphi : (M, g) \to (Y, d_Y)$ of finite energy (in the sense of [KS 1993] or Definition 9.1 above) into complete geodesic spaces of nonpositive curvature. Take a Lip vector field Z on the closure $\overline{M} = M \cup bM$ of M. In analogy with the construction of $e(\varphi)$ earlier in this chapter, we follow Korevaar and Schoen in defining the directional energy ${}^{Z}e(\varphi)$, see Chapter 14, Theorem 1.9.6. And then the first fundamental form $\pi(\varphi)$ of φ by

$$2\,\pi(\varphi)(Z, W) = {}^{(Z+W)}e(\varphi) - {}^{(Z-W)}e(\varphi)$$

for Lip vector fields Z and W on $\overline{M}$. That definition is motivated by the parallelogram law, and requires the (Y, d_Y) have nonpositive curvature [KS 1993, §2.3]. And $\pi(\varphi)$ is a symmetric, nonnegative, continuous 2-covariant tensor field on (M, g) [KS 1993, Theorems 1.11 and 2.3.2]. If (Y, d_Y) is a Riemannian manifold, then in local charts

$$|d\varphi|^2 = g^{ij}\pi_{ij}(\varphi).$$

Extend the definition of volume density to a map $\varphi : (M, g) \to (Y, d_Y)$ by

$$dv(\varphi) = \sqrt{\frac{\det \pi(\varphi)}{\det g}}\, d\mu_g;$$

and finally, to maps $\varphi : X \to Y$ of an admissible Riemannian m-dimensional polyhedron X to a complete geodesic space Y of nonpositive curvature, as in Remark 9.1. Here

$$V(\varphi) = \int_X dv(\varphi) = \sum_{s\in\mathcal{S}^m(X)} \int_{s^\circ} dv(\varphi).$$

We shall not pursue further the variational theory of V, and its relationship to the energy functional E, in spite of their manifest interest (especially in various Plateau problems).

10

Hölder continuity of energy minimizers

In this chapter and the next we establish, in several different contexts, the existence and regularity of energy minimizing maps between certain Riemannian polyhedra.

For existence, the direct method of variational theory is used. For Hölder continuity of E-minimizers (optimal, in general), we employ results from Chapters 6 and 7 together with the weak Poincaré inequality for maps (Proposition 9.1), and adapt and complete the argument of Jost [J 1997a], which is based on a different notion of energy. (See Chapter 1 above for references to earlier work underlying [J 1997a].)

We suppose that (X, g) and (Y, h) are Riemannian polyhedra of dimensions m and n, respectively, with X admissible and g *simplexwise smooth*, and Y *complete* (hence a geodesic space), *simply connected and of nonpositive curvature*, cf. Chapter 2. In Theorem 10.2, however, the last two conditions are replaced by $\operatorname{curv} Y \leqslant K$ for some constant $0 < K < \infty$, and the locally E-minimizing map φ is required to have a suitably small range.

The polyhedral assumption on the space Y merely serves to ensure the validity of the weak Poincaré inequality for locally bounded maps into Y, and could therefore be replaced in both theorems by Y being, more generally, a complete geodesic space which is (locally) Lip homeomorphic to a closed subset of a Euclidean space.

When saying that a map $\varphi : X \to Y$ is Hölder continuous, we mean that there is a Hölder continuous map which equals φ almost everywhere in X.

When $m = 1$ every map $\varphi \in W^{1,2}_{\text{loc}}(X, Y)$ is Hölder continuous with exponent $\frac{1}{2}$ (see the text preceding Proposition 12.4).

Definition 10.1. For any metric space target Y, a map $\varphi \in W^{1,2}_{\text{loc}}(X, Y)$ is said to be *locally E-minimizing* if X can be covered by relatively compact domains $U \subset X$ for which $E(\varphi_{|U}) \leqslant E(\psi_{|U})$ for every map $\psi \in W^{1,2}_{\text{loc}}(X, Y)$ such that $\psi = \varphi$ a.e. in $X \setminus U$.

Clearly, every subdomain of U has the same property. If $E(\varphi) < \infty$, the inequality in the definition is of course equivalent to $E(\varphi) \leqslant E(\psi)$.

In Definition 10.1, as well as in Theorems 10.1 and 10.2 below, the concept of energy can be either that of Definition 9.1 or Definition 9.2. The proof of the versions of Theorems 10.1 and 10.2 for the energy concept of Definition 9.2 (in which Y is required to be a manifold, but g need not be simplexwise smooth) will be given in Chapter 12 (Propositions 12.1 and 12.2).

The case of a target of nonpositive curvature

Theorem 10.1. *If Y is a simply connected complete Riemannian polyhedron of nonpositive curvature, every locally E-minimizing map $\varphi : X \to Y$ is Hölder continuous.*

Remark 10.1. In the special case that the Riemannian polyhedron Y is isometrically embedded in some Euclidean space, [Chen 1995] has stated Theorem 10.1 (using the energy functional of [GS 1992]), but his proof is faulty, as noted in Remark 6.4.

In [J 1997a] Jost has made the same assertion as in the above theorem for more general spaces X and Y, with a different notion of energy defined as a certain limit of integrals. However, the existence of this limit is left undecided and seems dubious in the stated generality. Nonetheless, our proof of Theorem 10.1 will mostly follow the arguments in [J 1997a], adapted to our frame of Riemannian polyhedra X, Y as stated above.

By way of preparation we first establish a kind of Harnack inequality for weakly superharmonic functions in the spirit of [Mos 1961], and obtained by Sturm [St 1995a] in the framework of [BM 1995]. We give a similar proof based on our Moser-style results from Chapter 6 above.

Let $B^e(a,r)$ denote the closed ball with respect to the Euclidean Riemannian structure on X. For $u \in L^\infty(X)$ write

$$u_{+,r} = \operatorname*{ess\,sup}_{B^e(a,r)} u, \qquad u_{-,r} = \operatorname*{ess\,inf}_{B^e(a,r)} u,$$

and

$$u_r = \fint_{B^e(a,r)} u\, d\mu_g = \frac{1}{\mu_g(B^e(a,r))} \int_{B^e(a,r)} u\, d\mu_g.$$

Lemma 10.1. *For any point $a \in X$ there are positive constants $c_0 > 1$ and $\sigma, \delta < 1$ depending only on X, Λ_X, a such that $B^e(a,\sigma) \Subset X$ and the following holds for $u \in W^{1,2}(X) \cap L^\infty(X)$ and $r \leqslant \sigma/6$:*

(a) *If u is weakly superharmonic $\geqslant 0$ in $B^e(a,6r)^\circ$ then*

$$u_{5r} \leqslant c_0 u_{-,r}.$$

(b) *If u is weakly subharmonic $\geqslant 0$ in $B^e(a,6r)^\circ$ then*

$$u_{+,r} \leqslant \delta u_{+,6r} + (1-\delta) u_{5r}.$$

Proof. We may assume that a is a vertex (in a suitable triangulation of X). From Lemma 4.3 we have a bi-Lip homeomorphism θ of the compact star $S := \overline{\mathrm{st}}_X(a)$ onto a cell complex $P \subset [0,1]^k$ whose cells are unit cubes. Take $\sigma = |\theta|^{-1}$, where $|\theta|$ denotes the bi-Lip constant of θ with respect to the Euclidean Riemannian structure on S and on P; then $B^e(a,\sigma) \subset S$.

Ad (a). For $6|\theta|r \leqslant 1$ we have

$$\theta(B^e(a,5r)) \subset 5|\theta|rP \subset B^e_P(0, 5\sqrt{m}|\theta|r) \subset \theta(B^e(a, 5\sqrt{m}|\theta|^2 r)).$$

Now apply (6.8) and Remark 6.3 with $p = 1$, replacing $r/5$, $2r/5$ and r there by the present r, $5r$ and $6r$ (cf. the text at end of the paragraph following Proposition 6.1). Transporting u from S to P we then get, after some estimations,

$$u_{5r} \leqslant \Lambda_X^{2m} (\sqrt{m}|\theta|^2)^m c_2 (\kappa - 1)^{-2} u_{-,r}.$$

Ad (b). Apply (a) to the weakly superharmonic function $z = u_{+,6r} - u \geqslant 0$ in $B^e(a,6r)^\circ$:

$$u_{+,6r} - u_{5r} = z_{5r} \leqslant c_0 z_{-,r} \leqslant c_0 (u_{+,6r} - u_{+,r}).$$

This implies (b) with $\delta = 1 - 1/c_0$. □

Following the proof of [GT 1998, Lemma 8.23] we iterate the inequality in (b) $p-1$ times to get

$$u_{+,r/6^p} \leqslant \delta^p u_{+,r} + (1-\delta^p) \sum_{i=1}^{p} \tau_i u_{5r/6^i},$$

with

$$\tau_i = \frac{\delta^{p-i}(1-\delta)}{1-\delta^p}, \quad \sum_{i=1}^{p} \tau_i = 1.$$

Choose $p = p(X, \Lambda_X, a)$ so that $\delta^p \leqslant \varepsilon^2$ with $0 < \varepsilon < 1/6$ to be chosen later as $\varepsilon = 1/8$. Then, *for any function $u \in W^{1,2}(X) \cap L^\infty(X)$ such that u is weakly subharmonic $\geqslant 0$ in $B^e(a, 6r)^\circ$, $r \leqslant \sigma/6$,*

$$u_{+,\varepsilon^p r} \leqslant \varepsilon^2 u_{+,r} + \max_{1 \leqslant i \leqslant p} u_{5r/6^i}. \tag{10.1}$$

We now return to the study of maps $\varphi \in W^{1,2}_{\text{loc}}(X, Y)$. For any point $q \in Y$ we shall consider the function

$$u_q = d_Y^2(\varphi(\cdot), q). \tag{10.2}$$

For $u, v \in W^{1,2}(X)$ we shall write for brevity, guided by (9.8),

$$E(u, v) = c_m \int_X \langle \nabla u, \nabla v \rangle d\mu_g. \tag{10.3}$$

The key to the proof of Theorem 10.1 is the following lemma, of which Part (b) is due to [HKW 1975; 1976, Lemma 3] in the case where X and Y are Riemannian manifolds.

Lemma 10.2. *Let (Y, d_Y) be a simply connected complete geodesic space of nonpositive curvature, and let $\varphi \in W^{1,2}_{\text{loc}}(X, Y)$ be locally E-minimizing. Then φ is essentially locally bounded, and we have for any $q \in Y$:*

(a) *The function $\sqrt{u_q} = d_Y(\varphi(\cdot), q)$ of class $W^{1,2}_{\text{loc}}(X)$ is weakly subharmonic, in particular essentially locally bounded; and hence $u_q \in W^{1,2}_{\text{loc}}(X)$, and u_q is itself weakly subharmonic.*

(b) *$\Delta u_q \geqslant 2e(\varphi)$ in the weak sense:*

$$E(u_q, \lambda) \leqslant -2 \int_X e(\varphi) \lambda \, d\mu_g \tag{10.4}$$

holds for every $\lambda \in W^{1,2}_c(X) \cap L^\infty(X)$, $\lambda \geqslant 0$. If $E(\varphi) < \infty$ then (10.4) holds, more generally, for every $\lambda \in W^{1,2}_0(X) \cap L^\infty(X)$, $\lambda \geqslant 0$.

Proof. We adapt and complete the proof from [J 1997a], in which φ is compared with maps obtained by pulling $\varphi(x)$ towards a given point q of Y, an idea used in [JK 1979]. For brevity write $u_q = u$, $d_Y(\varphi(\cdot), q) = \sqrt{u_q} = v$. In proving (a) we may suppose that X is compact. Consider first the case that $\lambda \in \mathrm{Lip}(X)$, $0 \leqslant \lambda \leqslant 1$, and that $\operatorname{supp} \lambda \subset U$ for some domain U as in Definition 10.1. For $x \in X$ denote $\gamma_x : [0,1] \to Y$ the constant-speed geodesic in Y from $\gamma_x(0) = \varphi(x)$ to $\gamma_x(1) = q$. Define a map $\varphi_\lambda : X \to Y$ by

$$\varphi_\lambda(x) = \gamma_x(\lambda(x)), \qquad x \in X,$$

hence $\varphi_\lambda = \varphi$ off U. We shall compare the energy of φ with that of φ_λ.

To see that $\varphi_\lambda \in L^2(X,Y)$ (cf. Chapter 9), note that φ and hence φ_λ is μ_g-measurable because γ_x varies continuously in x with its endpoints $\varphi(x)$ and q; that is a consequence of Y being globally squared-convex (see the properties following Definition 2.8). Furthermore,

$$d_Y(\varphi_\lambda(x), q) \leqslant d_Y(\varphi(x), q) = v(x)$$

with $v \in L^2(X)$ since $\varphi \in L^2(X,Y)$, cf. Chapter 9. We show that $\varphi_\lambda \in W^{1,2}(X,Y)$. Triangle comparison [KS 1993, (2.1 ii)] gives, for $x, x' \in X$,

$$\begin{aligned} d_Y^2(\varphi(x), \varphi_\lambda(x')) \leqslant (1-\lambda(x'))d_Y^2(\varphi(x),\varphi(x')) + \lambda(x')d_Y^2(\varphi(x),q) - \\ \lambda(x')(1-\lambda(x'))d_Y^2(\varphi(x'),q) \end{aligned}$$

and

$$\begin{aligned} d_Y^2(\varphi_\lambda(x), \varphi_\lambda(x')) \leqslant (1-\lambda(x))d_Y^2(\varphi(x),\varphi_\lambda(x')) + \lambda(x)d_Y^2(\varphi_\lambda(x'),q) - \\ \lambda(x)(1-\lambda(x))d_Y^2(\varphi(x),q). \end{aligned}$$

Inserting $d_Y(\varphi_\lambda(x'), q) = (1-\lambda(x'))d_Y(\varphi(x'), q)$, $d_Y(\varphi(\cdot), q) = v$ gives

$$\begin{aligned} &d_Y^2(\varphi_\lambda(x), \varphi_\lambda(x')) - d_Y^2(\varphi(x), \varphi(x')) \\ &\quad \leqslant - \left[\lambda(x) + \lambda(x') - \lambda(x)\lambda(x')\right] d_Y^2(\varphi(x), \varphi(x')) \\ &\qquad - (\lambda(x) - \lambda(x'))[(1-\lambda(x))v^2(x) - (1-\lambda(x'))v^2(x')], \end{aligned}$$

or equivalently (essentially by a discrete version of the "polarization identity" (2.1)):

$$\begin{aligned} &d_Y^2(\varphi_\lambda(x), \varphi_\lambda(x')) - d_Y^2(\varphi(x), \varphi(x')) \\ &\quad \leqslant - \left[\lambda(x) + \lambda(x') - \lambda(x)\lambda(x')\right] \left[d_Y^2(\varphi(x), \varphi(x')) - (v(x) - v(x'))^2\right] \\ &\qquad - 2(v(x) - v(x'))[\lambda(x)v(x) - \lambda(x')v(x')] \\ &\qquad + [\lambda(x)v(x) - \lambda(x')v(x')]^2. \end{aligned}$$

Here $|\lambda(x') - \lambda(x)| \leqslant \text{const.}\, \varepsilon$ when $x' \in B_X(x, \varepsilon)$, and so

$$\lambda(x) + \lambda(x') - \lambda(x)\lambda(x') = 2\lambda(x) - \lambda(x)^2 + O(\varepsilon).$$

By Definition 9.1, Theorem 9.1 and Corollary 9.1 we infer, as $\varepsilon \to 0$, that φ_λ (viewed on U) has finite energy, like φ, and that

$$\begin{aligned} 0 &\leqslant E(\varphi_{\lambda|U}) - E(\varphi_{|U}) \\ &= \lim_{f \in C_c(U,[0,1])} \limsup_{\varepsilon \to 0} \int_U (e_\varepsilon(\varphi_\lambda) - e_\varepsilon(\varphi)) f \, d\mu_g \\ &\leqslant - \int_U (2\lambda - \lambda^2)(e(\varphi) - c_m |\nabla v|^2) \, d\mu_g \\ &\quad - 2c_m \int_U \langle \nabla v, \nabla(\lambda v) \rangle d\mu_g + c_m \int_U |\nabla(\lambda v)|^2 d\mu_g. \end{aligned}$$

For the equation here we use the interpretation of the supremum in (9.3) (now with U in place of X) as the limit of the corresponding increasing net with the upper directed index set $C_c(U, [0,1])$ (ordered pointwise).

For given $\lambda \in \mathrm{Lip}_c^+(U)$ we may replace λ above by $t\lambda$ for any number $t > 0$, divide by t, and let $t \to 0$. This shows that we may delete the terms above of second order in λ and thereby obtain

$$-c_m \int_U \langle \nabla v, \nabla(\lambda v) \rangle d\mu_g \geqslant \int_U \lambda(e(\varphi) - c_m |\nabla v|^2) \, d\mu_g \tag{10.5}$$

for every $\lambda \in \mathrm{Lip}_c^+(U)$.

By the triangle inequality, $|v(x) - v(x')|^2 \leqslant d_Y^2(\varphi(x), \varphi(x'))$, and so, by Definition 9.1 and Corollary 9.2, $c_m |\nabla v|^2 \leqslant e(\varphi)$. It therefore follows from (10.5) in particular that

$$\int_U \langle \nabla v, \nabla(\lambda v) \rangle d\mu_g \leqslant 0, \qquad \lambda \in \mathrm{Lip}_c^+(U). \tag{10.6}$$

This extends to functions $\lambda \in W_c^{1,2}(U) \cap L^\infty(U)$, $\lambda \geqslant 0$, because any such λ can be approximated in $W_c^{1,2}(U)$ by uniformly bounded functions in $\mathrm{Lip}_c^+(U)$ supported by a fixed compact subset of U, as shown by truncation.

To verify (a) we first show that $v = \sqrt{u_q}$ is weakly subharmonic. Let $T_n : \mathbb{R} \to \mathbb{R}$ denote the nearest point projection of $\mathbb{R}$ onto $[1/n, n]$. The bounded functions $t_n := T_n \circ d_Y(\cdot, q)$ and $1/t_n$ are of class $\mathrm{Lip}(Y)$ since $d_Y(\cdot, q)$ is so and $T_n, 1/T_n \in \mathrm{Lip}(\mathbb{R})$. It therefore follows from Corollary 9.1

that the bounded functions $v_n := t_n \circ \varphi_{|U}$ and $1/v_n$ (truncations of v) have finite energy because $\varphi_{|U}$ has so. The set

$$U_n = \{x \in U : 1/n < v(x) < n\}$$

is *quasiopen* when we use a quasicontinuous version of $v \in W^{1,2}_{\text{loc}}(X)$ (Proposition 7.7). Consider any function $\psi \in W^{1,2}_c(X) \cap L^\infty(X)$, $\psi \geqslant 0$, with $\operatorname{supp}\psi \subset U_n$. Then $\lambda_n := \psi/v_n$ has the same properties, and $\nabla\psi = \nabla(\lambda_n v)$ in U_n, so that by Remark 5.2

$$\int_{U_n} \langle \nabla v, \nabla\psi \rangle d\mu_g = \int_{U_n} \langle \nabla v, \nabla(\lambda_n v) \rangle d\mu_g \leqslant 0$$

by the extended version of (10.6). Varying ψ, we infer from [Fug 1982a, Théorème 10] (valid also for our admissible Riemannian polyhedron X in place of a Euclidean space) that v after modification on a null set becomes *finely subharmonic* off a polar subset of U_n, and hence q.e. in $U = \bigcup_n U_n$, every countable union of polar sets being polar, in particular finely closed, and the concept of fine subharmonicity being a local one in the fine topology, see [Fug 1972]. By the quoted result in [Fug 1982a], this shows that the function $v = d_Y(\varphi(\cdot), q)$ is weakly subharmonic in U and hence in X, in particular essentially locally bounded.

Alternatively, the passage from (10.6) to the weak subharmonicity of v can be easily made by use of the substitute for a partition of unity obtained in [KilM 1992, Lemma 2.4] for coverings by quasiopen sets (such as the above set U), again valid for our polyhedron X; cf. Chapter 7, Sobolev functions on quasiopen sets. In the same way it is shown that (10.5) holds with U replaced by X for any $\lambda \in \operatorname{Lip}_c^+(X)$.

To verify (b) with $\lambda \in \operatorname{Lip}_c^+(X)$ take for v in (10.3) the present λv, insert it in the version of (10.5) with U replaced by X, and use (2.1) while exploiting the essential local boundedness of v which ensures that $u = u_q = v^2 \in W^{1,2}_{\text{loc}}(X)$:

$$\begin{aligned} E(u_q, \lambda) = E(v^2, \lambda) &= 2E(v, \lambda v) - 2c_m \int_X \lambda |\nabla v|^2 d\mu_g \\ &\leqslant -2 \int_X \lambda e(\varphi)\, d\mu_g, \qquad \lambda \in \operatorname{Lip}_c^+(X). \end{aligned}$$

This extends immediately to functions $\lambda \in W^{1,2}_c(X)^+ \cap L^\infty(X)$.

If $E(\varphi) < \infty$ then, by truncation, any function $\lambda \in W^{1,2}_0(X)^+ \cap L^\infty(X)$ can be approximated in $W^{1,2}_0(X)$ by a uniformly bounded sequence of functions $\lambda_\nu \in W^{1,2}_c(X)^+ \cap L^\infty(X)$; and so $\int \lambda_\nu e(\varphi)\, d\mu_g \to \int \lambda e(\varphi)\, d\mu_g$, by the dominated convergence theorem, because $\lambda_\nu \to \lambda$ in $L^2(X)$, and so (λ_ν) has a subsequence which converges to λ pointwise a.e. □

Remark 10.2. The functions $u_q = d_Y^2(\varphi(\cdot), q)$ and $v = \sqrt{u}_q$ being weakly subharmonic, by part (a) of the above lemma, they can be redefined uniquely on a null set so as to become subharmonic (Definition 2.2), see Remark 7.4.

Corollary 10.1. *Under the hypotheses of Lemma 10.2, every locally E-minimizing map $\varphi : X \to Y$ is locally essentially bounded.*

Indeed, for given $q \in Y$, the weakly subharmonic function $v = \sqrt{u_q}$, cf. (10.2), coincides outside some null set Z with a subharmonic function $\widetilde{v}$. In particular (assuming X compact), $\widetilde{v}$ is bounded on X, say $\widetilde{v} \leqslant R$, and so $\varphi(X \setminus Z) \subset B_Y(q, R)$. Defining $\widetilde{\varphi} = \varphi$ on $X \setminus Z$ and e.g., $\widetilde{\varphi} = q$ in Z, we have $\widetilde{\varphi}(X) \subset B_Y(q, R)$ and $\widetilde{\varphi} = \varphi$ a.e. in X.

We shall need the following estimates of mollified Green functions $G_U \mu_r$, cf. [GH 1982, p. 134]. As in Lemma 10.1, a denotes a given point of X. For given $r > 0$ such that $r \leqslant \sigma/6$ write

$$U = B^{\mathrm{e}}(a, 5r)^\circ, \quad \mu_r = (\mu_g)_{|B^{\mathrm{e}}(a,r)}.$$

Sublemma 10.1. *There are constants $\gamma_1 > \gamma_2 > 0$ and $\varrho > 0$, depending only on X, Λ_X, a, such that*

$$\begin{aligned} G_U \mu_r &\leqslant \gamma_1 r^2 \qquad \text{on } U = B^{\mathrm{e}}(a, 5r)^\circ, \\ G_U \mu_r &\geqslant \gamma_2 r^2 \qquad \text{on } B^{\mathrm{e}}(a, r) \end{aligned}$$

for any $0 < r \leqslant \varrho$.

Proof. Let G_U^e denote the Green kernel, μ_e the volume measure and ∇_e the gradient on U, all with respect to the Euclidean Riemannian structure g^{e} on U, and write $\mu_r^e = (\mu_e)_{|B^{\mathrm{e}}(a,r)}$. By Theorem 7.4, $G_U^e(x, y)$ is majorized for $x, y \in B^{\mathrm{e}}(a, r)$ by a constant times $d_X^e(x, y)^{2-m}$ (when $m \neq 2$, the case $m = 2$ being similar). By classical potential theory this implies (since $B^{\mathrm{e}}(a, r) \subset S = \overline{\mathrm{st}}_X(a)$) that the potential $G_U^e \mu_r^e$ is bounded on $B^{\mathrm{e}}(a, r)$, hence of finite energy: $\int_U G_U^e \mu_r^e \, d\mu_r^e < \infty$.

Fix $\varrho \leqslant \sigma/6$ and c in accordance with Lemma 7.2, and apply the above, taking $r = \varrho$. Writing $V = B^e(a, 5\varrho)^\circ$, we thus have constants $c_1 > c_2 > 0$ depending only on X, Λ_X, a (after fixing ϱ) such that

$$c_2 \leqslant G_V^e \mu_\varrho^e \leqslant c_1 \qquad \text{on } B^e(a, \varrho).$$

Applying Lemma 7.3 to $\alpha = r/\varrho$ ($\leqslant 1$) leads to

$$c_2 r^2/\varrho^2 \leqslant G_U^e \mu_r^e \leqslant c_1 r^2/\varrho^2 \qquad \text{on } B^e(a, r)$$

when $r \leqslant \varrho$; and hence by Lemma 7.2

$$c^{-1} c_2 r^2/\varrho^2 \leqslant G_U \mu_r \leqslant c c_1 r^2/\varrho^2 \qquad \text{on } B^e(a, r).$$

The latter inequality actually holds in all of U by Frostman's maximum principle (see property (5) in the text following Theorem 7.3). □

Henceforth, $\varrho = \varrho(X, \Lambda_X, a)$ will remain as in Sublemma 10.1. Because $\int G_U \mu_r \, d\mu_r < \infty$ for $r \leqslant \varrho$ it follows from Theorem 7.3 (3) and Remark 7.2 that $G_U \mu_r \in L_0^{1,2}(U) = W_0^{1,2}(U)$ and

$$\int_U \langle \nabla G_U \mu_r, \nabla \psi \rangle d\mu_g = \int_U \psi \, d\mu_r, \qquad \psi \in W_0^{1,2}(U). \tag{10.7}$$

For the following inequality recall the notation $u_{+,r}$ and u_r preceding Lemma 10.1, to be applied now to the weakly subharmonic function $u = u_q = d_Y^2(\varphi(\cdot), q)$, $q \in Y$, see (10.2).

Lemma 10.3. *Suppose that X is compact, (Y, d_Y) is a simply connected complete geodesic space of nonpositive curvature, and that $\varphi : X \to Y$ is locally E-minimizing. For given $a \in X$ there is a constant c depending only on X, Λ_X, a such that, for $r \leqslant \varrho$, $q \in Y$, and corresponding u_q from (10.2),*

$$\frac{r^2}{\mu_g(B^e(a, r))} \int_{B^e(a,r)} e(\varphi) \, d\mu_g \leqslant c(u_{q,+,6r} - u_{q,+,r/5}).$$

Proof. With $u = u_q$ from (10.2) and $0 < r \leqslant \varrho$ from Sublemma 10.1 set

$$z = u_{+,6r} - u,$$

a superharmonic function of class $W^{1,2}(X) \cap L^\infty(X)$ satisfying $z \geqslant 0$ in $B^e(a, 6r)^\circ$. With $U = B^e(a, 5r)^\circ$ as in Sublemma 10.1,

$$g_r := G_U \mu_r$$

is bounded and of class $W_0^{1,2}(U) = L_0^{1,2}(U)$, hence so are g_r^2 and zg_r. (Indeed, g_r can be approximated in $W_0^{1,2}(U)$ by a uniformly bounded sequence of functions of class $\mathrm{Lip}_c(U)$, as seen by truncation.) Because $E(\varphi) < \infty$ (X being compact), this further implies by the last assertion of Lemma 10.2 and by polarization (cf. (2.1) and (10.3)),

$$\begin{aligned}2\int_U e(\varphi)g_r^2 d\mu_g &\leqslant -E(u, g_r^2) = E(z, g_r^2)\\ &= 2E(g_r, zg_r) - 2c_m \int_U |\nabla g_r|^2 z\, d\mu_g \leqslant 2E(g_r, zg_r)\end{aligned} \tag{10.8}$$

since $z \geqslant 0$ in $U \subset B^e(a, 6r)^\circ$. From (10.8) and from (10.7) applied to $\psi = zg_r \in W_0^{1,2}(U)$ we obtain, by Sublemma 10.1, (recall the abbreviation $\mu_r = (\mu_g)_{B^e(a,r)}$):

$$\begin{aligned}\int_{B^e(a,r)} e(\varphi)\, d\mu_g &\leqslant \frac{1}{\gamma_2^2 r^4} \int_U e(\varphi)g_r^2 d\mu_g\\ &\leqslant \frac{c_m}{\gamma_2^2 r^4} \int_U \langle\nabla(zg_r), \nabla g_r\rangle d\mu_g = \frac{c_m}{\gamma_2^2 r^4} \int_{B^e(a,r)} zg_r d\mu_g\\ &\leqslant \frac{c_m\gamma_1}{\gamma_2^2 r^2} \int_{B^e(a,r)} z\, d\mu_g = \frac{c_m\gamma_1}{\gamma_2^2 r^2} \mu_g(B^e(a,r))z_r\\ &\leqslant \frac{c_0 c_m\gamma_1}{\gamma_2^2 r^2} \mu_g(B^e(a,r))z_{-,r/5}\\ &= \frac{c}{r^2}\mu_g(B^e(a,r))(u_{+,6r} - u_{+,r/5}).\end{aligned}$$

For the last inequality we have applied Lemma 10.1 (a) with r replaced by $r/5$, noting that $z = u_{+,6r} - u \in W^{1,2}(X) \cap L^\infty(X)$ is weakly *superharmonic* in X by Lemma 10.2 (since $u = u_q$ from (10.2)). □

For the following lemma see [KS 1993, Proposition 2.5.4]; the easy proof is included for completeness.

Lemma 10.4. *Suppose that X is compact and that Y is a simply connected complete geodesic space of nonpositive curvature. For every measurable set $A \subset X$ with $\mu_g(A) > 0$, the meanvalue $\overline{\varphi}_A$ over A of a map $\varphi \in W^{1,2}(X, Y)$ lies in the closed convex hull of the essential image $\varphi(A)$.*

The *essential image* $\varphi(A)$ is defined as the closed set of all points $q \in Y$ such that $A \cap \varphi^{-1}(V)$ has positive measure for any neighbourhood V of q

in Y. The *closed convex hull* of a set $B \subset Y$ is defined as the intersection of all closed convex subsets of Y containing B.

By the *meanvalue* $\overline{\varphi}_A$ of φ over A is meant the unique point $\overline{y} \in Y$ for which

$$\int_A d_Y^2(\varphi(\cdot), \overline{y})\, d\mu_g = \min_{y \in Y} \int_A d_Y^2(\varphi(\cdot), y)\, d\mu_g. \tag{10.9}$$

See [KS 1993, Lemma 2.5.1] for the existence and uniqueness of $\overline{y}$. (By triangle comparison it is shown that every sequence which is minimizing for $\int_A d_Y^2(\varphi(\cdot), y)\, d\mu_g$, is a Cauchy sequence in (Y, d_Y).)

Proof of Lemma 10.4. Let C denote any closed convex set containing $\varphi(A)$. We shall rule out the possibility that a point $y \in Y \setminus C$ can be the meanvalue of φ over A. Let y^* denote the point of C nearest to y. (Any sequence $(y_j) \subset C$, minimizing for $d_Y(y, \cdot)$ on C, is Cauchy in (Y, d_Y), by triangle comparison, and hence has a unique limit y^*, lying in C.) It suffices to show that $d_Y(\varphi(\cdot), y) > d_Y(\varphi(\cdot), y^*)$ a.e. in A, and that will follow from

$$d_Y(z, y) > d_Y(z, y^*) \qquad \text{for every } z \in C.$$

Consider any point $w \in zy^*$ (the geodesic segment in C with endpoints z, y^*). In the Euclidean comparison triangle $\widetilde{z}\widetilde{y}\widetilde{y}^*$ let $\widetilde{w}$ correspond to w, i.e. $|\widetilde{w} - \widetilde{y}^*| = d_Y(w, y^*)$. Because Y has globally nonpositive curvature and $w \in C$,

$$|\widetilde{w} - \widetilde{y}| \geqslant d_Y(w, y) \geqslant d_Y(y^*, y) = |\widetilde{y}^* - \widetilde{y}|.$$

For w near y^*, hence $\widetilde{w}$ near $\widetilde{y}^*$, this shows that the Euclidean angle at $\widetilde{y}^*$ is $\geqslant \pi/2$, and so indeed

$$d_Y(z, y) = |\widetilde{z} - \widetilde{y}| > |\widetilde{z} - \widetilde{y}^*| = d_Y(z, y^*).$$

□

Corollary 10.2. *Under the hypotheses of Lemma 10.4 we have for every $q \in Y$ and every measurable set $B \supset A$*

$$\operatorname*{ess\,sup}_B d_Y(\varphi(\cdot), \overline{\varphi}_A) \leqslant 2 \operatorname*{ess\,sup}_B d_Y(\varphi(\cdot), q).$$

In fact, this reduces, by the triangle inequality, to

$$d_Y(\overline{\varphi}_A, q) \leqslant \operatorname*{ess\,sup}_A d_Y(\varphi(\cdot), q) \ \ (\leqslant \operatorname*{ess\,sup}_B d_Y(\varphi(\cdot), q)),$$

which in turn follows from Lemma 10.4 because $\varphi(A) \subset B_Y(q,R)$, where $R = \operatorname{ess\,sup}_A d_Y(\varphi(\cdot),q)$; and every ball $B_Y(q,R)$ is closed and convex.

Proof of Theorem 10.1

The proof will be given in the more general setting where φ maps into a simply connected complete geodesic space Y of nonpositive curvature and such that Y *is Lip homeomorphic to a closed subset of a Euclidean space* (cf. Lemma 4.1), this being sufficient for the Poincaré inequality (Proposition 9.1 and Remark 9.3) in view of Corollary 10.1 above.

The assertion of the theorem being local, we may suppose that X is compact. Fix a point $a \in X$. As in Lemma 10.3, suppose $r \leqslant \varrho$ (cf. Sublemma 10.1). For brevity, write $\overline{\varphi}_{5r/6}$ for the meanvalue of φ over $B^e(a,5r/6)$, and consider on X the function

$$w = d_Y^2(\varphi(\cdot), \overline{\varphi}_{5r/6}).$$

Returning to (10.1) and the text preceding it, we now fix $\varepsilon = 1/8$ and $p = p(X,\Lambda_X,a)$ so that $\delta^p \leqslant \varepsilon^2$. By Corollary 4.1 (ball doubling, now with "Euclidean" balls) there is a constant $c_3 = c_3(X,\Lambda_X,a)$ such that, for $i = 1,\cdots,p$,

$$w_{5r/6^i} \leqslant \frac{\mu_g(B^e(a,5r/6))}{\mu_g(B^e(a,5r/6^p))} w_{5r/6} \leqslant c_3 w_{5r/6}. \tag{10.10}$$

By (10.9) we have,

$$\begin{aligned} w_{5r/6} &= \fint_{B^e(a,5r/6)} d_Y^2(\varphi(\cdot),\overline{\varphi}_{5r/6})\, d\mu_g \\ &= \min_{y\in Y} \fint_{B^e(a,5r/6)} d_Y^2(\varphi(\cdot),y)\, d\mu_g. \end{aligned} \tag{10.11}$$

According to Corollary 10.1 and Remark 9.3 we may apply the weak Poincaré inequality (Proposition 9.1, again with "Euclidean" balls). For a suitable constant $\kappa = \kappa(X,\Lambda_X,a) \geqslant 1$, we have from (10.10) and (10.11) when $5\kappa r < 6\varrho$

$$\begin{aligned} \max_{1\leqslant i\leqslant p} w_{5r/6^i} \leqslant c_3 w_{5r/6} &\leqslant \frac{c_4 r^2}{\mu_g(B^e(a,5r/6))} \int_{B^e(a,5\kappa r/6)} e(\varphi)\, d\mu_g \\ &\leqslant c_5 (u_{q,+,5\kappa r} - u_{q,+,\kappa r/6}) \end{aligned} \tag{10.12}$$

by Lemma 10.3 for any $q \in Y$, with u_q from (10.2). Here c_4 and hence c_5 depend on X, Λ_X, a and $\widetilde{\Lambda}_Y$.

Combine that with (10.1) applied to the function $w = d_Y^2(\varphi(\cdot), \overline{\varphi}_{5r/6})$, which is weakly subharmonic by Lemma 10.2. Taking q in the essential image $\varphi(B^e(a, \varepsilon^p r))$, cf. Lemma 10.4, we have for any $\eta > 0$

$$\mu_g(\{x \in B^e(a, \varepsilon^p r) : d_Y(\varphi(x), q) < \eta\}) > 0,$$

and hence

$$\operatorname*{ess\,sup}_{B^e(a,\varepsilon^p r)} d_Y(\varphi(\cdot), \overline{\varphi}_{5r/6}) \geqslant d_Y(q, \overline{\varphi}_{5r/6}) - \eta.$$

By the triangle inequality this implies

$$\begin{aligned}\operatorname*{ess\,sup}_{B^e(a,\varepsilon^p r)} d_Y(\varphi(\cdot), q) &\leqslant \operatorname*{ess\,sup}_{B^e(a,\varepsilon^p r)} d_Y(\varphi(\cdot), \overline{\varphi}_{5r/6}) + d_Y(q, \overline{\varphi}_{5r/6}) \\ &\leqslant 2 \operatorname*{ess\,sup}_{B^e(a,\varepsilon^p r)} d_Y(\varphi(\cdot), \overline{\varphi}_{5r/6}) + \eta.\end{aligned}$$

Here we may replace η by 0. Writing $u = u_q = d_Y^2(\varphi(\cdot), q)$ and $w = d_Y^2(\varphi(\cdot), \overline{\varphi}_{5r/6})$ we obtain from (10.1) (applied to w) and (10.12):

$$\begin{aligned}u_{+,\varepsilon^p r} &= \operatorname*{ess\,sup}_{B^e(a,\varepsilon^p r)} d_Y^2(\varphi(\cdot), q) \leqslant 4 \operatorname*{ess\,sup}_{B^e(a,\varepsilon^p r)} d_Y^2(\varphi(\cdot), \overline{\varphi}_{5r/6}) = 4w_{+,\varepsilon^p r} \\ &\leqslant 4\varepsilon^2 w_{+,r} + 4 \max_{1\leqslant i\leqslant p} w_{5r/6^i} \\ &\leqslant 4\varepsilon^2 \operatorname*{ess\,sup}_{B^e(a,r)} d_Y^2(\varphi(\cdot), \overline{\varphi}_{5r/6}) + 4c_5(u_{+,5\kappa r} - u_{+,\kappa r/6}) \\ &\leqslant 16\varepsilon^2 u_{+,r} + 4c_5(u_{+,5\kappa r} - u_{+,\kappa r/6}) \\ &\leqslant (16\varepsilon^2 + 4c_5)u_{+,5\kappa r} - 4c_5 u_{+,\kappa r/6}.\end{aligned}$$

Here the second last inequality follows from Corollary 10.2 applied to $A = B^e(a, 5r/6)$, $B = B^e(a, r)$. From the paragraph containing (10.1) we have $\varepsilon^p \leqslant \varepsilon < 1/6 \leqslant \kappa/6$, and altogether we obtain, having chosen $\varepsilon = 1/8$,

$$(1 + 4c_5)u_{+,\varepsilon^p r} \leqslant u_{+,\varepsilon^p r} + 4c_5 u_{+,\kappa r/6} \leqslant (\tfrac{1}{4} + 4c_5)u_{+,5\kappa r}. \tag{10.13}$$

Write $\beta = \varepsilon^p/(5\kappa) < 1/30$, and replace $5\kappa r$ by r, assuming henceforth that $r \leqslant r_0 := 5\varrho$ with ϱ as in Sublemma 10.1. We obtain from (10.13)

$$u_{q,+,\beta r} \leqslant \gamma^2 u_{q,+,r} \quad \text{for } q \in \varphi(B^e(a, \beta r)), \tag{10.14}$$

whereby $0 < \beta, \gamma < 1$ depend on X, Λ_X, a only, viz. $\gamma^2 = (\frac{1}{4}+4c_5)/(1+4c_5)$.

By the hypothesis that Y can be bi-Lip embedded as a closed subset of a Euclidean space, every bounded closed subset of Y is compact. By Corollary 10.1 this applies to the essential image $\varphi(X)$. In particular, $\varphi(B^e(a,r))$ is compact and nonvoid (for any $r > 0$), its closed convex hull containing the meanvalue $\overline{\varphi}_r$ of φ over $B^e(a,r)$, by Lemma 10.4.

In the rest of the proof we accordingly fix a point

$$q \in \bigcap_{r>0} \varphi(B^e(a,r)) \quad (\neq \emptyset).$$

Following De Giorgi [Gi 1957] and Moser [Mos 1961], cf. [GT 1998, Lemma 8.23], we iterate the Harnack type inequality (10.14) $j-1$ times:

$$u_{q,+,\beta^j r} \leqslant \gamma^{2j} u_{q,+,r}, \qquad j = 1, 2, \cdots . \tag{10.15}$$

For $r > 0$ let $\omega(r)$, resp. $\omega^e(r)$, denote the essential oscillation of φ over $B(a,r)$, resp. $B^e(a,r)$, i.e., the diameter of the (bounded) essential image $\varphi(B(a,r))$, resp. $\varphi(B^e(a,r))$. For any $y \in Y$, $u_{y,+,r}$ is the square of the maximal distance between y and points of $\varphi(B^e(a,r))$, and hence, for $r \leqslant r_0$,

$$u_{q,+,r} \leqslant \omega^e(r)^2 \leqslant 4u_{q,+,r} \tag{10.16}$$

because $q \in \varphi(B^e(a,r))$. From (10.15) and (10.16) we get for $r \leqslant r_0$

$$\omega^e(\beta^j r) \leqslant 2\gamma^j \omega^e(r), \qquad j = 1, 2, \cdots .$$

With $\alpha > 0$ defined by $\beta^\alpha = \gamma$ this leads to

$$\omega^e(r) \leqslant 2\Big(\frac{r}{\beta r_0}\Big)^\alpha \omega^e(r_0) \qquad \text{for } r \leqslant \beta r_0.$$

In terms of the ellipticity constant $\Lambda = \Lambda_X$ from (4.1) and (4.2) we have $\omega(r) \leqslant \omega^e(\Lambda_X r)$, and we conclude that

$$\omega(r) \leqslant A r^\alpha \qquad \text{for } r \leqslant R,$$

with $A = 2\Lambda_X^\alpha \beta^{-\alpha} r_0^{-\alpha} \omega^e(r_0)$, $R = \beta r_0/\Lambda_X$, and α all depending on X, Λ_X, a only. This implies for each $a \in X$ the existence of the essential limit

$$\varphi^*(a) = \operatorname*{ess\,lim}_{x \to a} \varphi(x),$$

the unique point of Y common to the compact essential images $\varphi(B^e(a,r))$ of diameter $\omega(r) \to 0$ as $r \to 0$. Using Lusin's theorem, as at the end of the proof of Theorem 6.2, we find that $\varphi^* = \varphi$ a.e. For any ball $B^e(a,r)$, $r \leqslant R$, and any inner point x thereof, $\varphi^*(x) \in \varphi(B^e(a,r))$, and so

$$\operatorname*{osc}_{B^e(a,r)^\circ} \varphi^* \leqslant \operatorname*{ess\,osc}_{B^e(a,r)} \varphi = \omega(r) \leqslant Ar^\alpha,$$

which shows that φ^* is indeed Hölder continuous. □

The case of a target of upper bounded curvature

Theorem 10.1 extends as follows to targets of *upper bounded curvature*, now with a restriction on the diameter of the range.

Theorem 10.2. *Let (Y, d_Y) be a complete Riemannian polyhedron with* $\operatorname{curv}(Y, d_Y) \leqslant K$ *for some $K > 0$. Let $\varphi : X \to Y$ be a locally E-minimizing map, and suppose that the essential range $\varphi(X)$ is contained in a convex compact set $V \subset Y$ of diameter $D < \pi/(2\sqrt{K})$ such that geodesics in V are uniquely determined by their endpoints. Then φ is Hölder continuous.*

Some limitation on the size of $\varphi(X)$ is necessary, see Example 12.3.

Proof. The proof of Theorem 10.2 will be given in the more general setting where φ maps into a compact geodesic space V which is Lip homeomorphic to a closed subset of a Euclidean space (cf. Remark 9.3) such that V has diameter $D < \pi/(2\sqrt{K})$ and satisfies bi-point uniqueness. We may assume that X is compact and $K = 1$ (after rescaling the metric d_Y).

We first establish Lemma 10.2, now for $q \in V$ and with a suitable constant factor > 0 on the right. Every geodesic triangle in V is 1-dominated, cf. Chapter 2. Consider in V a triangle $yy'q$ with sides

$$d_Y(y,q) = v, \quad d_Y(y',q) = v', \quad d_Y(y,y') = d$$

(all $\leqslant D$). For given $0 \leqslant \lambda, \lambda' < 1$ let $y_\lambda \in [y,q]$ and $y'_\lambda \in [y',q]$ be given by $d_Y(y,y_\lambda) = \lambda v$, $d_Y(y',y'_\lambda) = \lambda' v'$, and write $d_Y(y_\lambda, y'_\lambda) = d_\lambda$. We shall estimate d_λ from above in terms of d.

Triangle comparison for the triangle qyy'_λ in V gives, by spherical trigonometry after elimination of the angle θ between $y_\lambda y'_\lambda$ and $y_\lambda q$, writing $d_Y(y, y'_\lambda) = a$,

$$\sin v \cos d_\lambda \geqslant \sin(v - \lambda v)\cos a + \sin(\lambda v)\cos(v' - \lambda' v').$$

Similarly, for the triangle $qy'y$,

$$\sin v' \cos a \geqslant \sin(v' - \lambda' v') \cos d + \sin(\lambda' v') \cos v.$$

Eliminating a between these two inequalities leads to

$$\begin{aligned}\sin v \sin v' \cos d_\lambda \geqslant &\sin(v - \lambda v) \sin(v' - \lambda' v') \cos d \\ &+ \sin(v - \lambda v) \sin(\lambda' v') \cos v \\ &+ \sin v' \sin(\lambda v) \cos(v' - \lambda' v').\end{aligned}$$

Inserting $\cos d = 1 - 2\sin^2(\frac{1}{2}d)$ and similarly for $\cos d_\lambda$, and dividing by $\sin v \sin v'$, gives the following basic estimate (after some manipulations serving to render Theorem 9.1 and Corollary 9.2 applicable):

$$\frac{\sin D}{2D}\,(d_\lambda^2 - d^2) \leqslant 2\sin^2(\tfrac{1}{2}\, d_\lambda) - 2\sin^2(\tfrac{1}{2}d) \leqslant R = \sum_{i=1}^{4} R_i. \tag{10.17}$$

Here,

$$\begin{aligned}R_1 :&= -2\sin^2(\tfrac{1}{2}\, d)\Big(1 - \frac{\sin(v-\lambda v)}{\sin v}\,\frac{\sin(v' - \lambda' v')}{\sin v'}\Big) \\ &= -2\sin^2(\tfrac{1}{2}\, d)(\lambda v \cot v + \lambda' v' \cot v')(1 + O(\lambda^2 + \lambda'^2)) \\ &\leqslant -\frac{\sin^2(\frac{1}{2}D)}{D^2}\, 2d^2 D \cot D(2\lambda - (\lambda - \lambda'))(1 + O(\lambda^2 + \lambda'^2)),\end{aligned}$$

the functions $t^{-1}\sin(t/2)$ and $t\cot t$ being decreassing for $0 < t \leqslant D\ (< \pi/2)$;

$$\begin{aligned}R_2 :&= \cos(\lambda v)\cos(\lambda' v')(\cos v - \cos v')\Big(\frac{\tan(\lambda v)}{\sin v} - \frac{\tan(\lambda' v')}{\sin v'}\Big) \\ &= (\cos v - \cos v')\Big(\frac{\tan(\lambda v)}{\sin v} - \frac{\tan(\lambda' v')}{\sin v'}\Big)(1 + O(\lambda^2 + \lambda'^2)); \\ R_3 :&= (\cos v - \cos v')^2 \frac{\sin(\lambda v)}{\sin v}\,\frac{\sin(\lambda' v')}{\sin v'} \\ &= (\cos v - \cos v')^2 O(\lambda^2 + \lambda'^2); \\ R_4 :&= 2\sin^2\frac{\lambda v - \lambda' v'}{2} - 2\sin^2\frac{v - v'}{2}\,\frac{\sin(\lambda v)}{\sin v}\,\frac{\sin(\lambda' v')}{\sin v'} \\ &\leqslant \tfrac{1}{2}(\lambda v - \lambda' v')^2.\end{aligned}$$

With U as in Definition 10.1, consider a function $\lambda \in \mathrm{Lip}_c(U)$, $0 \leqslant \lambda < 1$, and apply the above to

$$y = \varphi(x), \quad y' = \varphi(x'), \quad \lambda = \lambda(x), \quad \lambda' = \lambda(x')$$

for $x \in X$ and $x' \in B_X(x, \varepsilon)$; and write $y_\lambda = \varphi_\lambda(x)$, $y'_\lambda = \varphi_\lambda(x')$. Then

$$v = v(x) = d_Y(\varphi(x), q), \quad v' = v(x') = d_Y(\varphi(x'), q),$$

$$d = d_Y(\varphi(x), \varphi(x')), \quad d_\lambda = d_Y(\varphi_\lambda(x), \varphi_\lambda(x')).$$

Clearly, $v \in W^{1,2}(U)$ because $d_Y(\cdot, q) \in \mathrm{Lip}(Y)$. It follows that the functions $\cos v$, λv, $\tan(\lambda v)/\sin v$ and $\sin(\lambda v)/\sin v$ are of class $W^{1,2}(U)$. Furthermore, $\varphi_\lambda = \varphi$ in $U \setminus \operatorname{supp} \lambda$. Note also that $|\lambda(x) - \lambda(x')| \leqslant$ const. ε since $x' \in B_X(x, \varepsilon)$.

Inserting all this in (10.17) we see from Definition 9.1, Theorem 9.1 and Corollary 9.2 that φ_λ (viewed on U) has finite energy, like $\varphi_{|U}$, and that

$$\begin{aligned}
0 &\leqslant \frac{\sin D}{2D}\,(E(\varphi_\lambda) - E(\varphi)) \\
&\leqslant \limsup_{f \in C_c(U,[0,1])}\ \limsup_{\varepsilon \to 0} \int_U f(x)\, d\mu_g(x) \int_{B_X(x,\varepsilon)} \frac{R}{\varepsilon^{m+2}}\, d\mu_g(x') \\
&\leqslant -(1 + O(\|\lambda\|^2_{L^\infty})) \left(\frac{\sin(\frac{1}{2}D)}{\frac{1}{2}D}\right)^2 D \cot D \int_U \lambda e(\varphi)\, d\mu_g \\
&\quad + c_m(1 + O(\|\lambda\|^2_{L^\infty})) \int_U \Big\langle \nabla \cos v, \nabla \frac{\tan(\lambda v)}{\sin v} \Big\rangle d\mu_g \\
&\quad + c_m O(\|\lambda\|^2_{L^\infty}) \int_U |\nabla \cos v|^2 d\mu_g \\
&\quad + c_m \frac{1}{2} \int_U |\nabla(\lambda v)|^2 d\mu_g.
\end{aligned}$$

For given $\lambda \in \mathrm{Lip}_c^+(U)$, $\lambda \not\equiv 0$, apply this with $t\lambda$ in place of λ above, $0 < t < 1/\|\lambda\|_{L^\infty}$; divide by t, and let $t \to 0$, noting that

$$\nabla \frac{\tan(t\lambda v)}{t \sin v} \to \nabla \frac{\lambda v}{\sin v} \qquad \text{in } L^2(U) \text{ as } t \to 0.$$

This leads to

$$\int_U \Big\langle \nabla \cos v, \nabla \frac{\lambda v}{\sin v} \Big\rangle d\mu_g \geqslant \frac{c'}{c_m} \int_U \lambda e(\varphi)\, d\mu_g, \qquad \lambda \in \mathrm{Lip}_c^+(U), \qquad (10.18)$$

where $c' = (\sin(\frac{1}{2}D)/\frac{1}{2}D)^2 D \cot D$. And here $\lambda v/\sin v$ can be replaced by any $\psi \in \mathrm{Lip}_c^+(U)$. Indeed,

$$\lambda := \psi v^{-1} \sin v \in W^{1,2}(U) \cap L^\infty(U),$$

and $\operatorname{supp} \lambda \subset \operatorname{supp} \psi$ (compact). It remains to choose a sequence of functions $\lambda_j \in \mathrm{Lip}_c^+(U)$, converging to λ in $W^{1,2}(U)$, supported by a fixed compact set, and satisfying $0 \leqslant \lambda_j \leqslant \|\lambda\|_{L^\infty}$ (the last two properties being achieved by truncation). Then

$$\frac{\lambda_j v}{\sin v} \to \frac{\lambda v}{\sin v} \qquad \text{boundedly in } W_c^{1,2}(U).$$

Since $\lambda = \psi v^{-1} \sin v \geqslant \psi D^{-1} \sin D$, it now follows from (10.18) that

$$\int_U \langle \nabla \cos v, \nabla \psi \rangle d\mu_g \geqslant \frac{c}{c_m} \int_U \psi e(\varphi)\, d\mu_g, \qquad \psi \in \mathrm{Lip}_c^+(U), \tag{10.19}$$

where $c = c' D^{-1} \sin D$ (cf. [JK 1979, Lemma 4]). In particular, $\cos v$ is weakly superharmonic and hence *superharmonic* on U (after correction on a null set, cf. Remark 7.4), and therefore actually on all of X.

The last step in the proof of the present version of Lemma 10.2 is to pass from $\cos v$ in (10.19) to v itself and to

$$u_q = v^2 = d_Y^2(\varphi(\cdot), q) = F \circ (1 - \cos v),$$

where

$$F(s) := \arccos^2(1 - s), \qquad 0 \leqslant s \leqslant 1 - \cos D,$$

is smooth, increasing and convex with $F'(s) \geqslant 2$. The auxiliary function

$$G(s) := F(s) - 2s, \qquad 0 \leqslant s \leqslant 1 - \cos D,$$

is therefore likewise smooth, increasing and convex. As noted above, the function $s := 1 - \cos v$ is subharmonic on U, in particular upper semicontinuous, and so therefore is the function

$$G \circ s = u_q - 2(1 - \cos v), \tag{10.20}$$

by Jensen's inequality applied in the definition of subharmonicity (cf. Definition 2.2):

$$G \circ s(x) \leqslant G\Big(\int_{\partial W} s\, d\omega_x^W\Big) \leqslant \int_{\partial W} (G \circ s)\, d\omega_x^W$$

for any regular neighbourhood W of x in U, the harmonic measure ω_x^W being a probability measure (see the text following Definition 2.1). Being of class $W^{1,2}(U)$, like v, $G \circ s$ is therefore weakly subharmonic on U:

$$\int_U \langle \nabla(G \circ s), \nabla\psi \rangle d\mu_g \leqslant 0, \qquad \psi \in \mathrm{Lip}_c^+(U).$$

Adding this to (10.19), multiplied by -2, and inserting (10.20) leads to

$$c_m \int_U \langle \nabla u_q, \nabla\psi \rangle d\mu_g \leqslant -2c \int_U \psi e(\varphi)\, d\mu_g, \qquad \psi \in \mathrm{Lip}_c^+(U), \tag{10.21}$$

with $c = (\sin\frac{1}{2}D/\frac{1}{2}D)^2 \cos D > 0$. By application of a partition of unity, U can be replaced here by all of X. This proves Lemma 10.2 for $q \in V$, except for the present constant factor $c > 0$ on the right. In particular, $u_q = d_Y^2(\varphi(\cdot), q)$ is subharmonic for any $q \in V$, and so Lemma 10.3 remains in force with the same proof (now with Y replaced by V).

Lemma 10.4 and hence Corollary 10.2 likewise carry over when the previous meanvalue $\overline{\varphi}_A$ of φ over $A \subset X$ is replaced by the unique point $\widetilde{\varphi}_A$ of V minimizing

$$I(y) = \int_A \sin^2[\tfrac{1}{2} d_Y(\varphi(x), y)]\, d\mu_g(x), \qquad y \in V.$$

In order to show that any I-minimizing sequence $(y_j) \subset V$ is Cauchy, let $z \in V$ denote the midpoint of $y_i y_j$, and write $d = d_Y(y_i, y_j)$ and

$$u_i = d_Y(\varphi(x), y_i), \quad u_j = d_Y(\varphi(x), y_j), \quad u = d_Y(\varphi(x), z)$$

for $x \in A$. From spherical trigonometry we then have by triangle comparison, noting that $d \leqslant D < \pi/2$ when we take $K = 1$:

$$2 \cos D \sin^2 \tfrac{1}{2} d \leqslant \sin^2 \tfrac{1}{2} u_i + \sin^2 \tfrac{1}{2} u_j - 2 \sin^2 \tfrac{1}{2} u. \tag{10.22}$$

After integration over $x \in A$ this shows that (y_j) is indeed a Cauchy sequence in V, hence has a unique limit $\widetilde{\varphi}_A \in V$, and that point minimizes $I(y)$, $y \in V$.

As in the proof of Lemma 10.4, let C be any closed convex subset of V containing the essential image $\varphi(A)$. To see that no point $y \in V \setminus C$ can equal $\widetilde{\varphi}_A$, we denote by y^* the point of C minimizing $d_Y(y, z)$, $z \in C$. This nearest point y^* exists (uniquely) because any minimizing sequence $(y_j) \subset C$ for $d_Y(y, \cdot)$ is Cauchy, by (10.22), now with y in place of $\varphi(x)$. For $z \in C$, triangle comparison for yy^*z with a point $w \in y^*z$ near y^* gives, by the spherical cosine relation, $d_Y(y,z) > d_Y(y^*,z)$, as in the case of nonpositive curvature. For $z = \varphi(x)$ this leads by integration over $x \in X$ to $I(y) > I(y^*)$, and so indeed $y \neq \widetilde{\varphi}_A$, i.e., $\widetilde{\varphi}_A \subset C$.

For the rest of the proof of Theorem 10.2 (cf. the preceding section), fix again $a \in X$. Now write $\widetilde{\varphi}_{5r/6} = \widetilde{\varphi}_A$ with $A = B^e(a, 5r/6)$, $r \leqslant \varrho$. Consider on X the functions

$$\begin{aligned} w &= d_Y^2(\varphi(\cdot), \widetilde{\varphi}_{5r/6}), \\ \widetilde{w} &= 4\sin^2[\tfrac{1}{2} d_Y(\varphi(\cdot), \widetilde{\varphi}_{5r/6})] = 4\sin^2(\tfrac{1}{2}\sqrt{w}), \end{aligned} \tag{10.23}$$

and note that

$$\widetilde{w} \leqslant w \leqslant \widetilde{D}\widetilde{w}, \qquad \widetilde{D} = \Big(\frac{D/2}{\sin(D/2)}\Big)^2 < \frac{\pi^2}{8}.$$

Instead of (10.10) and (10.11) (combined) we now get

$$w_{5r/6^i} \leqslant \widetilde{D} c_3 \widetilde{w}_{5r/6} \leqslant \widetilde{D} c_3 \min_{y \in V} \fint_{B^e(a,5r/6)} d_Y^2(\varphi(\cdot), y)\, d\mu_g$$

for $i = 1, \cdots, p$; and this leads to (10.12) as it stands (though with a new constant c_5). From here on the proof of Theorem 10.2 is completed in exactly the same way as that of Theorem 10.1, replacing throughout $\overline{\varphi}_{5r/6}$ by $\widetilde{\varphi}_{5r/6}$, and hence w by $d_Y^2(\varphi(\cdot), \widetilde{\varphi}_{5r/6})$ as in (10.23). □

Remark 10.3. The inequality (10.18), leading to the key inequality (10.21) above, follows alternatively from the proof of [Se 1995, Corollary 2.10], stated there for a Riemannian domain (X, g), but immediately extendible (in view of Theorem 9.1) to the present setting, in which (X, g) is an admissible Riemannian polyhedron with simplexwise smooth Riemannian metric.

For the case that X and Y are both Riemannian manifolds, (10.21) is due to [HKW 1976, Lemma 3] (see also Proposition 12.2).

Corollary 10.3. *Every continuous locally E-minimizing map φ of X into a complete Riemannian polyhedron Y with upper bounded curvature is Hölder continuous.*

Indeed, φ satisfies locally the hypothesis on the smallness of the range in Theorem 10.2.

11

Existence of energy minimizers

For the existence of minimizing maps in various homotopy classes, the following topological theorem is basic:

Homotopy classification. *The homotopy classes of maps* $\varphi : X \to Y$ *between locally finite polyhedra are in a natural bijective correspondence* $\varphi \mapsto \varphi_*$ *with the conjugacy classes of homomorphisms* $\varphi_* : \pi_1(X) \to \pi_1(Y)$ *of their fundamental groups, in either of these two cases:*

(a) *The universal covering space* $\widetilde{Y}$ *of* Y *is contractible, or*
(b) *the second homotopy group* $\pi_2(Y) = 0$, *and* $\dim X \leqslant 2$.

See [Sp 1966, the proof of Theorem 9 in Chapter 8.1].

As in Chapter 10 we consider two *Riemannian polyhedra* (X, g), (Y, h) with X admissible and g simplexwise smooth, and Y *complete* and (to begin with) *of nonpositive curvature.* The universal covers then have the same properties; see Example 4.5. For simplicity of exposition, X will be assumed to be *compact.*

We shall work primarily with the energy functional E from Definition 9.1, based on [KS 1993]. It will therefore be understood throughout this chapter that g *is simplexwise smooth.* See, however, the sections of Chapter 9 on energy of maps into a manifold or a polyhedron. Recall that a map $\varphi : X \to Y$ is of class $W^{1,2}(X, Y)$ iff $E(\varphi) < \infty$.

The results of the present chapter are also valid with mostly the same proofs in the case where the target Y is a complete geodesic space with the stated curvature property such that every bounded closed subset of Y can be bi-Lip embedded as a closed subset of a Euclidean space, cf. Remark 9.3 and Remark 11.4 below.

For a complete geodesic space Y of nonpositive curvature the universal cover $\pi_Y : \widetilde{Y} \to Y$ is contractible. The metric $d_{\widetilde{Y}}$ induced by π_Y makes $\widetilde{Y}$ a simply connected complete geodesic space of nonpositive curvature. Any two points of $(\widetilde{Y}, d_{\widetilde{Y}})$ can therefore be joined by a unique geodesic depending continuously on its endpoints, by the Hadamard–Cartan theorem [AB 1990, Theorem 1] (see also Chapter 2). Thus (a) above applies.

In a complete geodesic space Y (locally compact, as usual) without conjugate points (e.g., $\operatorname{curv} Y \leqslant 0$), *any continuous path $I \to Y$ is fixed-end homotopic to a unique constant-speed geodesic*, [AB 1990, Theorems 2 and 6]. This leads to

Proposition 11.1. *If Y has nonpositive curvature then homotopic maps $\varphi_0, \varphi_1 : X \to Y$ can be joined by a geodesic homotopy, i.e., by a homotopy $\varphi_I = (\varphi_t)_{t \in I}$ such that for any $x \in X$ the path $\varphi_I(x)$, given by $t \mapsto \varphi_t(x)$, is a constant-speed geodesic joining $\varphi_0(x)$ to $\varphi_1(x)$.*

Indeed, for any homotopy ψ_I joining $\varphi_0 = \psi_0$ to $\varphi_1 = \psi_1$ and any point $x \in X$, take for $\varphi_I(x)$ the unique constant-speed geodesic joining $\varphi_0(x)$ to $\varphi_1(x)$ and homotopic to $\psi_I(x)$.

Next, we derive certain convexity properties of E, assuming that Y is *simply connected* and has nonpositive curvature; Y then has globally nonpositive curvature (see Chapter 2). From [KS 1993, §2.1] we therefore have the following strong convexity property (which also entered in the proof of Lemma 10.2), basically due to Reshetnyak [Resh 1968]: For an open convex subset V of Y and constant-speed geodesic segments $\alpha, \beta : [0,1] \to V$,

$$\begin{aligned} d_Y^2(\alpha(t), \beta(t)) \leqslant (1-t) d_Y^2(\alpha(0), \beta(0)) + t\, d_Y^2(\alpha(1), \beta(1)) \\ - t(1-t)\big(d_Y(\alpha(0), \alpha(1)) - d_Y(\beta(0), \beta(1))\big)^2. \end{aligned} \tag{11.1}$$

Let φ_I be a geodesic homotopy between $\varphi_0, \varphi_1 \in W^{1,2}(X, Y)$. Then (11.1) implies for any $x, x' \in X$,

$$\begin{aligned} d_Y^2(\varphi_t(x), \varphi_t(x')) \leqslant (1-t) d_Y^2(\varphi_0(x), \varphi_0(x')) + t\, d_Y^2(\varphi_1(x), \varphi_1(x')) \\ - t(1-t)\big(d_Y(\varphi_0(x), \varphi_1(x)) - d_Y(\varphi_0(x')), \varphi_1(x'))\big)^2. \end{aligned}$$

By Theorem 9.1 (with the supremum in (9.3) interpreted as the limit of an increasing net) and Corollary 9.2 it follows that

$$E(\varphi_t) \leqslant (1-t) E(\varphi_0) + t E(\varphi_1) - c_m t(1-t) \int_X |\nabla d_Y(\varphi_0, \varphi_1)|^2. \tag{11.2}$$

In particular, we have obtained:

Proposition 11.2. *If Y is simply connected and has curvature* $\operatorname{curv} Y \leqslant 0$ *then the energy functional E is convex along geodesic homotopies.*

The case of free homotopy

Theorem 11.1. *Let X and Y be compact Riemannian polyhedra. Assume that*

(i) *X is admissible; and*
(ii) *Y has nonpositive curvature.*

Then every homotopy class $\mathcal{H}$ of continuous maps $X \to Y$ has an E-minimizer relative to $\mathcal{H}$, and any such is Hölder continuous.

Proof. Step 1. We show that $\mathcal{H}$ has members of class $\operatorname{Lip}(X,Y)$, hence of finite energy (see the text following Definition 9.1). Because X and Y are compact polyhedra, a given continuous map $\varphi : X \to Y$ can be approximated uniformly by simplicial maps φ^* relative to fine subdivisions of given triangulations of X and Y; and simplicial maps are Lipschitz. We show that φ^* can be taken to be homotopic to φ.

Because Y is compact and has nonpositive curvature, there exists $r > 0$ such that every ball $B_Y(y, 2r)$ is convex and that triangles therein are 0-dominated (cf. Chapter 2). Finitely many balls $B_Y(y_i, r)$, $i = 1, \cdots, k$, cover Y. We may assume that

$$|\varphi^*(x) - \varphi(x)| < r \qquad \text{for } x \in X.$$

(Henceforth, by abuse of notation, we write $|y - y'|$ for $d_Y(y, y')$, $y, y' \in Y$.)

For given $i = 1, \cdots, k$ and $x \in U_i := \varphi^{-1}(B_Y(y_i, r))$ there is a unique constant-speed geodesic $\gamma_x : I \to B_Y(y_i, 2r)$ joining $\varphi(x)$ to $\varphi^*(x)$ within $B_Y(y_i, 2r)$. The map $\varphi : [0,1] \times U_i \to B_Y(y_i, 2r)$ given by $\varphi(t, x) = \gamma_x(t)$ is continuous because geodesic segments in the convex ball $B_Y(y_i, 2r)$ vary continuously with their endpoints, Y being compact.

Step 2. Take an E-minimizing sequence of continuous $W^{1,2}$-maps in $\mathcal{H}$, and an L^2-convergent subsequence (φ_i), convergent also pointwise a.e. (This is possible by Lemma 9.2 because $d_Y(\varphi_i(\cdot), q) \leqslant \operatorname{diam} Y$ and $\int_X d\mu_g < \infty$.) Each of these maps lifts to a continuous map $\tilde{\varphi}_i \in W^{1,2}_{\mathrm{loc}}(\widetilde{X}, \widetilde{Y})$ between the indicated Riemannian universal covers:

$$\begin{array}{ccc} \widetilde{X} & \xrightarrow{\ \tilde{\varphi}_i\ } & \widetilde{Y} \ni \tilde{\varphi}_i(\tilde{x}_0) \\ {\scriptstyle \pi_X}\downarrow & & \downarrow{\scriptstyle \pi_Y} \\ x_0 \in X & \xrightarrow{\ \varphi_i\ } & Y \end{array}$$

Each of the maps $\tilde{\varphi}_i$ is equivariant with respect to the covering groups Γ_X, Γ_Y; i.e., letting $(\tilde{\varphi}_i)_* : \Gamma_X \to \Gamma_Y$ denote the homomorphism induced by the map φ_i,

$$\tilde{\varphi}_i \circ \gamma = (\tilde{\varphi}_i)_* (\gamma) \circ \tilde{\varphi}_i \qquad \text{for all } \gamma \in \Gamma_X. \tag{11.3}$$

Normalize these lifted maps by fixing a point $x_0 \in X$ and choosing an image point called $\tilde{\varphi}_i(\tilde{x}_0)$ belonging to ${\pi_Y}^{-1}(\varphi_i(x_0))$. Hypothesis (ii) insures that $\widetilde{Y}$ is contractible, and the classes $\mathcal{H}$ are identified with the conjugacy classes of homomorphisms $\Gamma_X \to \Gamma_Y$, independently of the choice of $\tilde{\varphi}_i(\tilde{x}_0)$ in the case of the minimizing sequence (φ_i); so choose all $\tilde{\varphi}_{i*} = \tilde{\varphi}_{1*}$. In view of (11.3) the pointwise limit $\tilde{\varphi} = \lim_{i\to\infty} \tilde{\varphi}_i$ satisfies

$$\tilde{\varphi} \circ \gamma = \tilde{\varphi}_{1*}(\gamma) \circ \tilde{\varphi} \qquad \text{a.e. in } \widetilde{X} \text{ for all } \gamma \in \Gamma_X. \tag{11.4}$$

Now Γ_X acts isometrically and simplicially on $\widetilde{X}$. We shall call a compact set $\widetilde{F} \subset \widetilde{X}$ a fundamental domain of Γ_X if $\partial \widetilde{F}$ has measure 0 and each point of $\widetilde{X}$ is Γ_X-equivalent either to exactly one point of the interior of $\widetilde{F}$ or to at least one point of $\partial \widetilde{F}$. Because X is compact, such a fundamental domain $\widetilde{F}$ can clearly be assembled as a suitable union of top-dimensional simplexes of $\widetilde{X}$, possibly after subdivision. Furthermore, $\widetilde{F}$ is contained in the interior $\widetilde{U}$ of a suitable union $\widetilde{G}$ of finitely many (say N) Γ_X-translates of $\widetilde{F}$.

Denote by $\mathcal{A}$ the class of all maps in $W^{1,2}_{\mathrm{loc}}(\widetilde{X}, \widetilde{Y})$ which are equivariant, as in (11.4). Then the above map $\tilde{\varphi} = \lim \tilde{\varphi}_i$ belongs to $\mathcal{A}$.

Now modify the above construction. Choose a minimizing sequence $(\tilde{\varphi}_i)$ in $\mathcal{A}$ for $\int_{\widetilde{F}} \tilde{e}(\tilde{\varphi})$, $\tilde{e}(\tilde{\varphi})$ denoting the energy density of a map $\tilde{\varphi}$ of class $W^{1,2}_{\mathrm{loc}}(\widetilde{X}, \widetilde{Y})$. By (11.4), $(\tilde{\varphi}_i)$ is likewise minimizing for

$$\widetilde{E}(\tilde{\varphi}_{|\widetilde{U}}) = \widetilde{E}(\tilde{\varphi}_{|\widetilde{G}}) = N \int_{\widetilde{F}} \tilde{e}(\tilde{\varphi}),$$

and we may therefore assume, by Lemma 9.2 applied to compact subsets of $\widetilde{U}$, that $(\tilde{\varphi}_i)_{|\widetilde{U}}$ converges in $L^2_{\mathrm{loc}}(\widetilde{U})$ and pointwise a.e. in $\widetilde{U}$ to some map $\tilde{\varphi}_{\widetilde{U}} \in W^{1,2}(\widetilde{U}, \widetilde{Y})$ which minimizes the energy of restrictions to $\widetilde{U}$ of maps of class $\mathcal{A}$, according to Lemma 9.1. Clearly $\tilde{\varphi}_i$ converges pointwise a.e. in $\widetilde{X}$, by (11.4), to an extension $\tilde{\varphi}$ of $\tilde{\varphi}_{\widetilde{U}}$, and $\tilde{\varphi}$ therefore likewise satisfies (11.4), so $\tilde{\varphi} \in \mathcal{A}$. Furthermore, $\int \tilde{e}(\tilde{\varphi}_{|\widetilde{F}}) = N^{-1} \widetilde{E}(\tilde{\varphi}_{\widetilde{U}})$ minimizes the corresponding integral among all restrictions to $\widetilde{F}$ of maps of class $\mathcal{A}$, again by (11.4).

Any minimizer $\tilde{\varphi}$ as above is also locally $\widetilde{E}$-minimizing on $\widetilde{X}$. Indeed, every point of $\widetilde{X}$ has a relatively compact neighbourhood $\tilde{\omega}$ small enough so that $\tilde{\omega} \cap \gamma(\tilde{\omega}) = \emptyset$ for all $\gamma \in \Gamma_X \setminus \{\mathrm{id}\}$. If $\tilde{\psi} \in W^{1,2}_{\mathrm{loc}}(\widetilde{X}, \widetilde{Y})$ satisfies $\tilde{\psi} = \tilde{\varphi}$ in $\widetilde{X} \setminus \tilde{\omega}$, then the map $\tilde{\psi}^* : \widetilde{X} \to \widetilde{Y}$ defined by $\tilde{\psi}^*(\gamma\tilde{x}) = \tilde{\psi}(\tilde{x})$ for every $\tilde{x} \in \tilde{\omega}$ and $\gamma \in \Gamma_X$, while $\tilde{\psi}^* = \tilde{\psi}$ elsewhere, belongs to $\mathcal{A}$, and

$$\widetilde{E}(\tilde{\psi}_{|\widetilde{U}}) \geqslant \widetilde{E}((\tilde{\psi}^*)_{|\widetilde{U}}) \geqslant \widetilde{E}(\tilde{\varphi}_{|\widetilde{U}}).$$

Because the universal cover $\widetilde{Y}$ satifies the hypotheses of Theorem 10.1, $\tilde{\varphi}$ is Hölder continuous, and so therefore is the map $\varphi : X \to Y$ covered by $\tilde{\varphi}$. Furthermore, φ is of class $\mathcal{H}$ according to (11.4) and minimizes E on $\mathcal{H}$. Indeed, any map $\psi \in \mathcal{H}$ lifts to a map $\tilde{\psi} \in \mathcal{A}$, and

$$E(\psi) = \int_{\widetilde{F}} \tilde{e}(\tilde{\psi}) \geqslant \int_{\widetilde{F}} \tilde{e}(\tilde{\varphi}) = E(\varphi)$$

because $\tilde{\varphi}$ is minimizing relative to $\mathcal{A}$. □

The following are some immediate consequences of the existence of E-minimizers in $\mathcal{H}$ in Theorem 11.1:

Let $\varphi_0, \varphi_1 : X \to Y$ be E-minimizers in a homotopy class $\mathcal{H}$; they are joined by a geodesic homotopy φ_I, cf. Proposition 11.1. From (11.2) we obtain

$$E(\varphi_0) = E(\varphi_t) = E(\varphi_1) \quad \text{for all} \quad t \in I.$$

Thus every map φ_t is an E-minimizer in $\mathcal{H}$. Furthermore, (11.2) also insures that

$$\int_X |\nabla d_Y(\varphi_0, \varphi_1)|^2 = 0,$$

so $d_Y(\varphi_0, \varphi_1)$ is constant:

Corollary 11.1. *If two E-minimizers $\varphi_0, \varphi_1 : X \to Y$ in a homotopy class agree at a point of X, then they are identical.*

Remark 11.1. The above proof of Theorem 11.1 follows the pattern of [Lem 1978], see also [Sch 1997, §6]. For the corollary, see [GS 1992, §4] and [Sch 1997, §3].

Remark 11.2. In Theorem 11.1, compactness of Y can be replaced by completeness, in the presence of a suitable growth at infinity, [ES 1964, §8 and Theorem 10D] and [EL 1978, §5.2].

Remark 11.3. Theorem 11.1 can be demonstrated in the framework of evolution: As noted in the text preceding equation (9.1), $L^2(X,Y)$ is a complete metric space with distance function D given by

$$D^2(\varphi,\psi) = \int_X d_Y^2(\varphi(x),\psi(x))\, d\mu_g(x).$$

It is elementary to verify that $L^2(X,Y)$ itself becomes a (generally not locally compact) geodesic space of nonpositive curvature.

A general method of Mayer [May 1998], see also [J 1997c], provides a discrete-time difference scheme converging to a flow which is a form of eternal evolution associated to E; that flow lies in $W^{1,2}(X,Y)$, and E decreases along it.

Remark 11.4. For the extensions of the above theorem and of Theorem 11.2 below to targets Y which are geodesic spaces as in Remark 9.3, rather than polyhedra, one may apply [KS 1993, Proposition 2.5.2] with $\alpha = 1$ to a single map $f = h$ $(= \varphi)$. Denoting by $\varphi(x,r)$ the meanvalue of φ (cf. Lemma 10.4) with respect to normalized "Euclidean" volume measure on the ball $B_X^e(x,r)$ for small r, the family $(\varphi(x,r))$ (extended by $\varphi(x,0) = \varphi(x)$) is continuous in (x,r), and each $\varphi(\cdot,r)$ is of class $\mathrm{Lip}(X,Y)$; this replaces Step 1 in the above proof. Similarly for Theorem 11.4, replacing any distance d between points of Y by $2\sin(d/2)$ in the definition of meanvalue of a map (now into a suitably small ball in a complete geodesic space Y with $\operatorname{curv} Y \leqslant 1$), see the paragraph containing (10.22).

Example 11.1. We mention some special cases of Theorem 11.1.

(a) Assume $\dim X = 1$. Then Theorem 11.1 is valid without the curvature hypothesis on Y, by elementary considerations: The piecewise geodesic maps form a compact deformation retract of $\mathcal{H}$, along which the restriction of E to $\mathcal{H} \cap W^{1,2}$ is continuous and decreasing.

If $X = S^1$, then every closed path on a compact Riemannian Lipschitz manifold (N,h) can be deformed to a closed geodesic which is a Lipschitz E-minimizer.

For any Riemannian polyhedron (X,g) with $\dim X = 1$ and any compact Riemannian manifold $Y = (N,h)$, Kotani and Sunada [KoSu 1998] have shown that every continuous map $(X,g) \to (N,h)$ can be deformed to a harmonic map (in the sense of Definition 12.1 below). And if (N,h) has nonpositive curvature, these harmonic maps are the E-minimizers in a given homotopy class. They apply these results to characterize the equilibrium configurations of crystals. For an extensive survey, see [KoSu 2000].

(b) Taking for domain a compact oriented Riemannian Lipschitz manifold (M, g), every continuous map $M \to S^1$ can be deformed to a Hölder continuous E-minimizer φ. Its differential $d\varphi = \omega$ can be interpreted as a harmonic 1-form on M with integral periods. Indeed, the method of Morrey and Eells [ME 1956] (see also [Te 1983]) provides a Hodge decomposition of L^2-differential forms ω on (M, g) for which ω, $d\omega$, $*\omega$, $d{*}\omega$ are all of class L^2. Then *ω is harmonic iff $d\omega = 0 = d{*}\omega$.*

(c) If $\dim X = 1 = \dim Y$, both Theorem 11.1 and Corollary 11.1 have been found by Lebeau [Leb 1996], and applied to matters of super-rigidity of discrete cocompact groups acting on locally compact trees.

(d) Taking X as in Example 8.8 gives a substantial generalization of the main theorem of Chiang and Ratto [CR 1992], and certainly a different proof.

(e) Taking X as a compact Riemannian orbifold as in Example 8.13 gives a generalization of the main existence theorem of Chiang [Chi 1990], again with a different proof.

(f) Theorem 11.1 applies to maps $X \to Y$ between compact Riemannian polyhedra if X is admissible and $\dim Y = 1$, again in view of Example 8.1.

Ratto [Ra 1987, Chapter III, Proposition 3.4.1] has observed that for homotopy classes of maps $X \to S^1$ of certain (rather special) closed Riemannian circuits X into the circle $S^1 = Y$, intersection cohomology and Cheeger's harmonic 1-forms combine to prove existence of a harmonic representative, as in Theorem 11.1 above. He also constructed continuous (and partially harmonic) maps between Riemannian circuits (having conical singularities).

(g) Theorem 11.1 applies when X is a compact normal complex analytic space with a Kähler structure, as in Example 8.9. There is related work in:

An unfinished manuscript by Li and Tian. Their domains are certain projective algebraic varieties X, and ranges are smooth Riemannian manifolds of nonpositive curvature. They start with an operator Laplacian whose closure is selfadjoint in $L^2(X)$, as in [LiT 1995].

An unpublished study by Napier and Ramachandran of harmonic and pluriharmonic maps of compact normal complex spaces with Kähler structures into smooth Riemannian manifolds of nonpositive curvature. Here a continuous map $\varphi : X \to Y$ is *pluriharmonic* if for every holomorphic curve (i.e., Riemann surface A and holomorphic map) $i : A \to X$, the composition $\varphi \circ i : A \to Y$ is harmonic. For instance, taking $Y = \mathbb{R}$, φ is pluriharmonic iff it is locally the real part of a holomorphic function [GRos 1965, p. 272]. A holomorphic function $\psi : X \to \mathbb{C}$ is pluriharmonic.

Proposition 11.3. *Let X be a normal complex analytic subspace of a Kähler manifold and $\varphi : X \to (N, h)$ a holomorphic map to a Kähler manifold. Then φ is a local E-minimizer.*

Indeed, when $S = S(X)$ denotes the singular set of X as a normal circuit, $X \setminus S$ is a Kähler manifold, so $\varphi_{|X \setminus S}$ is harmonic; and an E-minimizer on every small ball in $X \setminus S$ [EL 1983, Corollary 8.15]. To show that φ is a local E-minimizer on X, take any small ball $D \subset X$ and map $\psi \in W^{1,2}(X, N)$ such that $\psi = \varphi$ in $X \setminus D$. Then

$$\int_D e(\varphi) = \int_{D \setminus S} e(\varphi) \leqslant \int_{D \setminus S} e(\psi) = \int_D e(\psi).$$

Such maps φ occur when blowing down the singularities of X.

Example 11.2. Let M and N be compact Riemann surfaces of the same genus $\geqslant 2$, and $\mathcal{H}$ a homotopy class of orientation preserving continuous maps $\varphi : M \to N$; assume that $\mathcal{H}$ contains no holomorphic map. Fix a holomorphic quadratic differential $\beta \not\equiv 0$ on N, as in Example 3.4 (and maintain its notation). Define the energy $E(\varphi)$ of a map $\varphi : M \to (N, h^\beta)$, and denote

$$\mathcal{H}_E = \{\psi \in \mathcal{H} : E(\psi) < \infty\}.$$

Then, as a consequence of [GS 1992, Theorem 4.4], *there is a map φ minimizing E on $\mathcal{H}_E$; φ is Lipschitz continuous as a map $\varphi : M \to (N, d^\beta)$.*

Furthermore, as shown by Kuwert [Ku 1996], following earlier contributions by Leite [Lei 1988] and Miyahara [Miy 1968]:

(a) φ is smooth off the preimage of the zero set of β, and satisfies $\tau(\varphi) = 0$ there;

(b) The $(2,0)$-part $(\varphi^* h^\beta)^{2,0}$ of the pull-back by φ of the singular Hermitian metric h^β determined by β, cf. Example 3.4, is a holomorphic quadratic differential on M;

(c) *If $\mathcal{H}_E$ contains a diffeomorphism, then the minimizer φ is unique, and is the Teichmüller map*, that is, it minimizes the supremum of the dilatation $(|\partial_z \varphi| + |\partial_{\bar{z}} \varphi|)/(|\partial_z \varphi| - |\partial_{\bar{z}} \varphi|)$ in the family of quasiconformal maps homotopic to it. It has constant dilatation.

The Dirichlet problem relative to a homotopy class

Assume now that the admissible Riemannian polyhedron X has a boundary $bX \neq \emptyset$, cf. Chapter 4. For any map $\psi \in W^{1,2}(X,Y)$ (Y a geodesic space) let

$$W^{1,2}_{\psi}(X,Y) = \{\varphi \in W^{1,2}(X,Y) : \varphi_{|bX} = \psi_{|bX} \text{ a.e.}\}$$

(almost everywhere with respect to the $(m-1)$-dimensional measure on bX, with φ and ψ chosen to be quasicontinuous (cf. Theorem 9.1); or equivalently: $\mathrm{tr}_{bX}\,\varphi = \mathrm{tr}_{bX}\,\psi$ in terms of the trace operator of [KS 1993, §1.12] applied to each m-simplex s of X having an $(m-1)$-face contained in bX).

If ψ is also continuous, let $C_{\psi}(X,Y)$ denote the class of all continuous maps of class $W^{1,2}_{\psi}(X,Y)$. The components of $C_{\psi}(X,Y)$ are the homotopy classes of maps holding the boundary map $\psi_{|bX}$ pointwise fixed.

Theorem 11.2. *Let X and Y be compact Riemannian polyhedra. Assume that*

(i) *X is admissible and has a boundary $bX \neq \emptyset$, and*
(ii) *Y has nonpositive curvature.*

Fix a continuous map $\psi \in W^{1,2}(X,Y)$. Then every class $\mathcal{H}$ of homotopic maps $\varphi \in C_{\psi}(X,Y)$ has a unique E-minimizer relative to $\mathcal{H}$, and that minimizer is Hölder continuous in $X \setminus bX$.

Proof. The uniqueness follows from Corollary 11.1. The rest of the proof follows the lines of that of Theorem 11.1 once it has been established that every map $\varphi \in C_{\psi}(X,Y)$ can be deformed within $C_{\psi}(X,Y)$ to a map of finite energy.

Choose $r > 0$ so that every ball $B_Y(y, 3r)$ is convex and its triangles are 0-dominated. Again, let $B_Y(y_i, r)$, $i = 1, \cdots, k$, cover Y, and write

$$U_i = \varphi^{-1}(B_Y(y_i, r)), \qquad i = 1, \cdots, k. \tag{11.5}$$

In the beginning of the proof of Theorem 11.1 we found a map φ^* of class $\mathrm{Lip}(X,Y)$ such that

$$|\varphi^*(x) - \varphi(x)| < r, \qquad x \in X. \tag{11.6}$$

It remains to replace φ^* by a map $\psi^* \in C_{\psi}(X,Y)$ such that

$$|\psi^*(x) - \varphi(x)| < 2r, \qquad x \in X; \tag{11.7}$$

indeed, such a map ψ^* will be homotopic to φ relative to $C_\psi(X,Y)$ by the argument in Step 1 of the proof of Theorem 11.1.

For $x \in X$ write

$$\xi(x) = \min_{x' \in bX} d_X(x,x').$$

Choose $\delta > 0$ so that, for $x, x' \in X$,

$$|\psi(x') - \psi(x)|,\ |\varphi^*(x') - \varphi^*(x)| < r \ \text{ when } d_X(x,x') \leqslant 3\delta, \tag{11.8}$$

and, for $x \in X$,

$$|\varphi(x) - \psi(x)| < r \ \text{ when } \xi(x) \leqslant 3\delta. \tag{11.9}$$

Now define a continuous map $\psi^* : X \to Y$ by

$$\psi^*(x) = (1 - \delta^{-1}\xi(x))\psi(x) + \delta^{-1}\xi(x)\varphi^*(x) \qquad \text{for } \xi(x) \leqslant \delta, \tag{11.10}$$

and $\psi^*(x) = \varphi^*(x)$ otherwise. This abuse of notation means that $\psi^*(x) = \gamma(t)$ for $t = \delta^{-1}\xi(x)$ when $\gamma : [0,1] \to Y$ denotes the constant-speed geodesic in Y joining $\psi(x)$ to $\varphi^*(x)$. Thus $\psi^*(x)$ is well defined and belongs to $B_Y(y_i, 2r)$ if $x \in U_i$ from (11.5) (i.e., if $\varphi(x) \in B_Y(y_i, r)$), for then $\varphi^*(x)$ and $\psi(x)$ lie in $B_Y(y_i, 2r)$ by (11.6) and (11.9), which also implies (11.7). Clearly, $\psi^* = \psi$ on bX, and it remains to prove that ψ^* has finite energy.

Fix for a while $0 < \varepsilon \leqslant \delta$. For any $x \in X$ with $\xi(x) \geqslant 2\delta$ we have, by (9.2), $e_\varepsilon(\psi^*)(x) = e_\varepsilon(\varphi^*)(x)$ because $\xi(\cdot) \geqslant 2\delta - \delta$ on $B_X(x,\varepsilon)$ and hence $\psi^* = \varphi^*$ there. Because $\varphi^* \in \mathrm{Lip}\,(X,Y)$ has finite energy, to show that also ψ^* has finite energy amounts, by Definition 9.1, to proving that

$$\limsup_{\varepsilon \to 0} \int_{\xi(\cdot) \leqslant 2\delta} e_\varepsilon(\psi^*) d\mu_g < \infty. \tag{11.11}$$

And to verify this, fix $x \in X$ with $\xi(x) \leqslant 2\delta$; we may suppose that $x \in U_i$ from (11.5). For any $\varepsilon \leqslant \delta$ and $x' \in B_X(x,\varepsilon)$ we have $\xi(x') \leqslant 2\delta + \varepsilon \leqslant 3\delta$, and since $\varphi(x) \in B_Y(y_i, r)$ it follows from (11.6), (11.8) and (11.9) that

$$\psi(x), \psi(x'), \varphi^*(x), \varphi^*(x') \in B_Y(y_i, 3r).$$

Now estimate $|\psi^*(x) - \psi^*(x')|$ via the auxiliary point

$$z = (1 - \delta^{-1}\xi(x))\psi(x') + \delta^{-1}\xi(x)\varphi^*(x'), \tag{11.12}$$

to be compared with $\psi^*(x')$, cf. (11.10). Thus

$$|z - \psi^*(x')| = \delta^{-1}|\xi(x) - \xi(x')|\,|\varphi^*(x') - \psi(x')| \leqslant 2r\varepsilon/\delta \tag{11.13}$$

since $\xi(\cdot)$ has Lip constant 1, and $|\varphi^*(x') - \psi(x')| < 2r$ by (11.6) and (11.9) together with $\xi(x') \leqslant 3\delta$. Because $B_Y(y_i, 3r)$ is locally squared-convex (cf. Chapter 2),

$$|\psi^*(x) - z| \leqslant \tfrac{1}{2}|\psi(x) - \psi(x')| + \tfrac{1}{2}|\varphi^*(x) - \varphi^*(x')|. \tag{11.14}$$

When combined with (11.13) this leads to an estimate of $|\psi^*(x) - \psi^*(x')|$, from which it follows that

$$e_\varepsilon(\psi^*)(x) \leqslant \tfrac{3}{4}e_\varepsilon(\psi)(x) + \tfrac{3}{4}e_\varepsilon(\varphi^*)(x) + 12r^2\delta^{-2}\varepsilon^{-m}\mu_g(B_X(x,\varepsilon)).$$

The last term on the right being bounded in ε, by Lemma 4.4, we have verified (11.11) because ψ and φ^* have finite energy. □

Query. Suppose furthermore that ψ is Hölder continuous on X. Is the E-minimizer also Hölder continuous there? That is true if X is a smooth Riemannian manifold [Se 1994].

Example 11.3. Taking X to be a Riemannian graph with k boundary points and any $\psi : bX \to Y$, there is a Hölder continuous E-minimizer

$$\varphi : X \to Y$$

with $\varphi_{|bX} = \psi$. Refinement of this leads to a solution of the Plateau problem for minimal networks in Y: Given k points of Y, find a spanning network of minimal length.

Remark 11.5. In [GS 1992, Theorems 2.3 and 4.4] and [KS 1993, Theorems 2.4.6 and 2.7.1] it is shown that the E-minimizers in Theorems 11.1 and 11.2 are Lipschitz continuous, provided that the domain X is a smooth Riemannian manifold and the boundary map ψ is Lipschitz continuous. In Remark 6.4 it has been noted that no such improvement is valid in general for Riemannian polyhedra X (not even when $Y = \mathbb{R}$).

The ordinary Dirichlet problem

For a simply connected target Y we deal with the ordinary Dirichlet problem (no homotopy requirement, and φ, ψ not necessarily continuous). Assuming that $\operatorname{curv} Y \leqslant K$ for some $K \geqslant 0$ we have the following two theorems, for the cases $K = 0$ and $K > 0$, respectively.

Theorem 11.3. *Take X as in Theorem 11.2 and Y a simply connected complete geodesic space of nonpositive curvature. For any map $\psi \in W^{1,2}(X,Y)$ there is a unique E-minimizer relative to $W^{1,2}_\psi(X,Y)$. That minimizer is Hölder continuous in $X \setminus bX$, provided that Y is a Riemannian polyhedron.*

Again, for the case that X is a Riemannian domain with Lipschitz boundary (cf. Chapter 14), this theorem, even with Lipschitz continuity, is due to Korevaar-Schoen [KS 1993, Theorems 2.2 and 2.4.6]. Their proof of existence and uniqueness, even for Y not locally compact, carries over immediately to the setting of our Theorem 11.3, in view of Theorem 9.1 and the inequality (11.2); for details, see also the proof of Theorem 11.4 below. Hölder continuity in $X \setminus bX$ follows from Theorem 10.1. If bX is smooth and ψ is Lip on X, then the E-minimizer is Hölder continuous on all of X [Se 1994].

Next consider the case where the target has *upper bounded curvature.* In a complete geodesic space (Y, d_Y) with $\operatorname{curv} Y \leqslant K$ for some $K > 0$, let $B = B_Y(q,R)$ be a closed ball in Y of radius $R < \pi/(2\sqrt{K}\,)$ such that B satisfies *radial uniqueness* with respect to q in the sense of [AB 1996, §4], i.e., B is convex, and geodesics in B are uniquely determined by their endpoints and vary continuously with these. It suffices to require this when one of the endpoints is q. Since we have assumed that Y is locally compact the continuous dependence is automatic, and the requirement as to the convex closed ball B is therefore equivalent to every triangle in B with perimeter $< 2\pi/\sqrt{K}$ being K-dominated, cf. Chapter 2. Clearly, B is *simply connected* (as seen by use of geodesic homotopy, noting that B has no conjugate points, and so any continuous path $I \to Y$ is fixed-end homotopic to a unique constant-speed geodesic, by [AB 1990, Theorems 3 and 6]).

With $W^{1,2}_\psi(X,Y)$ as defined in the text preceding Theorem 11.2 we have the following counterpart to Theorem 11.3:

Theorem 11.4. *Take X as in Theorem 11.2 and Y a complete geodesic space with $\operatorname{curv} Y \leqslant K$ for some constant $K > 0$. Let $B = B_Y(q,R)$ be a closed ball in Y of radius $R < \pi/(2\sqrt{K})$ satisfying radial uniqueness with respect to q.*

For any map $\psi \in W^{1,2}(X,B)$ there is a unique E-minimizer relative to $W^{1,2}_\psi(X,B)$. That minimizer is Hölder continuous in $X \setminus bX$ if $R < \pi/(4\sqrt{K})$ and Y is a Riemannian polyhedron.

For the case where X is a Riemannian domain with Lip boundary, this theorem is due to Serbinowski [Se 1995], who even established Lipschitz continuity of the solution up to the boundary when ψ is Lipschitz continuous, see also Chapter 15 below.

Proof of Theorem 11.4. We may suppose that $K = 1$. The basic inequality (11.2) with $t = \frac{1}{2}$, valid in the case $K = 0$ when Y is simply connected and complete, is now replaced by

$$c_m \cos^8 R \int_X \left|\nabla \frac{\tan(\frac{1}{2}d)}{\cos \varrho}\right|^2 d\mu_g \leqslant \tfrac{1}{2}E(\varphi_0) + \tfrac{1}{2}E(\varphi_1) - E(\varphi_{\frac{1}{2}}), \qquad (11.15)$$

valid for maps $\varphi_0, \varphi_1 \in W^{1,2}(X, B)$, writing $d(x) = d_Y(\varphi_0(x), \varphi_1(x))$ and $\varrho(x) = d_Y(\varphi_{\frac{1}{2}}(x), q)$ ($\leqslant R < \pi/2$), $\varphi_{\frac{1}{2}}(x)$ being, as before, the midpoint of the geodesic segment joining $\varphi_0(x)$ to $\varphi_1(x)$.

This deep inequality is due to Serbinowski in his Thesis [Se 1995] for the case mentioned above. In our setting, (11.15) is derived simply by applying Serbinowski's inequality to each top dimensional simplex of X, and adding up.

The basic idea in the proof of this inequality in [Se 1995] (with (X, g) a Riemannian domain (M, g)) is to pull the midpoint map $\varphi_{\frac{1}{2}}$ towards q so as to obtain a map $\hat{\varphi}_{\frac{1}{2}}$ defined (with the standard abuse of notation) by

$$\hat{\varphi}_{\frac{1}{2}}(x) = (1 - \eta(x))\varphi_{\frac{1}{2}}(x) + \eta(x)q$$

in terms of a function $\eta \in C_c^1(M, [0, 1])$. Choosing η so that

$$\frac{\sin[(1 - \eta)\varrho]}{\sin \varrho} = \cos \frac{d}{2},$$

this leads to (11.15) after careful estimations involving extensive use of the directional energies of Korevaar-Schoen [KS 1993], see also Chapter 14 below. (Using total energies alone would suffice provided that $R < \pi/(4\sqrt{K})$.)

As in [Se 1995], uniqueness in Theorem 11.4 follows immediately from (11.15) applied to two solutions φ_0, φ_1, noting that the right hand member of (11.15) is then nonpositive, and so $\tan(\frac{1}{2}d)/\cos \varrho$ is constant, and indeed 0 because $d = 0$ on bX.

As to existence, it follows from Lemma 9.2 that any minimizing sequence (φ_i) for Theorem 11.4 subconverges in $L^2(X, B)$ to a solution φ, and $\varphi = \psi$ on bX according to [KS 1993, Theorem 12.6.2]. Serbinowski's existence proof, valid without local compactness of Y, amounts to deducing from (11.15), in a standard way, that (φ_i) is in fact a Cauchy sequence.

When Y is a Riemannian polyhedron and $\varphi(X) \subset B_Y(q, R/(4\pi\sqrt{K}))$, Hölder continuity in $X \setminus bX$ follows from Theorem 10.2 because φ is locally E-minimizing there. □

Remark 11.6. For the Hölder continuity of φ in Theorems 11.3 and 11.4 above one may likewise allow Y to be a geodesic space (rather than a polyhedron) provided that Y be Lip homeomorphic to a closed subset of a Euclidean space (cf. the beginning of the proof of Theorem 10.1).

The case where the target is a Riemannian manifold

In this case we may consider instead the compatible setting of Definition 9.2, so that the Riemannian metric g on the compact admissible domain polyhedron X is allowed to be just measurable with local elliptic bounds, while the target Y is a smooth Riemannian manifold without boundary. Again, every Lipschitz map $X \to Y$ has finite energy (now because its components are Lipschitz). The content of the present chapter so far carries over to that alternative setting with unchanged proofs in view of Propositions 12.1 and 12.2 and Remark 9.6. This version of Theorem 11.4 above should be compared with [HKW 1977], where also X is a manifold; and with [Sch 1997, Remark 4.6], where X is a Riemannian Lipschitz manifold (as in Example 3.6), and results of [HW 1977] are used.

The case of 2-dimensional manifold domains

As an analogue of a theorem of Lemaire [Lem 1978] and Sacks–Uhlenbeck [SaU 1981], we formulate:

Theorem 11.5. *Let M be a closed Riemann surface without boundary, and Y a compact Riemannian polyhedron with curvature bounded above and second homotopy group $\pi_2(Y) = 0$. Then every homotopy class of continuous maps $M \to Y$ contains a Lipschitz continuous E-minimizer.*

Proof. These homotopy classes are classified by the conjugacy classes of homomorphisms $\pi_1(M) \to \pi_1(Y)$ by part (b) in the homotopy classification at the beginning of this chapter.

Step 1. The direct method of variational theory produces an E-minimizer in $W^{1,2}(M, Y)$, as in [Lem 1978]. Take such a map φ.

Step 2. There exists $\varepsilon_0 > 0$ depending only on φ and M such that for any ball $B(x,\sigma) \subset M$ with $E_{B(x,\sigma)}(\varphi) \leqslant \varepsilon_0$, the restriction $\varphi_{|B(x,\sigma/2)}$ is Lipschitz [Se 1995, Theorem 3.1], cf. Chapter 15.

Step 3. Morrey's method. See [Lem 1978, §12]: There exists $\varepsilon > 0$ such that for any point $x \in M$, there is an $r_x > 0$ such that $E_{B(x,r_x)}(\varphi) \leqslant \varepsilon$.

To prove this, take a small ball $B \subset M$. For any $f \in W^{1,2}(B,Y)$ let φ denote the weakly harmonic extension of $f_{|\partial B}$. Then

$$E_B(\varphi) \leqslant \frac{1}{2}\int_0^{2\pi} |\partial f/\partial\theta|^2 d\theta$$

cf. [Lem 1978, Lemma 12.5].

Write (using the conical embedding space of [Se 1995, §3.1.1], cf. Chapter 15)

$$f^\alpha(\theta) = \frac{1}{2}a_0^\alpha + \sum_{k=1}^{\infty}(a_k^\alpha \cos k\theta + b_k^\alpha \sin k\theta).$$

Then, when s denotes the radius of B,

$$\varphi^\alpha(r,\theta) = \frac{1}{2}a_0^\alpha + \sum_k \frac{r^k}{s^k}(a_k^\alpha \cos k\theta + b_k^\alpha \sin k\theta).$$

Consequently,

$$\begin{aligned}2E_B(\varphi) &= \sum_\alpha\sum_k k\pi\left((a_k^\alpha)^2 + (b_k^\alpha)^2\right)\\ &\leqslant \sum_\alpha\sum_k k^2\pi\left((a_k^\alpha)^2 + (b_k^\alpha)^2\right) = \sum_\alpha\int_0^{2\pi}\left(\frac{\partial f^\alpha}{\partial\theta}\right)^2 d\theta.\end{aligned}$$

Next, by [Mor 1966, Theorem 3.1.8], every $\varphi \in W^{1,2}$ can be represented locally by a map in a ball $B(x) \subset M$ with polar coordinates (r,θ) such that φ is absolutely continuous on the boundary $\partial B(x,r)$ a.e. (with respect to r). Take $f = \varphi_{|\partial B(x,r)}$. Then, as in [Lem 1978, §12], $\int_0^{2\pi}|\partial f/\partial\theta|^2 d\theta \leqslant \text{const.}\varrho_x$; and ϱ_x can be made arbitrarily small. Thus Step 3 is proved. □

Example 11.4. (Application of [EL 1980, §3].) As a closed orientable 2-manifold, the 2-sphere S has an essentially unique holomorphic structure, and therefore a harmonic space structure (using the elementary correspondence on simply connected domains in S between holomorphic functions and pairs of conjugate harmonic functions). The harmonic functions on S are the local solutions to the Laplace–Beltrami equation on S (cf. Example 3.1). Denoting by $S \to P$ the 2-leaved cover of the real projective plane, then

(a) *any harmonic map* $S \to S$ *is conformal* (i.e., $\pm$holomorphic); and

(b) *any harmonic map* $P \to S$ *is constant.* In particular, the single nontrivial homotopy class of maps $P \to S$ has no harmonic representative.

Query. Is there a harmonic space structure on the 2-torus and a harmonic map $T \to S$ of degree 1?

Questions and remarks

1. With reference to Theorem 11.4, in the case where both X and Y are Riemannian manifolds, smooth existence plus boundary regularity was proved in [HKW 1977], and uniqueness in [JK 1979]. The assertion of Theorem 11.2 is also valid in the following context (Baldes [B 1984; 1986]): X is a smooth manifold with an admissible degenerate (in Baldes' sense) Riemannian metric with discrete singular set; and Y a smooth Riemannian manifold.

If X is a smooth Riemannian manifold and Y a geodesic space with $\operatorname{curv} Y \leqslant K$, then a version of Theorem 11.4 has been established in [Se 1995, Theorem 1.16], cf. Chapter 15, with an argument based on a convexity property like (11.1), supplemented with the condition

$$d_Y(\alpha(0),\alpha(1)) + d_Y(\alpha(0),\beta(0)) + d_Y(\beta(0),\beta(1)) + d_Y(\alpha(1),\beta(1)) < \frac{2\pi}{\sqrt{K}}.$$

It is also concluded there that the unique E-minimizer is Lipschitz continuous up to the boundary, provided ψ is. See [J 1996] for an alternative proof of existence relative to $W^{1,2}_{\psi}$.

A study of existence of solutions to the Plateau problem for area-minimizing discs in geodesic spaces of curvature at most K has been made by Nikolaev [N 1979]. See also Chapter 9, The volume of a map.

2. Let (M,g), (N,h) denote smooth complete Riemannian manifolds (N without boundary); and $\mathfrak{X} = (X_1, \cdots, X_p)$ a system of smooth vector fields on M satisfying Hörmander's condition (for some r) on iterated brackets. As in Example 3.5, define the $\mathfrak{X}$-*energy density* $e^{\mathfrak{X}}(\varphi)$ *of a map* $\varphi : (M,g) \to (N,h)$ by

$$e^{\mathfrak{X}}(\varphi)(x) = \tfrac{1}{2} g^{ij}(x) \mathcal{L}_{X_i}\varphi^{\alpha}(x) \mathcal{L}_{X_j}\varphi^{\beta}(x) h_{\alpha\beta}(\varphi(x)).$$

The $\mathfrak{X}$-*energy* of φ is

$$E^{\mathfrak{X}}(\varphi) = \int_M e^{\mathfrak{X}}(\varphi)(x)\, d\mu_g(x),$$

whose Euler-Lagrange operator $\tau^{\mathfrak{X}}$ is an hypoelliptic system.

These can be effectively studied via the perturbation of $\tau^{\mathfrak{X}}$ to the tension operator τ^{ε} on maps (M, g^{ε}), where $g^{\varepsilon} = a + \varepsilon g$ is the smooth nondegenerate Riemannian metric of Example 3.5. The limit of metrics d^{ε} in (3.1) insures that

$$E^{\mathfrak{X}}(\varphi) = \lim_{\varepsilon \to 0} \frac{1}{2} \int_M \operatorname{trace}_{g^{\varepsilon}}(\varphi^* h)\, d\mu_{g^{\varepsilon}}.$$

As a representative special case, take for (M, g) a smoothly bounded Euclidean domain U in $\mathbb{R}^m$ on whose closure $\overline{U}$ the vector fields of $\mathfrak{X}$ are defined and smooth. In addition to Hörmander's condition on $\mathfrak{X}$ assume that at no point of the boundary of U do all the vector fields of $\mathfrak{X}$ vanish in the direction of bU. Consider the Dirichlet problem for maps of $\overline{U}$ into a chart V of (N, h). Jost and Xu [JX 1998] have established the existence of an $E^{\mathfrak{X}}$-minimizer φ with prescribed boundary values and range in a sufficiently small ball in a chart of (N, h). The map φ is Hölder continuous in U and up to its boundary. In this framework, full interior regularity is given in [XZ 1997]; and uniqueness in [Z 1999].

3. Theorem 11.5 remains valid if M is a compact Riemann surface with boundary bM, and $\psi : M \to Y$ a Lip map: *Every component of* $C_{\psi}(M, Y)$ *contains a Lip continuous E-minimizer.* For the relevant matters of relative homotopy, see [Lem 1982, Lemma 2.2].

The existence of a continuous E-minimizer in Theorem 11.5 has been established in [J 1994, §5], using [HW 1975, p. 14].

It seems that Theorem 11.5 admits an analogue for the case where the domain X is an admissible Riemannian 2-dimensional polyhedron without boundary, obtaining a Hölder continuous E-minimizer. Start by expressing X as a connected sum of closed Riemann surfaces and surfaces with boundary, using the construction in Example 8.5.

4. Lipschitz maps $\varphi : (M, g) \to (N, h)$ between Riemannian Lipschitz manifolds have a good energy functional with an elementary definition: Let $\mathcal{V} = (\chi_r, V_r)$ be a countable locally finite atlas of (N, h). Its transition maps $\chi_s^{-1}\chi_r$ are locally Lipschitz, so are differentiable off a null set $D_{rs} \subset V_r \cap V_s$; then $D(\mathcal{V}) = \bigcup_{r,s} D_{rs}$ is a null set of N. For any null set $Z \subset N$, say that a map $\varphi \in \operatorname{Lip}(M, N)$ *is transverse to* Z (written $\varphi \pitchfork Z$) if $\varphi^{-1}(Z)$ is null in

M. Set $\mathrm{Lip}^Z(M,N) = \{\varphi \in \mathrm{Lip}(M,N) : \varphi \pitchfork Z\}$. Note that if $\varphi \pitchfork D(\mathcal{V})$, then $|\nabla\varphi| : M \to \mathbb{R}$ is locally bounded measurable a.e. In analogy with [CP 1990], define

$$E^{\mathcal{V}}(\varphi) = \frac{1}{2}\int_M |\nabla\varphi|^2 \, d\mu_g \qquad \text{for } \varphi \in \mathrm{Lip}^{D(\mathcal{V})}(M,N).$$

Then the *energy* $E(\varphi)$ of $\varphi \in \mathrm{Lip}(M,N)$ is defined by

$$E(\varphi) = \sup_Z \{\liminf_{\theta\to\varphi} E^{\mathcal{V}}(\theta) : \theta \pitchfork Z\},$$

over all null sets $Z \subset N$. (The convergence is uniform convergence.)

It is not clear to us how to define a suitable concept of energy for maps $\varphi : (M,g) \to (N,h)$ of class $W^{1,2}_{\mathrm{loc}}$ between Riemannian–Lipschitz manifolds, in general. That has been done (Theorem 9.1) if (M,g) is a triangulable manifold with simplexwise smooth Riemannian metric. Or (Definition 9.2) if (N,h) is a C^1-Riemannian manifold. In both cases Theorem 11.1 remains valid (both M and N without boundaries).

5. Theorem 11.1 applies when X and Y are Riemann–Finsler manifolds (without boundary), as in Example 3.3. We do not know whether that is true using various other Laplacians on X. See [Ce 2000].

6. In [KK 1994; 1996] the authors have introduced the notion of spectral distance between isometry classes of connected compact Riemannian manifolds (M,g), defined and developed through their Dirichlet space structures. A Cauchy sequence (M_k, g_k) of such manifolds of dimesion m with Ricci curvature bounded below by $-(m-1)$ and of diameter bounded above has a limit (called the spectral limit) X, a connected compact geodesic measure space with natural Dirichlet space structure.

Kasue [Ka 2000] has established Theorem 11.1 (and various relative versions) for homotopy classes of maps of such a spectral limit space into a compact Riemannian manifold of nonpositive curvature. His E-minimizers are Lipschitz.

7. Following ideas of Thom and Whitney, we define a *stratification of a space* X to be a disjoint union of a locally finite set of C^1-manifolds such that:

(i) Each of these sets – called strata – has constant dimension; and
(ii) the frontier of each stratum is a union of a set of lower dimensional strata. Often further regularity properties are imposed, [Thom

1966; 1969], [W 1965a, §8], [W 1965b, §19]. For instance, *if a stratified Riemannian space X satisfies Whitney's regularity condition* (b), *then X is a Riemannian polyhedron*, [Pf 2000, §3.4.2], [Ver 1984, Theorem 7.8].

The methods of this monograph can be broadened to include

(1) domains X which are suitably stratified by Riemannian manifolds; and

(2) targets Y which are complete geodesic spaces (not necessarily locally compact).

On the other hand, we do not know how to define a useful energy $E(\varphi)$ for maps $X \to Y$ of an admissible Dirichlet space X into a complete geodesic space Y. However, in the presence of a weak Poincaré inequality for maps, here are two situations where we can, and where furthermore Hölder continuity holds for local E-minimizers:

(3) Y is a Riemannian polyhedron with nonpositive curvature embedded isometrically in a Euclidean space V;

(4) Y is a separable (locally compact) geodesic space of lower bounded curvature (Sturm [St 1997; 1999a]). For Hölder continuity, see [Lin 1997].

In [St 1999b], [St 1999c] Sturm develops a theory of maps $X \to Y$, where X is a measure space with (symmetric) Markov kernel and Y a simply connected, complete geodesic space of nonpositive curvature. His notions of "harmonic", "weakly harmonic", and "energy minimizer" are generally different from those in this book, even for $Y = \mathbb{R}$.

8. Jost has published various attempts to define an energy functional in rather general frameworks, and to prove existence of a minimizer of class L^2: [J 1994, Theorem 3.1], [J 1995, Theorem 2], [J 1996, Theorem 5.1], [J 1997a, Lemma 4], [J 1997b, Chapter 4.1]. The present state of these efforts and the suitability of his conclusions remain unclear to us. We have therefore chosen to base our exposition of the existence theory on [GS 1992, Theorem 4.4] and [KS 1993], adapted to our case of a polyhedral domain.

12

Harmonic maps. Totally geodesic maps

We propose a definition of *harmonic maps* from an admissible Riemannian polyhedron X to some metric space Y as continuous local E-minimizers in a slightly weakened sense suitable for continuous maps (Definition 12.1). This is motivated in part by the Hölder continuity of local E-minimizers when Y has nonpositive or upper bounded curvature (Theorems 10.1, 10.2, and Propositions 12.1, 12.2), and also by some characterizations of harmonic maps, given in this chapter under suitable restrictions on X or Y. In particular, we show that the harmonic maps from X to a Riemannian *manifold* N are the same as those continuous maps φ which are *E-critical*, or equivalently: *weakly harmonic*, i.e., of tension field $\tau(\varphi) = 0$ in the weak sense (Definition 12.2). We also extend the pull-back characterizations of harmonic and of totally geodesic maps due to Ishihara [Ish 1979] for maps between Riemannian manifolds (Theorems 12.1, 12.2), cf. also Proposition 12.4. As a tool we establish Jensen's inequality on a simply connected complete geodesic space of nonpositive curvature (Proposition 12.3).

A concept of harmonic map

Consider an admissible Riemannian polyhedron (X, g), $\dim X = m$, and a metric space (Y, d_Y) (possibly subject to restrictions). From Chapter 9 we have three mutually consistent (Theorems 9.2 and 9.3) concepts of energy of a continuous map $\varphi : X \to Y$, the first for the case that g is simplexwise smooth (Definition 9.1), the second for Y a Riemannian C^1-manifold without boundary (Definition 9.2), and the third for Y a Riemannian polyhedron with continuous Riemannian metric. The following definition and lemma pertain to each of these three cases.

Definition 12.1. A *harmonic map* $\varphi : X \to Y$ is a continuous map of class $W^{1,2}_{\text{loc}}(X, Y)$ which is *bi-locally E-minimizing* in the sense that X can be covered by relatively compact subdomains U for each of which there is an open set $V \supset \varphi(U)$ in Y such that

$$E(\varphi_{|U}) \leqslant E(\psi_{|U})$$

for every continuous map $\psi \in W^{1,2}_{\mathrm{loc}}(X,Y)$ with $\psi(U) \subset V$ and $\psi = \varphi$ in $X \setminus U$.

It would clearly amount to the same to require that $\psi = \varphi$ off a compact subset of U (rather than just in $X \setminus U$). If $E(\varphi) < \infty$, the inequality in the definition is equivalent to $E(\varphi) \leqslant E(\psi)$.

Every continuous, locally E-minimizing map $\varphi : X \to Y$ (Definition 10.1) *is harmonic* (because one may take $V = Y$ in Definition 12.1). The converse holds e.g., for $Y = \mathbb{R}^n$, by the proof of Proposition 5.2 (where $n = 1$). More generally:

Lemma 12.1. *Let Y be a simply connected complete Riemannian polyhedron of nonpositive curvature. A harmonic map $\varphi : X \to Y$ is the same as a continuous locally E-minimizing map.*

Proof. Suppose that φ is harmonic. For given $a \in X$ let $U \ni a$ and V be as in Definition 12.1. We may assume that a is a vertex of X, that the open star $\mathrm{st}_X(a)$ has compact closure S in U, and that $\mathrm{st}_X(a) = S^\circ$, i.e., $bS = \partial S$, cf. Remark 4.2. By Theorem 11.3, $W^{1,2}_\varphi(S,Y)$ has an E-minimizer φ^*. Let $\varphi_I : S \to Y$ be the geodesic homotopy joining $\varphi_0 = \varphi$ to $\varphi_1 = \varphi^*$, cf. Proposition 11.1. Because $\varphi(S)$ and $\varphi^*(S)$ are compact, and geodesics in Y vary continuously with their endpoints, it follows from $\varphi(S) \subset \varphi(U) \subset V$ that $\varphi^*(S) \subset V$ for small t, and hence

$$E(\varphi) \leqslant E(\varphi_t) \leqslant (1-t)E(\varphi) + tE(\varphi^*),$$

by Definition 12.1 and Proposition 11.2. This implies $E(\varphi) \leqslant E(\varphi^*) \leqslant E(\psi)$ for any map $\psi \in W^{1,2}_\varphi(S,Y)$, and so φ is indeed locally E-minimizing. □

Remark 12.1. If instead $\mathrm{curv}\, Y \leqslant K$ for some constant $K > 0$, a similar result is established in the same way in the setting of Theorem 11.4 for maps φ of X into a closed ball of radius $< \pi/(2\sqrt{K})$ satisfying radial uniqueness.

Say that a geodesic space Y has *locally upper bounded curvature* if Y can be covered by convex subdomains of upper bounded curvature.

For harmonicity of continuous maps into such a geodesic space Y, a certain infinitesimal type of bi-local E-minimization suffices:

Lemma 12.2. *Let Y be a locally compact geodesic space with locally upper bounded curvature. A continuous map $\varphi \in W^{1,2}_{\mathrm{loc}}(X,Y)$ is harmonic iff X can be covered by relatively compact subdomains U for each of which there is a compact convex set $V \subset Y$ satisfying bi-point uniqueness such that $\varphi(U) \subset V$ and, when $E(\varphi) < \infty$,*

$$\limsup_{t \to 0+} \frac{1}{t}(E(\varphi_t) - E(\varphi)) \geqslant 0 \tag{12.1}$$

for any geodesic homotopy $(\varphi_t)_{t \in I}$ within $W^{1,2}_{\mathrm{loc}}(X,V)$ with $\varphi_1 = \varphi$ in $X \setminus U$ and $\varphi_0 = \varphi$.

If $E(\varphi) = \infty$, replace φ and φ_t in (12.1) by their restrictions to U.

Proof. By continuity of φ we may suppose that $\operatorname{curv} Y \leqslant K$ and, after rescaling, that $K = 1$. Fix $p \in X$, and write $\varphi(p) = q$. Choose a compact convex ball $B = B_Y(q,R)$ of radius $R < \pi/2$ so that B satisfies bi-point uniqueness (see Chapter 2). Then B is simply connected (cf. the text preceding Theorem 11.4).

Suppose first that φ is harmonic, and choose U and V as in Definition 12.1. By diminishing R and U we arrange that $B \subset V$ and $\varphi(U) \subset B$. Clearly, (12.1) holds for any geodesic homotopy (cf. Proposition 11.1) as stated in the lemma (now with V replaced by B).

Conversely, with U and V as in the lemma and with a continuous map $\psi \in W^{1,2}_{\mathrm{loc}}(X,Y)$ such that $\psi(U) \subset V$ and $\psi = \varphi$ off a compact subset of U, there is a geodesic homotopy $(\varphi_t)_{t \in I}$ within $W^{1,2}_{\mathrm{loc}}(U,V)$ joining $\varphi_0 = \varphi_{|U}$ to $\varphi_1 = \psi_{|U}$, cf. the proof of Proposition 11.1, applicable to B in place of Y in view of [AB 1990, Theorems 3 and 6]. Then

$$E(\varphi_t) \leqslant (1-t)E(\varphi_0) + tE(\varphi_1), \qquad 0 \leqslant t \leqslant 1,$$

first for $t = \frac{1}{2}$ by Serbinowski's inequality (11.15). By repeated application the same follows for any dyadic number $t \in [0,1]$; and finally it holds for any $t \in [0,1]$ by lower semicontinuity of the energy functional (Lemma 9.1 and Remark 9.6) because the distance function $d_Y(\varphi_t(\cdot), \varphi(\cdot))$ converges locally uniformly, and hence in $L^2_{\mathrm{loc}}(X)$ to 0 on X as $t \to 0$. Hence,

$$0 \leqslant \limsup_{t \to 0+} \frac{1}{t}(E(\varphi_t) - E(\varphi_0)) \leqslant E(\varphi_1) - E(\varphi_0),$$

and so indeed $E(\varphi_{|U}) \leqslant E(\psi_{|U})$. □

Remark 12.2. There is equality in (12.1) when Y is a Riemannian manifold (see Theorem 12.1 below), but not always for polyhedral Y, as shown by the following example. Take $X = \mathbb{R}$, $Y =$ the union of the x-axis and the positive y-axis in the xy-plane (Cornea's tripod, cf. Remark 5.6). The inclusion map $\varphi : X \to Y$ given by $\varphi(x) = (x, 0)$ is clearly locally E-minimizing and hence harmonic (this also follows from Lemma 12.5, φ being a speed 1 geodesic). For any $\alpha > 0$ consider the geodesic homotopy $(\varphi_t)_{t \geqslant 0}$ within $W^{1,2}_{\text{loc}}(X, Y)$, given by $\varphi_t(-x) = -\varphi_t(x)$ and by $\varphi_t(x) = (x, 0)$ $(= \varphi(x))$ for $|x| \geqslant \alpha$, while for $0 \leqslant x \leqslant \alpha$

$$\varphi_t(x) = \big([(1+t)x - t\alpha]^+, [(1+t)x - t\alpha]^-\big).$$

In particular, $\varphi_0 = \varphi$. The paths $0 \leqslant t \mapsto \varphi_t(x)$ and $[0, \alpha] \ni x \mapsto \varphi_t(x)$ are geodesics in Y of constant speed $(\alpha - |x|)^+$ and $1 + t$, respectively. It follows that $E((\varphi_t)_{[-\alpha,\alpha]}) = 2\alpha(1+t)^2$; and so

$$\frac{d}{dt}\big(E(\varphi_t) - E(\varphi)\big)\Big|_{t=0+} = 4\alpha,$$

which shows that the harmonic map φ cannot be regarded as E-critical.

A similar example can be made with $X = \mathbb{R}$, $Y =$ the Riemann surface of $\sqrt{z}$ (see Example 4.4), into which the tripod can be isometrically embedded.

Remark 12.3. In the setting of Definition 12.1, if X is *compact* and Y is simply connected complete with nonpositive curvature, then every harmonic map $\varphi : X \to Y$ is *constant.* (We omit here the energy concept of Definition 9.3.) Indeed, for fixed $p \in X$, $q = \varphi(p)$, the continuous function $d_Y^2(\varphi(\cdot), q)$ is subharmonic on X, by Lemma 10.2, resp. the proof of Proposition 12.1; and hence constant, by the minimum principle (see the text following Definition 2.2), X being compact. It follows that $d_Y(\varphi(x), q) = d_Y(\varphi(p), q) = 0$, i.e., $\varphi(x) = q$ for all $x \in X$.

Here "simply connected" cannot be dropped, not even if X is a circle, and hence the harmonic maps $X \to Y$ are the constant-speed closed geodesics, by Corollary 12.3 below. (For example: Y is a circle, an annulus in the plane, or a cylinder.) Likewise, nonpositive curvature cannot be weakened to upper bounded curvature, except when the range of φ is suitably small. (For example: Y is a round 2-sphere).

Example 12.1. Let $\pi : M \to (N, h)$ be a Riemannian branched cover of a Riemannian manifold (N, h), as in Example 8.11. Then according to Example 13.4, π is a harmonic map. A similar situation occurs for an analytic branched covering $\pi : X \to (N, h)$ of a Kähler manifold by a normal complex analytic space, using Proposition 11.3.

Example 12.2. As in Example 8.12, let K be a compact group of isometries of smooth Riemannian manifolds $(M, g), (N, h)$; and $\psi : M \to N$ a smooth K-equivariant map (i.e., $\psi(k \cdot x) = k \cdot \psi(x)$ for all $x \in M, k \in K$). Then ψ induces a continuous map $\varphi : X = M/K \to Y = N/K$:

$$\begin{array}{ccc} M & \xrightarrow{\psi} & N \\ {\scriptstyle \pi_M}\downarrow & & \downarrow{\scriptstyle \pi_N} \\ X & \xrightarrow{\varphi} & Y \end{array}$$

Under various circumstances, harmonicity of ψ insures that of φ (in the sense of Definition 12.1), and conversely. A model framework is described in [EL 1978, §4.17].

Subexample 12.2 (i). Example 13.5, with K acting trivially on N.

Subexample 12.2 (ii). Starting with a smooth map $\theta : (M, g) \to (N, h)$ between smooth Riemannian manifolds, form the r-th power map

$$\theta^{(r)} = \theta \times \cdots \times \theta : \quad M^{(r)} = M \times \cdots \times M \longrightarrow N^{(r)} = N \times \cdots \times N.$$

As in Example 8.13 (ii), the symmetric group $\mathfrak{S}_r$ acts isometrically by permuting these factors. Then $\theta^{(r)}$ is an $\mathfrak{S}_r$-equivariant map, determining the r-th symmetric power map $\mathfrak{S}^r(\theta)$:

$$\begin{array}{ccc} M^{(r)} & \xrightarrow{\theta^{(r)}} & N^{(r)} \\ \downarrow & & \downarrow \\ M^{(r)}/\mathfrak{S}_r = \mathfrak{S}^r M & \xrightarrow{\mathfrak{S}^r(\theta)} & \mathfrak{S}^r N = N^{(r)}/\mathfrak{S}_r \end{array}$$

If $\theta : M \to N$ is harmonic, then $\mathfrak{S}^r(\theta)$ is a harmonic map between Riemannian orbifolds.

Weakly harmonic maps into a Riemannian manifold

In this section (X, g) is an arbitrary admissible Riemannian polyhedron (g just bounded measurable with local elliptic bounds), $\dim X = m$, while

(N, h) is a smooth Riemannian manifold without boundary, $\dim N = n$. In particular, N has locally upper bounded curvature.

Recall from Definition 9.2 that a map $\varphi \in W^{1,2}_{\text{loc}}(X, N)$ has a quasicontinuous version (to be used throughout), and the components $\varphi^1, \cdots, \varphi^n$ of φ in any chart $V \to \mathbb{R}^n$ are of class $W^{1,2}_{\text{loc}}(U)$ on every quasiopen set $U \subset \varphi^{-1}(V)$ of compact closure in X, see Chapter 7, Sobolev functions on quasiopen sets.

We proceed to characterize harmonic maps as continuous, weakly harmonic maps, denoting by $\Gamma^k_{\alpha\beta}$ the Christoffel symbol on N.

Definition 12.2. A *weakly harmonic map* $\varphi : X \to N$ is a quasicontinuous map of class $W^{1,2}_{\text{loc}}(X, N)$ with the following property:

For any chart $\eta : V \to \mathbb{R}^n$ on N and any quasiopen set $U \subset \varphi^{-1}(V)$ of compact closure in X, the equation

$$\int_U \langle \nabla\lambda, \nabla\varphi^k \rangle d\mu_g = \int_U \lambda(\Gamma^k_{\alpha\beta} \circ \varphi)\langle \nabla\varphi^\alpha, \nabla\varphi^\beta \rangle d\mu_g \tag{12.2}$$

holds for every $k = 1, \cdots, n$ and every bounded function $\lambda \in W^{1,2}_0(U)$.

It is enough to verify (12.2) for every test function $\lambda \in D(U)$ (Definition 7.5). Also, by covariance, it suffices that (12.2) holds for some sequence of couples (U_j, V_j) like (U, V) above such that $\bigcup_j U_j = X$; this is seen by using the substitute for a partition of unity given in [KilM 1992, Lemma 2.4]. The concept of a weakly harmonic map is therefore a local one.

In Definition 12.2, equation (12.2) can of course be replaced equivalently by

$$\int_U \langle \nabla\lambda_k, \nabla\varphi^k \rangle d\mu_g = \int_U \lambda_k(\Gamma^k_{\alpha\beta} \circ \varphi)\langle \nabla\varphi^\alpha, \nabla\varphi^\beta \rangle d\mu_g, \tag{12.3}$$

valid for any couple (U, V) as above and any vector field λ on U along φ of class $W^{1,2}_0(U) \cap L^\infty(U)$ (or just of class $D(U)$) with covariant components λ_k. Cf. [EL 1983, Proposition 2.4 and equation (2.5)] for the case that (X, g) is also a Riemannian manifold.

Weak harmonicity of a map φ can be interpreted as *vanishing tension field* $\tau(\varphi) = 0$, cf. (1.1), again in the weak sense; or as vanishing first variation $\delta E(\varphi) = 0$ (cf. Lemma 12.3), i.e., as φ being *E-critical.*

The concept of a weakly harmonic map simplifies in the following two situations (perhaps the only cases considered in the literature on harmonic maps between manifolds): (i) φ is *continuous*, or (ii) N is diffeomorphic to a domain in $\mathbb{R}^n$ (i.e., N has an atlas with *only one chart*). In either case, each

$\varphi^{-1}(V)$ is *open* (not just quasiopen), and it is clearly enough to consider relatively compact open sets $U \subset \varphi^{-1}(V)$ in Definition 12.2 and subsequent paragraphs.

Remark 12.4. We sketch a coordinate-free version of Definition 12.2 for smooth maps and smooth vector fields between smooth Riemannian manifolds. It is based on the differential geometry of vector bundles, as described in [EL 1983, Part I, §§1, 2]; we refer to that for terminology, background and further details.

Consider a (smooth) map $\varphi : (M, g) \to (N, h)$ between Riemannian manifolds. Its differential at any point $x \in M$ is a linear map $T_x(M) \to T_{\varphi(x)}(N)$, which is treated as an element of the tensor product $T_x^*(M) \otimes T_{\varphi(x)}(N)$, and denoted here by $\nabla\varphi(x)$. The map φ pulls the tangent bundle of N back to the *induced vector bundle* $\varphi^{-1}T(N) \to M$. Its sections are called *vector fields along* φ (or sometimes *infinitesimal variations* of φ).

The Riemannian metric g on M determines its Levi Civita connection on $T(M)$, as well as on the cotangent bundle $T^*(M)$. Similarly for h on N; and on $\varphi^{-1}(N)$. Taken together, these produce a tensor product connection ∇ and a covariant constant metric on the tensor product *Riemannian vector bundle* $T^*(M) \otimes \varphi^{-1}(T(M)) \to M$.

If λ is a vector field along φ, it has a covariant differential $\nabla\lambda$ along φ which (like $\nabla\varphi$) is a section of $T^*(M) \otimes \varphi^{-1}T(M)$. Say that a map φ is *weakly harmonic* if

$$\int_M \langle \nabla\lambda, \nabla\varphi \rangle d\mu_g = 0 \tag{12.4}$$

for all compactly supported vector fields λ along φ.

Remark 12.5. An alternative, coordinate characterization of weak harmonicity, again for smooth maps between smooth Riemannian manifolds, can be derived using an isometric embedding $\iota : N \hookrightarrow V$ of N into some Euclidean space. (More elementary embeddings suffice as well; see [EP 1984, §3].) Start with a geometric form of Lagrange's method of multipliers: For any map $\varphi : M \to N$ let $\Phi = \iota \circ \varphi$. Then at any point $x \in M$

$$\Delta\Phi(x) = \iota_*\tau(\varphi)(x) + \nu(\Phi)(x),$$

where Δ denotes the Laplace–Beltrami operator on V-valued maps, $\tau(\varphi)$ the tension field of φ, and $\nu(\Phi)$ is a normal vector along Φ. In particular,

φ is harmonic iff at every point x of M, $\Delta\Phi(x)$ is orthogonal to $T_{\varphi(x)}(N)$.
Then in the present context (12.4) takes the form

$$-\int_M g^{ij}\frac{\partial\varphi}{\partial x^i}\nabla_{x_j}\lambda\, d\mu_g + \int_M g^{ij}\pi^k_{\alpha\beta}\frac{\partial\varphi^\alpha}{\partial x^i}\frac{\partial\varphi^\beta}{\partial x^j}\lambda\, d\mu_g = 0$$

($k = 1, \cdots, \dim V$), where ($\pi^k_{\alpha\beta}$) is essentially the second fundamental form of the embedding $\iota : N \hookrightarrow V$, and $\nabla_{x^j}\lambda$ is the covariant differential of λ in the x^j-direction.

Lemma 12.3. *Let $\varphi : X \to N$ be a continuous map of class $W^{1,2}_{\mathrm{loc}}(X, N)$. For any chart $\eta : W \to \mathbb{R}^n$ of N, any domain $V \Subset W$ such that $\eta(V)$ is convex (in $\mathbb{R}^n$), and any domain $U \Subset X$ such that $\varphi(U) \subset V$, the following estimate holds for any continuous map $\psi \in W^{1,2}_{\mathrm{loc}}(X, N)$ with $\psi(U) \subset V$ and $\psi = \varphi$ off a compact subset of U:*

$$E(\psi_{|U}) = E(\varphi_{|U}) + 2\delta E(\varphi)(\lambda) + R(\varphi, \lambda), \tag{12.5}$$

whereby $\lambda = \psi - \varphi$ (in the sense of contravariant components $\lambda^k = \psi^k - \varphi^k$ in the above chart). Here $\delta E(\varphi)$ (the first variation of $E(\varphi)$) is linear in λ and given by

$$\delta E(\varphi)(\lambda) = \int_U \big(\langle\nabla\lambda_k, \nabla\varphi^k\rangle - \lambda_k(\Gamma^k_{\alpha\beta}\circ\varphi)\langle\nabla\varphi^\alpha, \nabla\varphi^\beta\rangle\big)\, d\mu_g, \tag{12.6}$$

where $\lambda_k = (h_{kl}\circ\varphi)\lambda^l$. Furthermore,

$$|R(\varphi, \lambda)| \leqslant C(\|\lambda\|^2_\infty + \|\nabla_e\lambda\|^2_e), \tag{12.7}$$

where the constant $C > 0$ is independent of λ, and

$$\|\lambda\|^2_\infty := \|\sum_1^n (\lambda^k)^2\|_{L^\infty(U)}, \quad \|\nabla_e\lambda\|^2_e := \sum_1^n \int_U |\nabla_e\lambda^k|^2_e\, d\mu_g,$$

the subscript e referring to the Euclidean Riemannian metric g^e on X.

Proof. For any map ψ as stated, the corresponding map $\lambda = \eta\circ\psi - \eta\circ\varphi \in W^{1,2}_c(U, \mathbb{R}^n)$ is bounded and has finite energy, by Lemma 9.3, and

$$\begin{aligned} E(\psi_{|U}) - E(\varphi_{|U}) &= \int_U (h_{\alpha\beta}\circ\psi)\langle\nabla\psi^\alpha, \nabla\psi^\beta\rangle - \int_U (h_{\alpha\beta}\circ\varphi)\langle\nabla\varphi^\alpha, \nabla\varphi^\beta\rangle \\ &= 2\delta E(\varphi)(\lambda) + R(\varphi, \lambda), \end{aligned}$$

where $\delta E(\varphi)$ is defined as a function of $\lambda \in W_c^{1,2}(U,\mathbb{R}^n) \cap L^\infty(U,\mathbb{R}^n)$ by

$$\delta E(\varphi)(\lambda) = \int_U (h_{\alpha\beta} \circ \varphi)\langle \nabla\varphi^\alpha, \nabla\lambda^\beta\rangle + \frac{1}{2}\int_U (\partial_k h_{\alpha\beta} \circ \varphi)\lambda^k \langle \nabla\varphi^\alpha, \nabla\varphi^\beta\rangle.$$

Here, and elsewhere in the proof, integrations are with respect to μ_g.

The functional $\delta E(\varphi)$ is linear in λ and can be transformed into (12.6) by elementary manipulations, invoking the chain rule in Remark 9.7 and inserting

$$(h_{\alpha\beta} \circ \varphi)\nabla\lambda^\beta = \nabla\left[(h_{\alpha\beta} \circ \varphi)\lambda^\beta\right] - \lambda^\beta\left[(\partial_k h_{\alpha\beta}) \circ \varphi\right]\nabla\varphi^k$$

and $\Gamma^k_{\alpha\beta} = h^{kl}\Gamma_{l,\alpha\beta}$, where $2\Gamma_{k,\alpha\beta} = 2\Gamma_{k,\beta\alpha} = \partial_\alpha h_{k\beta} + \partial_\beta h_{k\alpha} - \partial_k h_{\alpha\beta}$. Next, (12.5) holds when we define, again for $\lambda \in W_c^{1,2}(U,\mathbb{R}^n) \cap L^\infty(U,\mathbb{R}^n)$,

$$R(\varphi,\lambda) = \int_U \left(R^{(0)}_{\alpha\beta}\langle \nabla\lambda^\alpha, \nabla\lambda^\beta\rangle + 2R^{(1)}_{\alpha\beta}\langle \nabla\varphi^\alpha, \nabla\lambda^\beta\rangle + R^{(2)}_{\alpha\beta}\langle \nabla\varphi^\alpha, \nabla\varphi^\beta\rangle\right),$$

where

$$\begin{aligned}
R^{(0)}_{\alpha\beta} &= h_{\alpha\beta} \circ \psi,\\
R^{(1)}_{\alpha\beta} &= h_{\alpha\beta} \circ \psi - h_{\alpha\beta} \circ \varphi,\\
R^{(2)}_{\alpha\beta} &= h_{\alpha\beta} \circ \psi - h_{\alpha\beta} \circ \varphi - \left[(\partial_k h_{\alpha\beta}) \circ \varphi\right]\lambda^k.
\end{aligned}$$

In terms of ellipticity constants Λ_U and Λ_V of $g_{|U}$ and $h_{|V}$ we have a.e. in U, by Taylor's formula, applied within the compact convex subset $\overline{\eta(V)}$ of $\mathbb{R}^n$, writing $c_0 = \Lambda_U^2\Lambda_V^2$; $|\lambda_e|^2 = \sum_1^n (\lambda^k)^2$; and $|\nabla_e\lambda|_e^2 = \sum_1^n |\nabla_e\lambda^k|_e^2$ and similarly with φ in place of λ:

$$\begin{aligned}
&|R^{(0)}_{\alpha\beta}\langle \nabla\lambda^\alpha, \nabla\lambda^\beta\rangle| \leqslant c_0|\nabla_e\lambda|_e^2,\\
&|R^{(1)}_{\alpha\beta}\langle \nabla\varphi^\alpha, \nabla\lambda^\beta\rangle| \leqslant c_1|\nabla_e\varphi|_e|\lambda|_e|\nabla_e\lambda|_e,\\
&|R^{(2)}_{\alpha\beta}\langle \nabla\varphi^\alpha, \nabla\varphi^\beta\rangle| \leqslant c_2|\nabla_e\varphi|_e^2|\lambda|_e^2,
\end{aligned}$$

as integrable functions of $x \in U$, whereby c_0, c_1, c_2 are positive constants. Inserting this in the definition of $R(\varphi,\lambda)$ leads to (12.7). □

In the present setting of maps into manifolds, Ishihara's pull-back characterization of harmonic maps between Riemannian manifolds [Ish 1979, Theorem 3.4] extends fully to polyhedral domains:

Theorem 12.1. *For a continuous map $\varphi \in W^{1,2}_{\text{loc}}(X,N)$ the following are equivalent:*

(a) *φ is harmonic, i.e., bi-locally E-minimizing* (Definition 12.1).
(b) *φ is weakly harmonic* (Definition 12.2).
(c) *φ pulls convex functions on open sets $V \subset N$ back to subharmonic functions on $\varphi^{-1}(V)$.*

It clearly suffices in (c) to consider *convex* open sets V.

As mentioned in Chapter 2, a function $v : V \to \mathbb{R}$, defined on an open subset V of a geodesic space Y, is said to be *convex* if $v \circ \gamma : I \to \mathbb{R}$ is convex for any constant-speed geodesic segment $\gamma : I \to Y$ with image in V. In the present case of a smooth Riemannian manifold (N,h) it is well known that a function $v \in C^2(V)$ is convex iff the Hessian $H(v)$ is positive semidefinite at each point of V. Recall that, in local coordinates y^α in N, the covariant components of $H(v)$ are the second order covariant derivatives $v_{\alpha\beta}$ of v:

$$v_{\alpha\beta} = \partial_\alpha \partial_\beta v - \Gamma^k_{\alpha\beta}\, \partial_k v, \qquad \alpha, \beta = 1, \cdots, n. \tag{12.8}$$

It is also known that, for any convex function v on an open subset V of N, every point of V has a convex neighbourhood on which v can be approximated uniformly by smooth, strictly convex functions on V. This follows from [GW 1976, Theorem 2 (a)] because every point $y_0 \in V$ has a convex neighbourhood in which the square of the distance function $y \mapsto d_Y(y, y_0)$ is strictly convex and bounded. Because a uniform limit of subharmonic functions is subharmonic, it suffices therefore in (c) of the above theorem to consider *smooth, strictly convex functions v on convex domains $V \subset N$* (see Chapter 2). Clearly, these domains may be taken to be coordinate patches.

Every convex function on an open subset of N is Lip continuous, as noted in [GW 1973, p. 642].

Proof of Theorem 12.1. (a) $\iff$ (b). For given $p \in X$ let $\eta : W \to \mathbb{R}^n$ be a chart of N such that $\varphi(p) \in W$.

Suppose first that φ is harmonic. Choose U and V as in Definition 12.1 and so that $p \in U$, $V \Subset W$, and $\eta(V)$ is convex. Given on U a continuous vector field λ along φ of class $\text{Lip}_c(U)$, define $\psi \in W^{1,2}_{\text{loc}}(X,N)$ by $\psi = \varphi$ off U and $\psi^k = \varphi^k + t\lambda^k$ in U, $t > 0$, whereby $\lambda^1, \cdots, \lambda^n$ denote the contravariant components of λ in the chosen chart; they are of class $W^{1,2}(U)$, by Definition 9.2. Furthermore, ψ is continuous, $\psi(U) \subset W$ for small t, and so $E(\psi_{|U}) \geqslant E(\varphi_{|U})$, by Definition 12.1. Inserting this in (12.5) and dividing by t leads for $t \to 0$ to $\delta E(\varphi) = 0$ in view of (12.7) and the linearity of $\delta E(\varphi)$. By (12.6) this implies (12.3) and therefore (12.2).

Conversely, suppose φ is weakly harmonic. This time choose U and V as in Definition 12.2, again so that $p \in U$, $V \Subset W$, and $\eta(V)$ is convex. Next, choose a ball B centred at $\varphi(p)$ and having bi-point uniqueness; and a domain U' with $p \in U' \Subset \varphi^{-1}(B)$. For any geodesic homotopy $(\varphi_t)_{t\in I}$ within $W^{1,2}_{\text{loc}}(U', B)$ with $\varphi_0 = \varphi_{|U'}$ it follows from (12.3) that $\delta E(\varphi) = 0$, and hence from Lemma 12.3 that

$$\lim_{t\to 0+} \frac{1}{t}(E(\varphi_t) - E(\varphi)) = 0$$

on account of (12.5) and (12.7) applied to $\lambda = \lambda_t := \varphi_t - \varphi_0$, which leads to $R(\varphi, \lambda_t) = O(t^2)$. Indeed, $\lambda_t(x) = O(t)$ and $|\nabla_e \lambda_t(x)|_e = O(t)$, uniformly for $x \in U'$, by smoothness of the exponential map $\exp : T(N) \to N$ and continuity of φ and ψ.

(b) $\iff$ (c). For any continuous map $\varphi \in W^{1,2}_{\text{loc}}(X, N)$, for any open sets $V \subset N$ and $U \subset \varphi^{-1}(V)$ with U of compact closure in X; and for any functions $v \in C^2(V)$ and $\lambda \in W^{1,2}_c(U) \cap L^\infty(U)$, we have by Remark 9.7,

$$\begin{aligned} -\int_U \langle \nabla\lambda, \nabla(v\circ\varphi)\rangle &= -\int_U \langle [(\partial_\alpha v)\circ\varphi]\nabla\lambda, \nabla\varphi^\alpha\rangle \\ &= -\int_U \langle \nabla(\lambda\,[(\partial_\alpha v)\circ\varphi]), \nabla\varphi^\alpha\rangle + \int_U \lambda\,[(\partial_\alpha\partial_\beta v)\circ\varphi]\langle\nabla\varphi^\alpha, \nabla\varphi^\beta\rangle. \end{aligned} \tag{12.9}$$

Inserting (12.8) in the last integral gives

$$\begin{aligned} &-\int_U \langle \nabla\lambda, \nabla(v\circ\varphi)\rangle \\ &\quad= -\int_U \langle \nabla(\lambda\,[(\partial_k v)\circ\varphi]), \nabla\varphi^k\rangle + \int_U \lambda(v_{\alpha\beta}\circ\varphi)\langle\nabla\varphi^\alpha, \nabla\varphi^\beta\rangle \\ &\quad\quad+ \int_U \lambda\,[(\partial_k v)\circ\varphi](\Gamma^k_{\alpha\beta}\circ\varphi)\langle\nabla\varphi^\alpha, \nabla\varphi^\beta\rangle. \end{aligned} \tag{12.10}$$

Suppose first that the continuous map φ is weakly harmonic. For any chart $\eta : V \to \mathbb{R}^n$ on N and any function $\lambda \in \text{Lip}_c(\varphi^{-1}(V))$, the first and the third integral on the right of (12.10) are then equal according to (12.2) applied with $\lambda\,[(\partial_k v)\circ\varphi] \in W^{1,2}_c(U) \cap L^\infty(U)$, $k = 1, \cdots, n$, in place of $\lambda \in W^{1,2}_c(U) \cap L^\infty(U)$.

If $v \in C^2(V)$ is convex, (12.10) therefore shows that $\int \langle\nabla\lambda, \nabla(v\circ\varphi)\rangle \leqslant 0$ for every $\lambda \in \text{Lip}^+_c(U)$, noting that the positive semidefinite matrix $v_{\alpha\beta}$ has the form $\sum_{k=1}^n c_{\alpha k}c_{\beta k}$, and hence

$$v_{\alpha\beta}\langle\nabla\varphi^\alpha, \nabla\varphi^\beta\rangle = \sum_{k=1}^n \|c_{\alpha k}\nabla\varphi^\alpha\|^2 \geqslant 0.$$

It follows that the continuous function $v\circ\varphi \in W^{1,2}_{\mathrm{loc}}(U)$ is indeed subharmonic in each U as above and hence in X, by Definition 5.2 and Remark 7.4.

Conversely, suppose that (c) holds. We may assume that (X, g) satisfies the Poincaré inequality (7.3), cf. Remark 7.2. By [Ish 1979, Lemma 3.1] or [AG 1991, §2.4], N has an atlas consisting of charts $\eta : V \to \mathbb{R}^n$ such that V has compact closure in N and the components $y^1, \cdots, y^n$ of η are bounded convex C^2-functions on V. By assumption, $y^k\circ\varphi = \varphi^k$ is subharmonic and of class $W^{1,2}$ on every domain $U \Subset \varphi^{-1}(V)$. Denoting by μ^k the positive Radon measure on U associated with the superharmonic function $-\varphi^k$ considered on U, we obtain, by use of Theorem 7.3 and the Riesz decomposition $-\varphi^k = G^U\mu^k+$ harmonic function on U (cf. Chapter 2),

$$-\int_U \langle\nabla\lambda, \nabla\varphi^k\rangle = \int_U \lambda\, d\mu^k, \qquad \lambda \in L_0^{1,2}(U). \tag{12.11}$$

Consider now any bounded convex C^2-function v on the above patch V. Then $v\circ\varphi \in W^{1,2}(U)$, by Remark 9.7; and $v\circ\varphi$ is subharmonic on U, by (c). It follows that $\int\langle\nabla\lambda, \nabla(v\circ\varphi)\rangle \leqslant 0$ for every $\lambda \in \mathrm{Lip}_c^+(U)$, by Remark 7.4 and Definition 5.2. Inserting this in (12.10) and applying (12.11) with λ replaced by $\lambda[(\partial_k v)\circ\varphi] \in L_0^{1,2}(U)$ while using again Remark 9.7, leads to

$$[(\partial_k v)\circ\varphi]\big(\mu^k + (\Gamma^k_{\alpha\beta}\circ\varphi)\langle\nabla\varphi^\alpha, \nabla\varphi^\beta\rangle\big) \geqslant -(v_{\alpha\beta}\circ\varphi)\langle\nabla\varphi^\alpha, \nabla\varphi^\beta\rangle \tag{12.12}$$

in U, the locally integrable function on the right being viewed as a locally absolutely continuous measure. Here we use that Lip_c is positively rich in C_c in the sense of [Bou 1965, Chapter III, §2, Proposition 2], cf. the beginning of Step 3 in the proof of Proposition 7.1. If $\partial_k v < -1$, say, on V for every $k = 1, \cdots, n$, it follows that each μ^k is locally absolutely continuous in U and therefore may be viewed as a function of class $L^1_{\mathrm{loc}}(U)$.

For given $x_0 \in U$ there exists, again by [Ish 1979, Lemma 3.1], a bounded convex C^2-function v near $y_0 := \varphi(x_0)$ such that $\partial_k v(y_0) < -1$, $k = 1, \cdots, n$, and hence $\partial_k v < -1$ in some open neighbourhood $V_0 \subset V$ of y_0. It follows therefore from (12.12) that each μ^k is locally *absolutely continuous* in each $U \cap \varphi^{-1}(V_0)$ and hence in U; we denote by f^k ($\in L^1_{\mathrm{loc}}(U)$) its derivative ($\geqslant 0$) with respect to the Riemannian volume measure on U:

$$d\mu^k = f^k d\mu_g \qquad \text{in } U. \tag{12.13}$$

Similarly, there exist, for each $l \in \{1, \cdots, n\}$ and $\varepsilon > 0$, two bounded convex C^2-functions v^+ and v^- in an open neighbourhood $V_\varepsilon \subset V$ of $y_0 = \varphi(x_0)$ such that

$$|\partial_k v^\pm \mp \delta_{kl}| < \varepsilon, \quad |v^\pm_{\alpha\beta}| < \varepsilon \qquad \text{in } V_\varepsilon \quad (k, \alpha, \beta = 1, \cdots, n).$$

Taking $v = v^+$, resp. v^-, in (12.12), we obtain, in view of (12.13),

$$|f^l + (\Gamma^l_{\alpha\beta} \circ \varphi)\langle\nabla\varphi^\alpha, \nabla\varphi^\beta\rangle| \leqslant \varepsilon F \qquad \text{a.e. in } U \cap \varphi^{-1}(V_\varepsilon), \tag{12.14}$$

where

$$F := \sum_{k=1}^{n}\Big(f^k + \sum_{\alpha,\beta=1}^{n}(1 + |\Gamma^k_{\alpha\beta} \circ \varphi|)|\langle\nabla\varphi^\alpha, \nabla\varphi^\beta\rangle|\Big) \in L^1_{\text{loc}}(U \cap \varphi^{-1}(V_\varepsilon)).$$

Now suppose that the above point x_0 lies in the interior s° of a top-dimensional simplex s of X, and that x_0 is a *Lebesgue point* for each of the locally integrable functions f^k, $\langle\nabla\varphi^\alpha, \nabla\varphi^\beta\rangle$, and F, considered in the open subset $s^\circ \cap U \cap \varphi^{-1}(V_\varepsilon)$ of s°. The concept of a Lebesgue point x may be taken e.g., with respect to Lebesgue measure and Euclidean balls in s°.

Integrating (12.14) over any ball $B^e_X(x, r) \subset s^\circ \cap U \cap \varphi^{-1}(V_\varepsilon)$ we find that the corresponding meanvalue of $|f^l + (\Gamma^l_{\alpha\beta} \circ \varphi)\langle\nabla\varphi^\alpha, \nabla\varphi^\beta\rangle|$ is majorized by that of εF, and letting first $r \to 0$ and then $\varepsilon \to 0$, we find that

$$f^k + (\Gamma^k_{\alpha\beta} \circ \varphi)\langle\nabla\varphi^\alpha, \nabla\varphi^\beta\rangle = 0$$

at any Lebesgue point x_0 as above, and hence a.e. in U.

Multiplying this by $\lambda \in \text{Lip}_c(U)$ $(\subset L^{1,2}_0(U))$ and integrating over U, we conclude by (12.11) and (12.13) that φ is indeed weakly harmonic in each U and hence in their union $\varphi^{-1}(V)$, see text following Definition 12.2. □

In view of Lemma 9.1 and Remark 9.6, Theorem 12.1, (a) $\Longleftrightarrow$ (c) implies easily the following:

Corollary 12.1. *The limit of a locally uniformly convergent E-bounded sequence of harmonic maps $X \to N$ is a harmonic map.*

Remark 12.6. If we drop the requirement that φ be continuous, the equivalence of (b) and (c) in Theorem 12.1 remains in force when, in (c), subharmonic functions (on open sets in X) are replaced by quasisubharmonic functions (on quasiopen subsets of X) in the following sense, cf. Definition 7.4:

A quasicontinuous function u on a quasiopen set $U \subset X$ is called *quasi-subharmonic* if U is the union of a sequence of quasiopen sets V such that $u_{|V} \in W^{1,2}(V)$ and

$$\int_V \langle\nabla u, \nabla\lambda\rangle d\mu_g \leqslant 0 \qquad \text{for every } \lambda \in [W^{1,2}_0(V)]^+,$$

cf. [KilM 1992, §4] (where such functions are termed quasilocal subsolutions of Δu in U). It suffices to take $\lambda \in D(U)$ (Definition 7.5).

A function u on a quasiopen set $U \subset X$ is quasisubharmonic iff (i) U is the union of a sequence of quasiopen sets V such that $u_{|V} \in W^{1,2}(V)$, and (ii) u is *finely subharmonic* q.e. in U in the sense of [Fug 1972, Definitions 8.2, 10.4]; this follows from [Fug 1972, Theorem 9.9] and [Fug 1982a, Théorème 11].

The proof that (c) (from Theorem 12.1), thus extended, is equivalent to the quasicontinuous map $\varphi \in W^{1,2}_{\text{loc}}(X, Y)$ being weakly harmonic (Definition 12.2) is quite similar to that of the equivalence of (b) and (c) for a continuous map (as in Theorem 12.1), mainly because we have the Riesz decomposition of bounded finely subharmonic functions, cf. [Fug 1982b, p. 67].

Hölder continuity revisited

We bring essentially equivalent versions of Theorems 10.1 and 10.2, respectively, in the present setting of maps from an arbitrary admissible Riemannian polyhedron (X, g) to a smooth Riemannian manifold (N, h) without boundary. First the case of nonpositive curvature:

Proposition 12.1. *If (N, h) is simply connected, complete and has nonpositive sectional curvature, every weakly harmonic map $\varphi : X \to N$ is Hölder continuous.*

Proof. As in the proof of Theorem 10.1 it suffices to consider the case where X is compact. For each $q \in N$ the exponential map $\exp_q T_q(N) \to N$ is a diffeomorphism, see e.g. [KN 1969, p. 102]. In terms of Euclidean coordinates $y^1, \cdots, y^n$ on $T_q(N)$ the inverse of $\exp_q$ is a diffeomorphism $\eta : N \to \mathbb{R}^n$. The Hessian $H(v)$ of the function

$$v(y) = d_N^2(y, q) = \sum_{k=1}^{n} (y^k)^2$$

has covariant components $v_{\alpha\beta}$ (the covariant second derivatives of v), and $H(y) - 2h(y)$ is positive semidefinite:

$$v_{\alpha\beta}(y)\xi^\alpha\xi^\beta \geqslant 2h_{\alpha\beta}(y)\xi^\alpha\xi^\beta, \qquad y \in N,\ (\xi^1, \cdots, \xi^n) \in \mathbb{R}^n. \tag{12.15}$$

In [HKW 1977, (2.2)] this is derived from Rauch's comparison theorem. Here is a short alternative proof, using the fact that N has global nonpositive curvature in the sense of Alexandrov, cf. Chapter 2. By continuity it suffices to consider the case $y \neq q$. For any geodesic $\gamma : \mathbb{R} \to N$ parametrized by path length and with $\gamma(0) = y$, comparison of the triangle $\gamma(t)\gamma(-t)q$, with median $\gamma(0)q$, gives

$$(v \circ \gamma)(t) + (v \circ \gamma)(-t) - 2(v \circ \gamma)(0) \geqslant 2t^2.$$

Dividing by t^2 and letting $t \to 0$ leads to $(v \circ \gamma)^{\cdot\cdot}(0) \geqslant 2$. By (12.8) and the differential equation for the geodesic γ of speed 1: $|\dot\gamma|^2 = (h_{\alpha\beta} \circ \gamma)\dot\gamma^\alpha\dot\gamma^\beta = 1$, we have

$$(v_{\alpha\beta} \circ \gamma)\dot\gamma^\alpha\dot\gamma^\beta \geqslant 2(h_{\alpha\beta} \circ \gamma)\dot\gamma^\alpha\dot\gamma^\beta \qquad \text{at } t = 0.$$

Inserting $\gamma(0) = y$ this establishes (12.15) because we may take $\dot\gamma^\alpha(0) = \xi^\alpha$ $(\alpha = 1, \cdots, n)$ when $(h_{\alpha\beta} \circ \gamma)\dot\gamma^\alpha\dot\gamma^\beta = 1$.

As in the proof of (b) $\Longleftrightarrow$ (c) in Theorem 12.1, we have from (12.2) and (12.10) (cf. Remark 12.6)

$$-\int_X \langle \nabla\lambda, \nabla(v \circ \varphi)\rangle = \int_X \lambda(v_{\alpha\beta} \circ \varphi)\langle \nabla\varphi^\alpha, \nabla\varphi^\beta\rangle$$

for $\lambda \in \mathrm{Lip}_c(X)$. Inserting (12.15) (evaluated at $y = \varphi(x)$, $x \in X$) gives

$$-\int_X \langle \nabla\lambda, \nabla(v \circ \varphi)\rangle \geqslant 2\int_X \lambda(h_{\alpha\beta} \circ \varphi)\langle \nabla\varphi^\alpha, \nabla\varphi^\beta\rangle = 2\int_X \lambda e(\varphi) \qquad (12.16)$$

by Definition 9.2, using the representation of a positive matrix (in this case $v_{\alpha\beta} - 2h_{\alpha\beta}$) in the paragraph following (12.10). And (12.16) extends to $\lambda \in W_0^{1,2}(X) \cap L^\infty(X)$ because such λ can be approximated in $W_0^{1,2}(X)$ by a uniformly bounded sequence of functions of class $\mathrm{Lip}_c(X)$, by truncation.

From Lemma 10.2 (b) thus established in the present setting follows that $v \circ \varphi = d_N^2(\varphi(\cdot), q)$ is weakly subharmonic, and the rest of the proof of Theorem 10.1 carries over immediately. □

Now pass to the case where N is a smooth Riemannian manifold with sectional curvature $\leqslant K$ for some $0 < K < \infty$.

Proposition 12.2. *Let $\varphi : X \to N$ be a weakly harmonic map, and suppose that the essential range $\varphi(X)$ is contained in an open convex set $V \Subset N$ of diameter $D < \pi/(2\sqrt{K})$ such that V lies within normal range of each of its points. Then φ is Hölder continuous.*

Proof. We may suppose that X is compact and $K = 1$. Because V lies within normal range of its points, geodesics in V are uniquely determined by their endpoints, and vary continuously with them, V being locally compact. Every triangle in V has perimeter $< 2\pi$ and is consequently K-dominated with $K = 1$ [A 1951; 1957], cf. Chapter 2.

For each $q \in V$ take normal coordinates in V centred at q, and write

$$u(y) = d_Y(y, q), \quad v(y) = u(y)^2, \qquad y \in V.$$

Consider a geodesic segment $\gamma : [-\varepsilon, \varepsilon] \to V$ with $\gamma(0) = y \neq q$. Triangle comparison of $\gamma(t)\gamma(-t)q$ with median $\gamma(0)q$ gives, writing $u(t)$ for $(u \circ \gamma)(t)$,

$$\cos u(t) + \cos u(-t) - 2 \cos u(0) \cos t \leqslant 0.$$

Subtracting $2 \cos u(0)(1 - \cos t)$ on both sides, and dividing by t^2, we obtain for $t \to 0$

$$\frac{d^2}{dt^2} \cos \sqrt{v(t)}\,\Big|_{t=0} \leqslant - \cos u(0),$$

and so

$$\ddot{v}(0) \geqslant \frac{2u(0)}{\tan u(0)} + \frac{1}{2u(0)} \left(\frac{1}{u(0)} - \frac{1}{\tan u(0)} \right) \dot{v}(0)^2 \geqslant 2c$$

with $c = D/\tan D$, noting that $D < \pi/2$. Thus $(v \circ \gamma)^{\cdot\cdot}(0) \geqslant 2c$, and hence $Hv - 2ch$ is positive semidefinite, as in the proof of Proposition 12.1 (also if $y = q$). Again, this is found in [HKW 1977, (2.2)] with the same constant c. The proof is completed in quite the same way as that of Proposition 12.1, cf. also the last three paragraphs of the proof of Theorem 10.2. □

Corollary 12.2. *Every harmonic map φ of (X, g) into an arbitrary smooth Riemannian manifold (N, h) without boundary is Hölder continuous.*

Remark 12.7. If N has *even* dimension n and is complete with sectional curvature satisfying $0 < \operatorname{curv} N \leqslant K$, then *every* ball in N of radius $R < \pi/(4\sqrt{K})$ is convex and within normal range of its points; similarly for odd n provided N is simply connected and $K/4 \leqslant \operatorname{curv} Y \leqslant K$, [GKM 1968, pp. 227–228], [Kar 1970, pp. 79–80].

Remark 12.8. When X is also a Riemannian manifold the hypothesis on the diameter of V in Proposition 12.2 can be weakened to $\varphi(X)$ being contained in a compact ball of radius $R < \pi/(2\sqrt{K})$, [HKW 1976; 1977]. When

$n > 2$ this is optimal, as shown by the following example. When $n = 2$, no restriction on the size of $\varphi(X)$ is needed, see [HW 1975, p. 67].

Example 12.3. [HKW 1977, §6]. Let S^{n-1}, resp. S^n, be the unit sphere in $\mathbb{R}^n$, resp. $\mathbb{R}^{n+1}$, $n > 2$. Let $p : \mathbb{R}^n \setminus \{0\} \to S^{n-1}$ denote central projection from 0 onto S^{n-1}; and $i : S^{n-1} \to S^n$ the inclusion map, which is clearly a totally geodesic map (Definition 12.3 below). It is known that p is a usual harmonic map (even a harmonic morphism, cf. Definition 13.1); this follows from [BE 1981, §5] because p is horizontally homothetic and has minimal fibres (straight lines in this case). The composite map $i \circ p : \mathbb{R}^n \setminus \{0\} \to S^n$ is therefore harmonic, by Theorem 12.2 and Theorem 12.1, (a) $\iff$ (c). For $n \geqslant 3$, $i \circ p$ is clearly of class $W^{1,2}_{\text{loc}}(\mathbb{R}^n, S^n)$, and the singularity at 0 is therefore removable, so that $i \circ p$ *extends to a weakly harmonic map* $\varphi \in W^{1,2}_{\text{loc}}(\mathbb{R}^n, S^n)$, according to [EP 1984]. The essential range $\varphi(\mathbb{R}^n)$ equals S^{n-1}, which is contained in a closed hemisphere in S^n, i.e., in a closed ball in S^n of radius $\pi/2 = \pi/(2\sqrt{K})$; but φ *does not agree a.e. with a continuous map* because of the essential discontinuity at 0.

Query. Can the assumption on $\varphi(X)$ in Proposition 12.2 be weakened to $\varphi(X)$ being contained in an open ball V in N of radius $< \pi/(2\sqrt{K})$ such that V lies within normal range of its points (cf. Remark 12.8)? What about the particular case when N is the standard n-sphere of radius $1/\sqrt{K}$? Similarly for Theorem 10.2?

Totally geodesic maps

We extend to suitable geodesic spaces X, Y the pull-back characterization of totally geodesic maps $\varphi : X \to Y$, due to Ishihara [Ish 1979, Theorem 3.2] for the case of maps between smooth Riemannian manifolds (without boundary). In the rest of this chapter, geodesics $\gamma : I \to X$ (with I an interval on $\mathbb{R}$ or a standard circle) are understood to have *constant speed.*

In the present section we do not require that geodesic spaces be locally compact. A geodesic space Y of *locally upper bounded curvature* (as defined in the text preceding Lemma 12.2) can be covered by convex subdomains V with $\operatorname{curv} V \leqslant K = K(V)$ and diameter $D \leqslant \pi/(2\sqrt{K})$ such that every geodesic triangle in V is K-dominated (cf. Chapter 2), and hence any two points of V can be joined by a unique geodesic in V.

It follows that the distance function $d_Y(\cdot, y)$ *is convex in* V *for every* $y \in V$. This means in fact that, in any triangle xyz in V, the median from y

has length $m \leqslant \frac{1}{2}(a+b)$, where $a = d_Y(x,y)$, $b = d_Y(z,y)$. And that follows by triangle comparison from the same inequality for the case when x, y, z are instead points of the standard sphere of radius $1/\sqrt{K}$. Taking $K = 1$, for simplicity of writing, we then have by spherical trigonometry, writing $c = d_Y(x,z)$:

$$\cos m \geqslant \cos m \, \cos \tfrac{1}{2}c \geqslant \tfrac{1}{2}(\cos a + \cos b),$$

and this implies indeed $m \leqslant \frac{1}{2}(a+b)$ because cosine is concave on $[0, \pi/2]$. (Cf. [AB 1996, Theorem 4.3], where y is fixed.)

Definition 12.3. A continuous map $\varphi : X \to Y$ between geodesic spaces (X, d_X) and (Y, d_Y) is said to be *totally geodesic* if $\varphi \circ \gamma : I \to Y$ is a (constant-speed) geodesic in Y for every geodesic $\gamma : I \to X$ in X.

Remark 12.9. (a) In this definition it suffices to consider (minimizing) geodesic segments $\gamma : [0,1] \to X$, even with images contained in prescribed open sets with the union X.

(b) The composite of two totally geodesic maps $\varphi : X \to Y$ and $\psi : Y \to Z$ is a totally geodesic map $\psi \circ \varphi : X \to Z$.

(c) If X and Y have locally upper bounded curvature, the inverse of a totally geodesic homeomorphism $\varphi : X \to Y$ is a totally geodesic homeomorphism $\varphi^{-1} : Y \to X$ ([ES 1964, §5] for the case of manifolds). To see this, cover X and Y by relatively compact convex domains $U \subset X$ and $V \subset Y$ such that $\varphi(U) \subset V$ and that geodesics in U, resp. V, are uniquely determined by their endpoints. For a geodesic $\eta : [0,1] \to \varphi(U)$ write $\varphi^{-1}(\eta(0)) = \xi_0$, $\varphi^{-1}(\eta(1)) = \xi_1$; and let $\xi : [0,1] \to U$ be the unique geodesic in U such that $\xi(0) = \xi_0$, $\xi(1) = \xi_1$. Then $\varphi \circ \xi : [0,1] \to V$ is the unique geodesic in V with endpoints $\eta(0), \eta(1)$; and so $\varphi \circ \xi = \eta$.

Remark 12.10. In the case of smooth Riemannian manifolds M^m, N^n without boundary it is known that a C^2-map $\varphi : M \to N$ is totally geodesic iff, for $i, j = 1, \cdots, m$ and $k = 1, \cdots, n$,

$$(\varphi^k)_{ij} + (\Gamma^k_{\alpha\beta} \circ \varphi)\, \partial_i \varphi^\alpha \partial_j \varphi^\beta = 0,$$

in terms of local coordinates x^i in M and y^α in N, whereby $\varphi^\alpha = y^\alpha \circ \varphi$, and $(\varphi^k)_{ij}$ denote the covariant second partials of φ^k.

In order to obtain a characterization of totally geodesic maps we first consider maps $\gamma : I \to Y$ of an interval $I \subset \mathbb{R}$ into a geodesic space (Y, d_Y) of *locally upper bounded curvature.*

Lemma 12.4. *A continuous map γ of an interval I into a geodesic space Y of locally upper bounded curvature is a constant-speed geodesic iff γ pulls convex functions $v : V \to \mathbb{R}$ on open sets $V \subset Y$ back to convex functions $v \circ \gamma$ on each component of $\gamma^{-1}(V) \subset I$.*

The proof will show that it suffices to verify this pull-back condition for convex functions $v : V \to \mathbb{R}$ with V convex and v of the form $v = d_Y(\cdot, y)$, $y \in V$.

Proof. If $\gamma : I \to Y$ is a (constant-speed) geodesic then $v \circ \gamma : J \to \mathbb{R}$ is convex for any convex function $v : V \to \mathbb{R}$, J being any component of $\gamma^{-1}(V) \subset I$; this expresses the definition of v being convex.

In the opposite direction we may assume, by continuity of φ, that the entire space Y has the properties of a convex subdomain V stated in the second paragraph of the present section. To prove that $\gamma : I \to \mathbb{R}$ is a geodesic, let $t_1, t_2 \in I$, $t_1 < t_2$, and write $y_i = \gamma(t_i)$, $i = 1, 2$, and further, on Y,

$$v_i = d_Y(\cdot, y_i), \quad v = v_1 + v_2.$$

Then v_i is convex, and so are therefore $v_i \circ \gamma$ and $v \circ \gamma$, by hypothesis. The triangle inequality shows, for any $y \in Y$, that $v(y) \geqslant \delta := d_Y(y_1, y_2)$, and that

$$v(y) = \delta \iff y \in [y_1, y_2],$$

the unique geodesic segment joining y_1 and y_2. In particular, $(v \circ \gamma)(t_i) = \delta$, hence $v \circ \gamma = \delta$ in $[t_1, t_2]$, by convexity of $v \circ \gamma \geqslant \delta$.

This shows that γ maps $[t_1, t_2]$ into (even onto) the geodesic segment $[y_1, y_2]$ in Y. The map $\gamma_{|[t_1,t_2]}$ is affine-linear (in terms of Riemannian path length on $[y_1, y_2]$) because the convex functions $v_i \circ \gamma = d_Y(\gamma(\cdot), \gamma(t_i))$ on $[t_1, t_2]$ are affine linear, their sum $v \circ \gamma$ being constant $= \delta$ on $[t_1, t_2]$. Thus γ is indeed a (constant-speed) geodesic. □

Remark 12.11. The curvature condition in the above lemma is clearly not needed for the "only if part", but cannot be dropped in the "if part". As an example, let Y be a standard *cone of revolution* such that Y, when cut along a generator, unfolds as the upper halfplane $\{(y_1, y_2\} \in \mathbb{R}^2 : y_2 > 0\}$, say. To the line $t \mapsto (t, \varepsilon)$ $(\varepsilon > 0)$ in $\mathbb{R}^2$ there corresponds on the cone Y a geodesic γ_ε: $\mathbb{R} \to Y$, converging as $\varepsilon \to 0$ to the "limit geodesic" $\gamma : \mathbb{R} \to Y$ formed by the generator g (directed towards the vertex of Y) followed by $-g$; and this path γ is clearly not a geodesic. But γ does pull germs of convex functions in Y back to germs of convex functions in $\mathbb{R}$. This shows that Y

does not have locally upper bounded curvature (this is also clear from the existence of two geodesic segments joining points of any standard cone of revolution Y on opposite generators). Every standard cone of revolution has nonnegative curvature (as mentioned in Example 4.3).

Theorem 12.2. *Let (X, d_X) and (Y, d_Y) be geodesic spaces, and suppose that Y has locally upper bounded curvature. A continuous map $\varphi : X \to Y$ is then totally geodesic if and only if φ pulls convex functions on open sets $V \subset Y$ back to convex functions on $\varphi^{-1}(V)$.*

This theorem is proved by a straightforward application of Lemma 12.4. Again it suffices to take $v = d_Y(\cdot, y)$ $(y \in V)$ on convex domains V, as in the beginning of the present section.

Corollary 12.3. *Let (X, g) be a complete admissible Riemannian polyhedron without boundary and with continuous and simplexwise smooth Riemannian metric g. Let (N, h) be a complete Riemannian manifold without boundary. Then every totally geodesic map $\varphi \in W^{1,2}_{\mathrm{loc}}(X, N)$ is harmonic.*

Indeed, X and N are geodesic spaces, and N has locally upper bounded curvature. By Theorem 12.2 and Proposition 7.9, φ has the pull-back property (c) of Theorem 12.1, and is therefore harmonic.

Geodesics as harmonic maps

In this section (Y, d_Y) is a locally compact geodesic space, and $\gamma : I \to Y$ a path, locally of finite energy in the sense of Korevaar-Schoen [KS 1993], or equivalently Definition 9.1. By Remark 9.2, γ has a continuous version; we show that this version is Hölder continuous with exponent $\frac{1}{2}$. We characterize geodesics without endpoints as harmonic maps (Definition 12.1), and also as continuous locally E-minimizing maps (Definition 10.1) of an open interval (or a standard circle) into Y.

First consider any continuous path $\gamma : I \to Y$, where $I = [a, b]$ is compact. For $0 < 2\varepsilon < b - a$ the restriction of γ to the Riemannian domain $I^\circ =]a, b[$ has the approximate energy density

$$e_\varepsilon(\gamma)(t) = \frac{d_Y^2(\gamma(t), \gamma(t+\varepsilon)) + d_Y^2(\gamma(t), \gamma(t-\varepsilon))}{2\varepsilon^2}, \qquad t \in I, \qquad (12.17)$$

understood as 0 for $t \notin [a + \varepsilon, b - \varepsilon]$. (We thus use the spherically averaged densities [KS 1993, p. 573] with $p = 2$. The factor 2 in the denominator serves to achieve full consistency with the case $(Y, d_Y) = \mathbb{R}$.)

Taking instead $p = 1$ leads to the *approximate length density*

$$l_\varepsilon(\gamma)(t) = \frac{d_Y(\gamma(t), \gamma(t+\varepsilon)) + d_Y(\gamma(t), \gamma(t-\varepsilon))}{2\varepsilon}, \qquad t \in I. \tag{12.18}$$

Clearly,

$$l_\varepsilon(\gamma)(t)^2 \leqslant e_\varepsilon(\gamma)(t), \qquad t \in I. \tag{12.19}$$

By definition, γ is of class $W^{1,2}(I, Y)$ iff the energy (i.e., the 2-energy)

$$E(\gamma) := \sup_{f \in C_c(I^\circ, [0,1])} \left(\limsup_{\varepsilon \to 0} \int_I e_\varepsilon(\gamma) f \, dt \right) < \infty. \tag{12.20}$$

Replacing $e_\varepsilon(\gamma)$ by $l_\varepsilon(\gamma)$ here leads to the definition of γ being of class $BV(I, Y)$, [KS 1993, §1.3], and of the 1-*energy*, which we denote by $L(\gamma)$:

$$L(\gamma) := \sup_{f \in C_c(I^\circ, [0,1])} \left(\limsup_{\varepsilon \to 0} \int_I l_\varepsilon(\gamma) f \, dt \right) < \infty. \tag{12.21}$$

Lemma 12.5. *In a metric space* (Y, d_Y), *let* $\gamma : [a, b] \to (Y, d_Y)$ *be a continuous path of finite energy* $E(\gamma)$. *Then* γ *is rectifiable of length* $L(\gamma)$ *from* (12.21), *and*

$$d_Y^2(\gamma(a), \gamma(b)) \leqslant L(\gamma)^2 \leqslant (b-a) E(\gamma). \tag{12.22}$$

Furthermore, γ *is uniformly Hölder continuous with exponent* $\frac{1}{2}$.

If Y *is a geodesic space, the following are equivalent:*

(a) γ *is a constant-speed geodesic,*
(b) γ *is* E*-minimizing for fixed endpoints* $\gamma(a)$, $\gamma(b)$,
(c) γ *is locally* E*-minimizing in* $]a, b[$ (Definition 10.1),
(d) γ *is harmonic in* $]a, b[$ (Definition 12.1),
(e) (12.22) *holds with equality signs.*

Proof. The latter inequality (12.22) follows from (12.19), (12.20), (12.21) by the Cauchy–Schwarz inequality because $\int f \, dt \leqslant b - a$. In particular, $L(\gamma) < \infty$.

Denote temporarily the usual length of γ by $\mathcal{L}(\gamma)$ ($\leqslant \infty$). To prove that $\mathcal{L}(\gamma) \leqslant L(\gamma)$ we first show, given $0 < \delta < \frac{1}{2}(b-a)$, that $\mathcal{L}(\gamma_{|[\alpha,\beta]}) \leqslant L(\gamma) + 2\delta$ for any $\alpha \in [a, a+\delta]$, $\beta \in [b-\delta, b]$. Choose $\varepsilon = (\beta - \alpha)/n$ for some integer $n > 2$ large enough so that (γ being uniformly continuous)

$$d_Y(\gamma(t), \gamma(t')) \leqslant \delta \qquad \text{when } |t - t'| \leqslant \varepsilon, \ t, t' \in I, \tag{12.23}$$

and, for given $q < \mathcal{L}(\gamma_{|[\alpha,\beta]})$, that

$$\sum_{i=1}^{m} d_Y(\gamma(t_{i-1}), \gamma(t_i)) > q \tag{12.24}$$

for any subdivision $\alpha = t_0 < \cdots < t_m = \beta$ of $[\alpha, \beta]$ with each $t_i - t_{i-1} \leqslant \varepsilon$. Writing $s_i = \alpha + i\varepsilon$ $(i = 0, \cdots, n)$, hence $s_n = \beta$, we obtain

$$\begin{aligned} &\int_\alpha^\beta d_Y(\gamma(t), \gamma(t+\varepsilon))\, dt \\ &\quad = \int_0^\varepsilon \sum_{i=0}^{n-1} d_Y(\gamma(s_i + s), \gamma(s_{i+1} + s))\, ds \geqslant (q - 2\delta)\varepsilon, \end{aligned}$$

as shown by considering for each $s \in \,]0, \varepsilon[$ the subdivision

$$\alpha < s_0 + s < s_1 + s < \cdots < s_{n-1} + s < \beta$$

of $[\alpha, \beta]$ into $n + 1$ subintervals of length $\leqslant \varepsilon$, the first and the last of which contribute, by (12.23), with at most δ each to the sum of the chord lengths. Similarly,

$$\int_\alpha^\beta d_Y(\gamma(t), \gamma(t-\varepsilon))\, dt \geqslant (q - 2\delta)\varepsilon,$$

and so from (12.18), for any $f \in C_c(I^\circ, [0,1])$ with $f(t) = 1$ for $t \in [\alpha, \beta]$,

$$\int_I l_\varepsilon(\gamma) f\, dt \geqslant \int_\alpha^\beta l_\varepsilon(\gamma)\, dt \geqslant q - 2\delta.$$

By (12.21) this implies for $n \to \infty$, hence $\varepsilon = (\beta - \alpha)/n \to 0$:

$$L(\gamma) \geqslant \limsup_{\varepsilon \to 0} \int_\alpha^\beta l_\varepsilon(\gamma)\, dt \geqslant q - 2\delta.$$

By varying q we infer that indeed

$$L(\gamma) \geqslant \mathcal{L}(\gamma_{|[\alpha,\beta]}) - 2\delta. \tag{12.25}$$

For any $q^* < \mathcal{L}(\gamma)$, (12.24) holds with q replaced by q^*, now for any sufficiently fine subdivision $a = t_0 < t_1 < \cdots < t_m = b$ of $I = [a, b]$. We may arrange that $t_1 < a + \delta$, $t_{m-1} > b - \delta$; and by (12.23) that $d_Y(a, t_1) < \delta$, $d_Y(t_{m-1}, b) < \delta$. Applying the above to $\alpha = t_1$, $\beta = t_{m-1}$, we obtain

$$\mathcal{L}(\gamma_{|[\alpha,\beta]}) \geqslant \sum_{i=2}^{m-1} d_Y(\gamma(t_{i-1}), \gamma(t_i)) > q^* - 2\delta,$$

and so, by (12.25), $L(\gamma) \geqslant q^* - 4\delta$. Making first $q^* \to \mathcal{L}(\gamma)$ and next $\delta \to 0$, we conclude that indeed $L(\gamma) \geqslant \mathcal{L}(\gamma)$.

In proving the opposite inequality $\mathcal{L}(\gamma) \geqslant L(\gamma)$, we may suppose that $\mathcal{L}(\gamma) < \infty$, i.e., γ is rectifiable. Consider the increasing bounded continuous function

$$\lambda(t) = \mathcal{L}(\gamma_{|[a,t]}), \qquad t \in I = [a, b],$$

with the approximate length density

$$l_\varepsilon(\lambda)(t) = \frac{\lambda(t+\varepsilon) - \lambda(t-\varepsilon)}{2\varepsilon} \geqslant l_\varepsilon(\gamma)(t), \tag{12.26}$$

cf. (12.18). By the consistency result [KS 1993, p. 582], the 1-energy $L(\lambda)$ of the function λ equals the usual 1-energy $\operatorname{Var}\lambda = \lambda(b) - \lambda(a) = \mathcal{L}(\gamma)$. Applying (12.21) to both maps γ and λ, and inserting (12.26), leads to $\mathcal{L}(\gamma) \geqslant L(\gamma)$.

Having thus proved that $L(\gamma)$ from (12.21) is the usual length of γ, the remaining former inequality (12.22) is obvious.

Replacing $[a, b]$ by a subinterval $[t, t']$ we derive from (12.22) the uniform Hölder continuity of γ:

$$d_Y^2(\gamma(t), \gamma(t')) \leqslant (t' - t)E(\gamma).$$

In the rest of the proof, (Y, d_Y) is a *geodesic space.*

(a) $\Longrightarrow$ (e) $\Longrightarrow$ (b). If γ is a geodesic of constant speed c then (12.18) and (12.17) show that $l_\varepsilon(\gamma)(t) = c$ and $e_\varepsilon(\gamma)(t) = c^2$, both for $t \in [a + \varepsilon, b - \varepsilon]$. By (12.21), (12.20) this implies

$$L(\gamma) = c(b - a), \quad E(\gamma) = c^2(b - a), \tag{12.27}$$

because $d_Y(\gamma(a), \gamma(b)) = c(b-a)$; and so equality prevails in both inequalities (12.22). This, in turn, implies that γ is E-minimizing for fixed endpoints.

(b) $\Longrightarrow$ (a). If γ is E-minimizing for fixed $\gamma(a), \gamma(b)$, then again (12.22) must hold with equality signs. It follows that γ is a minimizing geodesic segment, and (12.27) holds for a certain c (the average speed of γ). It remains to prove that γ has constant speed ($= c$). With $m = \frac{1}{2}(a+b)$, the paths

$$\gamma_1 = \gamma_{|[a,m]}, \quad \gamma_2 = \gamma_{|[m,b]}$$

in Y are likewise E-minimizing for fixed endpoints, and as in (12.27)

$$L(\gamma_i) = \tfrac{1}{2}c_i(b-a), \quad E(\gamma_i) = \tfrac{1}{2}c_i^2(b-a), \qquad i = 1, 2.$$

Inserting $L(\gamma) = L(\gamma_1) + L(\gamma_2)$, $E(\gamma) = E(\gamma_1) + E(\gamma_2)$, gives $c = \frac{1}{2}(c_1 + c_2)$, $c^2 = \frac{1}{2}(c_1^2 + c_2^2)$, and so $c_1 = c_2$. By repeated bisection it follows that $L(\gamma_{|[a,a+t]}) = ct$ for every t such that $t/(b-a)$ is dyadic in $[0,1]$, and by continuity for every $t \in [0, b-a]$; thus γ indeed has constant speed c.

(b) $\Longrightarrow$ (c) $\Longrightarrow$ (d). Obvious.

(d) $\Longrightarrow$ (c). Because γ is bi-locally E-minimizing in $]a,b[$ there are for every $t_0 \in\ ]a,b[$ an open interval $U =]t_0 - \delta, t_0 + \delta[\ \subset\ ]a,b[$ and an open ball $V = B_Y(\gamma(t_0), \varrho)^\circ$ serving as in Definition 12.1; in particular, $\gamma(U) \subset V$. By diminishing δ we arrange that $\overline{U} \subset\ ]a,b[$ and $\gamma(\overline{U}) \subset V' := B_Y(\gamma(t_0), \varrho/2)^\circ$, γ being continuous at t_0.

Consider any path γ' of finite energy which agrees with γ off some interval $U' :=]t_0 - \delta', t_0 + \delta'[$, where $0 < \delta' < \delta$. We may take γ' continuous. If $\gamma'(U') \subset V$ then $E(\gamma') \geqslant E(\gamma)$ according to Definition 12.1. Suppose therefore that $\gamma'(t') \notin V$ for some $t' \in U'$; then $d_Y(\gamma'(t_0 \pm \delta'), \gamma'(t')) \geqslant \varrho/2$ and e.g., $t' \geqslant t_0$. Applying (12.22) to the interval $[t_0 - \delta', t']$ of length $\geqslant \delta'$ gives

$$E(\gamma') \geqslant E(\gamma'_{|[t_0-\delta',t']}) \geqslant (\varrho/2)^2/\delta' > E(\gamma)$$

for sufficiently small δ'.

(c) $\Longrightarrow$ (a). From (c) it follows that every point $t_0 \in\ ]a,b[$ has a neighbourhood $U =]t_0 - \delta, t_0 + \delta[\ \Subset\ ]a,b[$ such that $\gamma_{|\overline{U}}$ is E-minimizing for fixed endpoints, and $\gamma_{\overline{U}}$ is therefore a constant-speed geodesic segment. When t_0 varies, this shows that γ is itself a geodesic, and has constant speed (because the speed must be the same in overlapping time intervals). □

Corollary 12.4. *In a geodesic space (Y, d_Y), a continuous path $\gamma : I \to Y$ from an open interval or a standard circle I is a constant-speed geodesic iff γ is harmonic, or equivalently iff γ is locally E-minimizing.*

Note that the "only if part" breaks down if I has a boundary, e.g., $I = [a,b]$, the only harmonic (or continuous locally E-minimizing) maps $I \to Y$ being then the constant maps; this should be compared with Lemma 12.5

(b), in which 'harmonic' is replaced by 'E-minimizing for fixed endpoints', which is weaker in the presence of a boundary.

Jensen's inequality for maps

This inequality, included here primarily on account of its use in Proposition 12.4, is stated without proof in [J 1994, Lemma 3.1]. Local compactness of geodesic spaces is not needed in this section.

We first characterize the (constant-speed) geodesics on the product of two geodesic spaces, invoking Lemma 12.5:

Lemma 12.6. *Let (X, d_X) and (Y, d_Y) be geodesic spaces, and define a metric d_Z on $Z = X \times Y$ by*

$$d_Z^2((x,y),(x',y')) = d_X^2(x,x') + d_Y^2(y,y').$$

Then

(a) *(Z, d_Z) is a geodesic space.*
(b) *A path $I \ni t \mapsto z(t) = (x(t), y(t)) \in Z$ is a constant-speed geodesic iff $t \mapsto x(t)$ and $t \mapsto y(t)$ are such.*
(c) *If X and Y have globally nonpositive curvature then so has Z.*

Proof. With $I = [0,1]$ let $I \ni t \mapsto x(t) \in X$ and $I \ni t \mapsto y(t) \in Y$ be continuous paths of finite energy. Then the path $I \ni t \mapsto z(t) := (x(t), y(t)) \in Z$ has

$$e_\varepsilon(z(\cdot)) = e_\varepsilon(x(\cdot)) + e_\varepsilon(y(\cdot)), \tag{12.28}$$

cf. (12.17), and hence $E(z(\cdot)) \leqslant E(x(\cdot)) + E(y(\cdot)) < \infty$, cf. (12.20). By [KS 1993, Theorem 1.5.1] applied to the Riemannian domain $]0,1[$ it therefore follows from (12.28) that actually

$$E(z(\cdot)) = E(x(\cdot)) + E(y(\cdot)). \tag{12.29}$$

Ad (a). Given two points $z_i = (x_i, y_i) \in Z$, $i = 0, 1$, let $I \ni t \mapsto x(t) \in X$ be the (constant-speed) geodesic segment in X with endpoints $x(0) = x_0$, $x(1) = x_1$, and similarly for $I \ni t \mapsto y(t) \in Y$. By Lemma 12.5 the continuous path $I \ni t \mapsto z(t) = (x(t), y(t)) \in Z$ of finite energy is rectifiable, and its length $L(z(\cdot))$ satisfies, by (12.29),

$$\begin{aligned} L(z(\cdot))^2 \leqslant E(z(\cdot)) &= E(x(\cdot)) + E(y(\cdot)) = L(x(\cdot))^2 + L(y(\cdot))^2 \\ &= d_X^2(x_0, x_1) + d_Y^2(y_0, y_1) = d_Z^2(z_0, z_1), \end{aligned}$$

so that (Z, d_Z) is indeed a geodesic space. And $z(\cdot)$ is a geodesic in X, by Lemma 12.5, (e) $\Longrightarrow$ (a).

Ad (b). The "if part" was shown in the proof of (a). Conversely, if $z(\cdot) = (x(\cdot), y(\cdot))$ is a geodesic in Z then the continuous paths $x(\cdot)$ and $y(\cdot)$ in X and Y, respectively, clearly have finite energy, and by (12.22), (12.29),

$$d_X^2(x_0, x_1) + d_Y^2(y_0, y_1) \leqslant E(x(\cdot)) + E(y(\cdot)) = E(z(\cdot)) = d_Z^2(z_0, z_1),$$

the last equation by (a) $\Longrightarrow$ (e) in Lemma 12.5. This shows that the equality sign prevails in the four inequalities $d_X^2(x_0, x_1) \leqslant L(x(\cdot))^2 \leqslant E(x(\cdot))$, $d_Y^2(y_0, y_1) \leqslant L(y(\cdot))^2 \leqslant E(y(\cdot))$, and so $x(\cdot)$ and $y(\cdot)$ are geodesics, again by Lemma 12.5.

Ad (c). Given triples $z_i = (x_i, y_i) \in Z$, $i = 1, 2, 3$, and comparison triples $\tilde{x}_i$ and $\tilde{y}_i$ in $\mathbb{R}^2$, the triple $\tilde{z}_i = (\tilde{x}_i, \tilde{y}_i)$ in $\mathbb{R}^4$ lies in a 2-dimensional Euclidean (affine) subspace V of $\mathbb{R}^4$ and is clearly a comparison triple for (z_i) satisfying the triangle comparison inequality; thus Z has globally nonpositive curvature. □

Remark 12.12. There is a well known, quite elementary analogue of (a) and (b) in Lemma 12.6, in which the exponent 2 is replaced by 1 in the definition of the distance on Z, and energy is replaced by length in (12.29). However, Lemma 12.6 as above, depending on Lemma 12.5, is indispensable for the proof of Jensen's inequality below.

Proposition 12.3. (Jensen's inequality.) *Let (Y, d_Y) be a simply connected, complete geodesic space of nonpositive curvature, v a lower semicontinuous convex function on Y, and μ a probability measure on Y. Then*

$$v(y_\mu) \leqslant \int v\, d\mu$$

when y_μ denotes the centre of mass of μ, i.e., the unique point of Y minimizing

$$I(y) := \int_Y d_Y^2(y, \eta)\, d\mu(\eta), \qquad y \in Y.$$

Proof. The centre of mass y_μ is the same as the meanvalue of the identity map id : $Y \to Y$, and so y_μ lies in the closed convex hull of $\operatorname{supp} \mu$, cf. Lemma 10.4.

Recall that Y has globally nonpositive curvature (cf. Chapter 2). By Lemma 12.6 the product space $Z = \mathbb{R} \times Y$ is therefore a complete geodesic space of globally nonpositive curvature (hence simply connected), with metric d_Z defined by

$$d_Z^2\left((t,t'),(y,y')\right) = (t-t')^2 + d_Y^2(y,y').$$

The convexity and lower semicontinuity of $v : Y \to \mathbb{R}$ amounts to its epigraph

$$G = \{(t,y) : y \in Y, t \geqslant v(y)\}$$

being a convex closed subset of Z. The image ν of μ under $y \mapsto (v(y), y)$ is a probability measure on G. The centre of mass of ν is $(\int v\,d\mu, y_\mu)$, as shown presently, and it belongs to G, so that indeed $v(y_\mu) \leqslant \int v\,d\mu$. For any $(t,y) \in \mathbb{R} \times Y$ we have in fact

$$\begin{aligned}
&\int_Z d_Z^2\left((t,y),(\tau,\eta)\right) d\nu(\tau,\eta)\\
&\quad = \int_Y d_Z^2\left((t,y),(v(\eta),\eta)\right) d\mu(\eta)\\
&\quad = \int_Y (t - v(\eta))^2 d\mu(\eta) + \int_Y d_Y^2(y,\eta) d\mu(\eta)\\
&\quad = \left(t - \int_Y v\,d\mu\right)^2 + \int_Y v^2 d\mu - \left(\int_Y v\,d\mu\right)^2 + \int_Y d_Y^2(y,\eta) d\mu(\eta),
\end{aligned}$$

which is smallest for $t = \int v\,d\mu$ and $y = y_\mu$. □

Harmonic maps from a 1-dimensional Riemannian polyhedron

We bring a version of Ishihara's characterization of harmonic maps, cf. Theorem 12.1, (a) $\Longleftrightarrow$ (c), for the case of a 1-dimensional Riemannian polyhedral domain (X,g). Every map $\varphi \in W^{1,2}_{\mathrm{loc}}(X,Y)$ of X into a geodesic space (Y,d_Y) has then a continuous version (Remark 9.2), and even a locally uniformly Hölder continuous version with exponent $\frac{1}{2}$, an easy consequence of Lemma 12.5 applied to the 1-simplexes of X containing a given vertex of X.

Proposition 12.4. *Let (X, g) be a 1-dimensional Riemannian polyhedron and (Y, d_Y) a locally compact geodesic space of nonpositive curvature. For a continuous map $\varphi \in W^{1,2}_{\text{loc}}(X, Y)$ the following are equivalent:*

(a) *φ is harmonic* (Definition 12.1),
(b) *φ is locally E-minimizing* (Definition 10.1),
(c) *φ pulls convex functions v on open sets $V \subset Y$ back to subharmonic functions $v \circ \varphi$ on $\varphi^{-1}(V)$.*

The proof will show that it suffices to consider convex functions v on Y of the form $v(y) = d_Y(y, q)$, $q \in Y$.

Proof. Because φ is continuous we may assume that X and Y are compact and that Y has globally nonpositive curvature, so that Y is simply connected. (Cf. the text at the beginning of Totally geodesic maps; and Chapter 2, Geodesic spaces.)

(a) $\Longleftrightarrow$ (b) follows therefore from Lemma 12.1.

(b) $\Longrightarrow$ (c). Suppose that φ is locally E-minimizing, and let v be a convex function on an open set $V \subset Y$. By Lemma 12.5, (c) $\Longrightarrow$ (a), and by Lemma 12.4, $v \circ \varphi$ is convex and hence subharmonic in the part in $\varphi^{-1}(V)$ of each open 1-simplex of X. It remains to show that $v \circ \varphi$ is subharmonic at each vertex $x_0 \in \varphi^{-1}(V)$ of X in a given triangulation; i.e., in terms of the 1-simplexes $s_1, \cdots, s_k$ of X having x_0 as a vertex,

$$\sum_{i=1}^{k} D_i(v \circ \varphi)(x_0) \geqslant 0, \tag{12.30}$$

where $D_i(v \circ \varphi)(x_0)$ denotes the derivative along s_i at x_0 with respect to path length from x_0, see Example 8.1. We may assume that the open ball $B_X(x_0, r)^\circ \subset \varphi^{-1}(V) \cap \mathrm{st}_X(x_0)$ can serve as U in Definition 10.1. Let $x_i \in s_i$ have $d_X(x_0, x_i) = r$, $i = 1, \cdots, k$. Writing $y_i = \varphi(x_i)$ for $i = 0, 1, \cdots, k$, we have from Lemma 12.5, (d) $\Longrightarrow$ (e), applied to each s_i°,

$$E(\varphi_{|U}) = \frac{1}{r} \sum_{i=1}^{k} d_Y^2(y_0, y_i), \tag{12.31}$$

and we conclude by varying y_0 in Y that y_0 must be the centre of mass of $y_1, \cdots, y_k$ (i.e., of the measure $\mu = k^{-1}(\varepsilon_{y_1} + \cdots + \varepsilon_{y_k})$).

The constant speed of the geodesic $\varphi_{|[x_0, x_i]}$ is $d_Y(y_0, y_i)/r$, and so

$$D_i(v \circ \varphi)(x_0) = \alpha_i \frac{d_Y(y_0, y_i)}{r} = \lim_{r \to 0} \frac{v(y_i) - v(y_0)}{r}, \tag{12.32}$$

where α_i denotes the derivative of the convex function $v_{|[y_0,y_i]}$ at y_0 with respect to path length $d_Y(y_0, y)$ on $[y_0, y_i]$. Adding over $i = 1, \cdots, k$ leads to (12.30) in view of Jensen's inequality (Proposition 12.3), according to which $v(y_0) \leqslant \frac{1}{k}\sum_1^k v(y_i)$.

(c) $\Longrightarrow$ (a). Suppose now that φ has the stated pull-back property with $v = d_Y(\cdot, q)$, and hence also with $v = d_Y^2(\cdot, q)$ on Y, $q \in Y$. Since subharmonic functions on an interval are convex it follows from Lemma 12.4 that the restriction of φ to any open 1-simplex of X is a (constant-speed) geodesic, i.e., locally E-minimizing by Lemma 12.5, (a) $\Longrightarrow$ (c). By Theorem 9.1, φ has finite energy, and it remains to prove that φ is E-minimizing for fixed restriction to $X \setminus U$ with $U = B_X(x_0, r)^\circ \subset \mathrm{st}_X(x_0)$.

Consider any map $\psi \in W^{1,2}_{\mathrm{loc}}(X, Y)$ with $\psi_{|\complement U} = \varphi_{|\complement U}$. By Remark 9.2 we may take ψ continuous. Write $q = \psi(x_0)$, $y_0 = \varphi(x_0)$, and $y_i = \varphi(x_i) = \psi(x_i)$ for $i = 1, \cdots, k$. From Lemma 12.5 we have

$$E(\psi_{|U}) \geqslant \frac{1}{r}\sum_{i=1}^{k} d_Y^2(q, y_i) \tag{12.33}$$

since $E(\psi_{|U})$ does not increase if we replace $\psi_{|[x_0,x_i]}$ by the unique constant-speed geodesic which maps $[x_0, x_i]$ onto $[q, y_i]$, in view of Lemma 12.5. By (12.31) and (12.33) it remains to prove that $\sum d_Y^2(q, y_i) \geqslant \sum d_Y^2(y_0, y_i)$.

For any $i = 1, \cdots, k$ let θ_i denote the angle at $\tilde{y}_0 = 0$ in the Euclidean comparison triangle $\tilde{y}_0\tilde{y}_i\tilde{q}$ for $y_0 y_i q$. Then $d_Y(y_0, q) = |\tilde{q}|$, $d_Y(y_0, y_i) = |\tilde{y}_i|$; and

$$\begin{aligned}\sum d_Y^2(q, y_i) &= \sum |\tilde{q} - \tilde{y}_i|^2 = \sum \left(|\tilde{q}|^2 + |\tilde{y}_i|^2 - 2\,|\tilde{q}|\,|\tilde{y}_i| \cos\theta_i\right) \\ &= k d_Y^2(y_0, q) + \sum d_Y^2(y_0, y_i) - 2 d_Y(y_0, q) \sum d_Y(y_0, y_i)\cos\theta_i.\end{aligned}$$

It suffices therefore to establish the inequality

$$\sum_{i=1}^{k} d_Y(y_0, y_i)\cos\theta_i \leqslant 0. \tag{12.34}$$

Take for v the convex function $v(y) = d_Y^2(q, y)$, $y \in Y$, and denote by $y_i(t)$, resp. $\tilde{y}_i(t)$, the point of $[y_0, y_i]$, resp. $[\tilde{y}_0, \tilde{y}_i]$, at distance t from y_0, resp. from $\tilde{y}_0 = 0$, $i = 1, \cdots, k$. Then $|\tilde{y}_i(t)| = t$, and the derivative α_i from (12.32) equals

$$\begin{aligned}\alpha_i &= \lim_{t\to 0} \frac{d_Y^2(q, y_i(t)) - d_Y^2(q, y_0)}{t} \\ &\leqslant \liminf_{t\to 0} \frac{|\tilde{q} - \tilde{y}_i(t)|^2 - |\tilde{q}|^2}{t} \leqslant -2|\tilde{q}|\cos\theta_i.\end{aligned}$$

From (12.30) (now valid because $v \circ \varphi$ is subharmonic at x_0) together with the former relation (12.32) (multiplied by r) we therefore obtain (12.34) as follows, again because $|\tilde{q}| = d_Y(y_0, q)$ and $|\tilde{y}_i| = d_Y(y_0, y_i)$,

$$0 \leqslant \sum_{i=1}^{k} \alpha_i d_Y(y_0, y_i) \leqslant -2d_Y(y_0, q) \sum_{i=1}^{k} d_Y(y_0, y_i) \cos \theta_i.$$

This completes the proof of Proposition 12.4. □

Queries. Can the requirement of nonpositive curvature for Y in the above proposition be weakened to locally upper bounded curvature?

Remark 12.13. It can be shown that Ishihara's characterizarion (a)⇔(c) in Theorem 12.1 and Proposition 12.4 also holds with X as in Theorem 12.1 and Y a 1-dimensional Riemannian polyhedron (hence $\operatorname{curv} Y \leqslant 0$). The case $\dim Y > 1$ remains open.

13

Harmonic morphisms

In this chapter we study harmonic morphisms and their relation to harmonic maps. For a survey on harmonic morphisms, see [EL 1978, §4], [EL 1988, §§2, 11], and [Fug 2000].

Harmonic morphisms are the morphisms for the category of Brelot harmonic spaces, and we start out in that generality.

Harmonic morphisms between harmonic spaces

These were introduced by Constantinescu and Cornea [CC 1965, §3] (under the name of harmonic maps) with the purpose of generalizing results about conformal maps between Riemann surfaces. In the present section X and Y are harmonic spaces (in the sense of Brelot), see Chapter 2.

Definition 13.1. A continuous map $\varphi : X \to Y$ is a *harmonic morphism* if, for every open set V in Y and for every harmonic function v on V, $v \circ \varphi$ is harmonic on $\varphi^{-1}(V)$.

Clearly, the composite of two harmonic morphisms is a harmonic morphism, and the identity map id : $X \to X$ is a harmonic morphism.

In the following theorem, properties (i) and (ii) were obtained in [CC 1965], and (iii) will be established below. As to (iv), obtained in [Fug 1979a;b], a point y of a Brelot space Y is called *strongly polar* if $s(y) = \infty$ holds for every superharmonic function s in a neighbourhood V of y such that s is harmonic in $V \setminus \{y\}$ but not at y. If Y has locally a Green kernel G (see Chapter 2, Harmonic spaces) then it is easily seen that y is strongly polar iff $G(y, y) = \infty$, [Fug 1979b].

Theorem 13.1. *Let $\varphi : X \to Y$ be a nonconstant harmonic morphism. Then:*

(i) *φ likewise pulls germs of superharmonic functions on Y back to germs of superharmonic functions on X (and this clearly characterizes the harmonic morphisms among all continuous maps).*

(ii) *The pre-image of any polar subset of Y is polar in X.*

(iii) *If φ is surjective and proper then a function $v : V \to [-\infty, \infty]$ (V open in Y) is superharmonic [resp. harmonic] if (and only if) $v \circ \varphi$ is superharmonic [resp. harmonic] on $\varphi^{-1}(V)$.*

(iv) *If every point of Y is strongly polar (defined above), then φ is an open map.*

Proof of (iii). We may assume that V is connected. Suppose that $v \circ \varphi$ is *superharmonic* on $\varphi^{-1}(V)$ (see Definition 2.2). Clearly, $v > -\infty$ and $v \not\equiv \infty$. To see that v is lower semicontinuous, let $y_0 \in V$ and $\alpha < v(y_0)$. According to [CC 1965, Lemma 3.1] the set $\{v > \alpha\} = \varphi(\{v \circ \varphi > \alpha\})$ is indeed a neighbourhood of y_0 because $\{v \circ \varphi > \alpha\}$ contains the compact fibre $\varphi^{-1}(y_0)$, hence also its components which are likewise compact.

Consider now a regular set W in Y such that $y_0 \in W$, $\overline{W} \subset V$. (If X is compact we further choose W small enough so that $\varphi^{-1}(W) \neq X$; this is possible because $\varphi^{-1}(y_0) \neq X$, φ being nonconstant.) For any $f \in C(\partial W)$ with $f \leqslant v_{|\partial W}$ denote by h the continuous extension of H_f^W (Definition 2.1) to $\overline{W}$ by $h = f$ on ∂W. Then $h \circ \varphi$ is harmonic on $\varphi^{-1}(W)$ (Definition 13.1) and continuous on $\overline{\varphi^{-1}(W)}$ ($\subset \varphi^{-1}(\overline{W})$); and $\overline{\varphi^{-1}(W)}$ is compact, φ being proper. By hypothesis, $v \circ \varphi$, hence also $v \circ \varphi - h \circ \varphi$, is superharmonic on $\varphi^{-1}(W)$ and lower semicontinuous on $\overline{\varphi^{-1}(W)}$ ($\subset \varphi^{-1}(V)$). Moreover,

$$v \circ \varphi - h \circ \varphi \geqslant 0 \qquad \text{on } \partial\varphi^{-1}(W) \ (\subset \varphi^{-1}(\partial W))$$

because $h = f \leqslant v$ on ∂W. Since $\varphi^{-1}(W) \neq X$ (connected), $\partial\varphi^{-1}(W) \neq \emptyset$, and we conclude from the boundary minimum principle (cf. Chapter 2) that $v \circ \varphi \geqslant h \circ \varphi$ in $\varphi^{-1}(W)$, and hence

$$v \geqslant h = H_f^W \qquad \text{in } W = \varphi(\varphi^{-1}(W)).$$

Because $h(y) = \int f \, d\omega_y^W$ for $y \in W$ (Definition 2.1), it follows by taking supremum over all continuous $f \leqslant v$ on ∂W that

$$v(y) \geqslant \int v \, d\omega_y^W \qquad \text{for every } y \in W,$$

and so v is indeed superharmonic on V. If $v \circ \varphi$ is *harmonic* on $\varphi^{-1}(V)$ then so is $(-v) \circ \varphi$, hence $\pm v$ is superharmonic, i.e., v is harmonic on V. □

In continuation of Theorem 12.1 we have the following:

Corollary 13.1. *([CC 1965].)*

(i) *The inverse of a bijective harmonic morphism between Brelot spaces is a harmonic morphism.*

(ii) *Every Brelot harmonic sheaf on a space is maximal under inclusion.*

(iii) *Every open harmonic morphism of a compact Brelot space X into a Brelot space Y is surjective, and hence Y must be compact as well.*

Here (i) follows from [CC 1965, Lemma 3.1], used in the above proof. As to (ii), if $\mathcal{H}$ and $\mathcal{H}'$ are Brelot harmonic sheaves on the same space X, and if $\mathcal{H} \subset \mathcal{H}'$ then $\mathcal{H} = \mathcal{H}'$ because $\mathrm{id} : (X, \mathcal{H}') \to (X, \mathcal{H})$ is a bijective harmonic morphism, and so is its inverse $\mathrm{id} : (X, \mathcal{H}) \to (X, \mathcal{H}')$, by (i). For (iii), $\varphi(X)$ is open and closed (compact) in Y (connected).

Harmonic morphisms between Riemannian polyhedra

Every admissible Riemannian polyhedron is a Brelot space and has, at least locally, a symmetric Green kernel (see Chapter 7). If $n := \dim Y > 1$ then $G(y, y) \equiv \infty$, by Theorem 7.4, and hence every point of Y is strongly polar, by [Fug 1979b] (cf. above), while every point of Y is nonpolar if $n = 1$. From Theorem 13.1 (iv) we therefore have case (i) of the following result, in which the hypothesis on the target in case (ii) serves to render Proposition 5.4 applicable:

Proposition 13.1. *Let φ be a nonconstant harmonic morphism between admissible Riemannian polyhedra (X, g) and (Y, h). Then φ is an open map in each of these two cases:*

(i) $n > 1$, *or*

(ii) *X is a (normal) circuit, g is Lipschitz continuous, and $bY = \emptyset$.*

Proof. In the remaining case (ii) with $n = 1$ we first show that, for any given $x_0 \in X$, either

(a) φ is constant in some neighbourhood of x_0, or else

(b) φ maps every neighbourhood of x_0 onto a full neighbourhood of $\varphi(x_0)$.

The proof of this dichotomy applies to any two admissible Riemannian polyhedra (X, g) and (Y, h) with $\dim Y = 1$. We may assume that the point $y_0 = \varphi(x_0)$ is a vertex of Y (otherwise add y_0 as a new vertex to the 1-dimensional polyhedron Y). Let V denote the (open) star of y_0 in Y.

Suppose that (b) fails to hold. Then there is a neighbourhood U of x_0 such that $\varphi(U)$ is *not* a neighbourhood of y_0. We may assume that $U \subset \varphi^{-1}(V)$ (a neighbourhood of x_0) and that U is connected and open. Then $\varphi(U)$ is connected, and there exists a 1-simplex t_1 of Y containing y_0 such that

$$t_1 \cap \varphi(U) = \{y_0\}. \tag{13.1}$$

If t_1 is the only 1-simplex of Y containing y_0 then $\varphi(U) \subset V \subset t_1$, and hence indeed $\varphi(U) = \{y_0\}$, by (13.1). Otherwise, let $t_2, \cdots, t_k$ $(k \geqslant 2)$ be the remaining 1-simplexes of Y containing y_0. Define a harmonic function v on V (cf. Example 8.1) by

$$v(y) = \begin{cases} (k-1)d_Y(y, y_0) & \text{for } x \in t_1 \cap V \\ -d_Y(y, y_0) & \text{for } x \in t_j \cap V,\ 2 \leqslant j \leqslant k. \end{cases} \tag{13.2}$$

Then $v \circ \varphi$ is harmonic in U, and $v \circ \varphi \leqslant 0$ by (13.1). Because $(v \circ \varphi)(x_0) = v(y_0) = 0$, it follows from the local maximum (or minimum) principle for harmonic functions (cf. Chapter 2) that $v \circ \varphi \equiv 0$ on U, and so again $\varphi(U) = \{y_0\}$ since $v(y) \neq 0$ for every $y \in V \setminus \{y_0\}$. This proves that indeed either (a) or (b) above holds (for any given point $x_0 \in X$).

Let A denote the open set of points of X with the property (a), i.e., having a neighbourhood in which φ is constant. Clearly $A \neq X$ because the nonconstant map φ is not even locally constant, X being connected. It remains to prove that $X \setminus A$ is open, like A, for then $A = \emptyset$, i.e., φ is an open map.

For given $x_0 \in X \setminus A$, let $t_1, \cdots, t_k$ be the 1-simplexes of Y containing $\varphi(x_0) = y_0$. Here $k \geqslant 2$ because $b(Y) = \emptyset$. We therefore have on $V = \mathrm{st}_Y(y_0)$ the harmonic function v from (13.2). As before, let U be a connected open neighbourhood of x_0 such that $U \subset \varphi^{-1}(V)$. We shall prove that $U \subset X \setminus A$.

If φ were constant in some neighbourhood U^* of a point $x^* \in U$ then $v \circ \varphi$ would be constant in U^* and hence in U, by Proposition 5.4. It would therefore follow that $v \circ \varphi = v(y_0) = 0$ in U because $v^{-1}(0) = \{y_0\}$. But this would contradict $x_0 \in X \setminus A$. We conclude that every point $x^* \in U$ is indeed in $X \setminus A$. □

Remark 13.1. In case (ii) with $n = 1$, Y cannot be allowed to have a boundary. (For example: $\varphi : \mathbb{R} \to [0, \infty]$ given by $\varphi(x) = \max\{x, 0\}$.) See Remark 5.6 as to the need for X to be circuit, and Remark 5.7 for g to be Lipschitz continuous if $m > 2$, even when $Y = \mathbb{R}$.

Harmonic morphisms into Riemannian manifolds

In this section we shall consider *maps from an admissible Riemannian polyhedron (X, g) to a smooth Riemannian n-manifold (N, h) without boundary.* We study the relation between harmonic morphisms and harmonic maps of X into N and extend known results from [Fug 1978] and [Ish 1979] for maps between smooth Riemannian manifolds.

A continuous map $\varphi : X \to N$ is of class $W^{1,2}_{\text{loc}}(X, N)$ iff the components $\varphi^\alpha = y^\alpha \circ \varphi$ of φ in terms of local coordinates (y^α) in N are of class $W^{1,2}_{\text{loc}}(X)$ (Definition 9.2). For every $v \in C^1(N)$ we then have $v \circ \varphi \in W^{1,2}_{\text{loc}}(X)$, by Remark 9.7.

Definition 13.2. A continuous map $\varphi : X \to N$ of class $W^{1,2}_{\text{loc}}(X, N)$ is called *horizontally weakly conformal* if there exists a scalar λ, defined a.e. in X, such that

$$\langle \nabla(v \circ \varphi), \nabla(w \circ \varphi)\rangle = \lambda[\langle \nabla_N v, \nabla_N w\rangle \circ \varphi] \qquad \text{a.e. in } X \tag{13.3}$$

for every pair of functions $v, w \in C^1(N)$.

Henceforth in this chapter the gradient operator ∇ without a subscript always refers to the domain space (X, g).

By polarization, it suffices in Definition 13.2 to consider pairs v, w with $v = w$. The property of horizontal weak conformality is a local one, and (13.3) reads, in terms of local coordinates (y^α) in N,

$$\langle \nabla\varphi^\alpha, \nabla\varphi^\beta\rangle = \lambda(h^{\alpha\beta} \circ \varphi) \qquad \text{a.e. in } X \tag{13.4}$$

for $\alpha, \beta = 1, \cdots, n$. Taking $\alpha = \beta$ this shows that λ is uniquely determined and $\geqslant 0$ a.e., and that $\lambda \in L^1_{\text{loc}}(X)$ because $\nabla\varphi^k \in L^2_{\text{loc}}(X)$. λ is called the *dilation* (or dilatation) of φ.

The term 'horizontally weakly conformal' comes from the case where also X is a smooth Riemannian manifold without boundary and $\varphi \in C^1(X, N)$, for then Definition 13.2 is equivalent to $d\varphi(x) : T_x(X) \to T_{\varphi(x)}(N)$ having, at each point $x \in X$ where $d\varphi(x) \neq 0$, a conformal restriction to the horizontal space H_x at x, H_x being the orthogonal complement of the kernel of $d\varphi(x)$ within $T_x(X)$; H_x is spanned by $\nabla\varphi^1(x), \cdots, \nabla\varphi^n(x)$, cf. [Fug 1978, p. 120].

A continuous map $\varphi \in W^{1,2}_{\text{loc}}(X, N)$ is *harmonic* (Definition 12.1) iff φ is *weakly harmonic* (Definition 12.2), i.e., in charts on N,

$$\int_X \langle \nabla\psi, \nabla\varphi^k\rangle d\mu_g = \int_X \psi(\Gamma^k_{\alpha\beta} \circ \varphi)\langle \nabla\varphi^\alpha, \nabla\varphi^\beta\rangle d\mu_g, \tag{13.5}$$

$k = 1, \cdots, n$, for every $\psi \in \mathrm{Lip}_c(X)$, and the same then also holds for any $\psi \in W_0^{1,2}(X) \cap L^\infty(X)$. (See Theorem 12.1.)

According to [GW 1975], N admits local coordinates (y^k) which are harmonic functions. For local questions we may therefore assume that N admits a diffeomorphism $y = (y^1, \cdots, y^n)$ onto an open set in $\mathbb{R}^n$ such that each y^k is harmonic.

Lemma 13.1. *Let $\varphi : X \to N$ be horizontally weakly conformal with dilation λ. Then φ is a harmonic map iff the components $\varphi^k = y^k \circ \varphi$ in terms of harmonic local coordinates y^k, $k = 1, \cdots, n$, are harmonic functions.*

Proof. The identity map $\mathrm{id} : N \to N$ is harmonic. This follows e.g., from Theorem 12.1 because every convex function on N is subharmonic, see [GW 1973]. For a direct proof, see [Fug 1978, p. 123]. Taking $\mathrm{id} : N \to N$ for φ in (13.5) we obtain in terms of harmonic local coordinates, i.e., $\Delta_N y^k = 0$:

$$h^{\alpha\beta}\Gamma^k_{\alpha\beta} = 0, \qquad k = 1, \cdots, n.$$

Inserting (13.4) in (13.5) we find that φ is a harmonic map iff

$$\int_X \langle \nabla\psi, \nabla\varphi^k \rangle = \int_X \psi\lambda\,[(h^{\alpha\beta}\Gamma^k_{\alpha\beta}) \circ \varphi] = 0$$

for every $\psi \in \mathrm{Lip}_c(X)$, $k \in \{1, \cdots, n\}$; that is, iff each component φ^k is harmonic. □

Theorem 13.2. *For a continuous map $\varphi : X \to N$ of class $W^{1,2}_{\mathrm{loc}}(X, N)$ the following are equivalent:*

(a) *φ is a harmonic morphism* (Definition 13.1).

(b) *φ is a horizontally weakly conformal, harmonic map* (Definitions 13.2 and 12.1).

(c) *There is a scalar $\lambda \in L^1_{\mathrm{loc}}(X)$ such that*

$$-\int_X \langle \nabla\psi, \nabla(v \circ \varphi) \rangle = \int_X \psi\lambda\,[(\Delta_N v) \circ \varphi]$$

for every $v \in C^2(N)$ and $\psi \in \mathrm{Lip}_c(X)$ (or $\psi \in W_0^{1,2}(X) \cap L^\infty(X)$).

In the affirmative case, λ from (c) *equals a.e. the dilation λ of φ (as a horizontally weakly conformal map).*

In the case where also X is a smooth Riemannian manifold (without boundary) and $\varphi \in C^2(X, N)$, this theorem was obtained in [Fug 1978] and independently in [Ish 1979]; then (c) reads as follows:

$$\Delta_X(v \circ \varphi) = \lambda\,[(\Delta_N v) \circ \varphi], \qquad v \in C^2(N).$$

The target manifold N cannot be replaced in Theorem 13.2 by an admissible Riemannian polyhedron, not even by a smooth Riemannian manifold with boundary (see Remarks 13.2 and 13.3 below).

Proof of Theorem 13.2. A constant map satisfies (a), (b) and (c), the last with $\lambda = 0$. Each of the properties (a), (b), (c) is a local one. (For (c), this is seen by multiplying $\psi \in \mathrm{Lip}_c(X)$ with the functions from a suitable partition of unity.) We may therefore assume that *φ is nonconstant* and that the whole of *N is diffeomorphic with a domain in $\mathbb{R}^n$*. In terms of the corresponding coordinates $y^1, \cdots, y^n$ on N we have for any $v \in C^2(N)$ (writing ∂_α for $\partial/\partial y^\alpha$, etc.)

$$\Delta_N v = h^{\alpha\beta}\partial_\alpha\partial_\beta v + (\Delta_N y^\alpha)\partial_\alpha v, \tag{13.6}$$

and for any $\psi \in \mathrm{Lip}_c(X)$, by partial integration (as in (12.9)),

$$\begin{aligned} &-\int_X \langle \nabla\psi, \nabla(v \circ \varphi)\rangle \\ &= -\int_X \langle \nabla(\psi\,[(\partial_\alpha v)\circ\varphi]), \nabla\varphi^\alpha\rangle + \int_X \psi\,[(\partial_\alpha\partial_\beta v)\circ\varphi]\langle\nabla\varphi^\alpha, \nabla\varphi^\beta\rangle. \end{aligned} \tag{13.7}$$

In the rest of the proof we choose the above coordinates y^α as *harmonic* functions, appealing to [GW 1975]. We may then assume that *the components $\varphi^\alpha = y^\alpha \circ \varphi$ are harmonic* on X, for this holds under any one of the hypotheses (a), (b) or (c). (Ad (a). Take $v = y^\alpha$ in Definition 13.1. Ad (b). Apply Lemma 13.1. Ad (c). Take $v = y^\alpha$ in (c); then Definition 5.1 is fulfilled with φ replaced by ψ and u by φ^α.) For $v \in C^2(N)$, the identities (13.6) and (13.7) now reduce to

$$\Delta_N v = h^{\alpha\beta}\partial_\alpha\partial_\beta v, \tag{13.8}$$

$$-\int_X \langle\nabla\psi, \nabla(v\circ\varphi)\rangle = \int_X \psi\,[(\partial_\alpha\partial_\beta v)\circ\varphi]\langle\nabla\varphi^\alpha, \nabla\varphi^\beta\rangle \tag{13.9}$$

for $\psi \in \mathrm{Lip}_c(X)$ since $(\partial_\alpha v) \circ \varphi \in W^{1,2}_{\mathrm{loc}}(X) \cap C(X)$ by Remark 9.7, and so $\psi[(\partial_\alpha v) \circ \varphi] \in W^{1,2}_c(X)$ by Remark 5.1. Consequently, (c) is equivalent to

$$[(\partial_\alpha \partial_\beta v) \circ \varphi]\langle \nabla \varphi^\alpha, \nabla \varphi^\beta \rangle = \lambda\,[(\Delta_N v) \circ \varphi] \qquad \text{a.e. in } X, \tag{13.10}$$

because two functions $g_1, g_2 \in L^1_{\mathrm{loc}}(X)$ are equal a.e. iff $\int_X \psi g_1 = \int_X \psi g_2$ for all $\psi \in \mathrm{Lip}_c(X)$; indeed, $\mathrm{Lip}_c(X)$ is positively rich in $C_c(X)$, as noted in the text following inequality (12.12).

(b) $\Longrightarrow$ (c). Multiplying (13.4) by $(\partial_\alpha \partial_\beta v) \circ \varphi$, and inserting (13.8) on the right, leads to (13.10), and hence to (c). This also proves the final assertion of the theorem.

(c) $\Longrightarrow$ (b). Inserting (13.8) in (13.10) yields

$$[(\partial_\alpha \partial_\beta v) \circ \varphi]\,\langle \nabla \varphi^\alpha, \nabla \varphi^\beta \rangle = \lambda\,[(h^{\alpha\beta} \partial_\alpha \partial_\beta v) \circ \varphi] \qquad \text{a.e. in } X$$

for every $v \in C^2(N)$. Taking $v = \frac{1}{2} y^\alpha y^\beta$ leads to (13.4), that is, φ is *horizontally weakly conformal* with dilation λ. By Lemma 13.1, φ is a harmonic map (its components being harmonic, as noted earlier).

(c) $\Longrightarrow$ (a). Clearly, (c) can be localized so as to apply to functions v defined merely in open subsets V of N. In this localized form, (c) obviously implies (a).

(a) $\Longrightarrow$ (c). If $n = 1$ we may assume that $N = \mathbb{R}$. The concepts harmonic morphism and harmonic map then both reduce to that of harmonic function; and every continuous map $\varphi \in W^{1,2}_{\mathrm{loc}}(X, \mathbb{R})$ is horizontally weakly conformal (with $\lambda = |\nabla \varphi|^2$). Thus (a) $\Longleftrightarrow$ (b) $\Longleftrightarrow$ (c).

In the remaining case $n > 1$, suppose φ is a nonconstant harmonic morphism. Using (13.4) with $\alpha = \beta = 1$ as a definition of λ, we put

$$\lambda = \frac{|\nabla \varphi^1|^2}{h^{11} \circ \varphi}, \tag{13.11}$$

and note that $\lambda \geqslant 0$ a.e. and $\lambda \in L^1_{\mathrm{loc}}(X)$. Consider the linear subspace of $C(X)$

$$P = \left\{ [(\Delta_N v) \circ \varphi]/(h^{11} \circ \varphi) : v \in C^2(N) \right\},$$

and the linear map $T : P \to L^1_{\mathrm{loc}}(X)$ defined by

$$T \frac{(\Delta_N v) \circ \varphi}{h^{11} \circ \varphi} = \frac{[(\partial_\alpha \partial_\beta v) \circ \varphi]\,\langle \nabla \varphi^\alpha, \nabla \varphi^\beta \rangle}{h^{11} \circ \varphi}, \tag{13.12}$$

cf. (13.10). This map is well defined. In fact, if $(\Delta_N v) \circ \varphi = 0$ then v is harmonic in the open set $\varphi(X) \subset N$ (the nonconstant harmonic morphism

φ being open, by Proposition 13.1, because $n > 1$). It follows that $v \circ \varphi$ is harmonic on X:

$$\int_X \langle \nabla\psi, \nabla(v \circ \varphi)\rangle = 0 \qquad \text{for every } \psi \in \mathrm{Lip}_c(X),$$

and so indeed $[(\partial_\alpha \partial_\beta v) \circ \varphi]\,\langle \nabla\varphi^\alpha, \nabla\varphi^\beta\rangle = 0$ a.e. by (13.9), again because $\mathrm{Lip}_c(X)$ is positively rich in $C_c(X)$.

The same argument shows that T is *local,* that is, for any open set $U \subset X$ and any $p \in P$, $p = 0$ in U implies $Tp = 0$ a.e. in U. Similarly, T is *positive*:

$$p \geqslant 0 \text{ in } U \quad \text{implies} \quad Tp \geqslant 0 \text{ a.e. in } U, \tag{13.13}$$

invoking (13.12), (13.9) and Theorem 13.1 (i). Note at this point that the subharmonic function $v \circ \varphi$ is of class $W^{1,2}_{\mathrm{loc}}(U)$ by Remark 9.7, and hence weakly subharmonic in U by Remark 7.4.

For $v = \frac{1}{2}(y^1)^2$ we obtain from (13.6): $\Delta_N v = h^{11}$, and so $1 \in P$. Furthermore, by (13.11), (13.12),

$$T1 = \lambda, \tag{13.14}$$

and (13.10) will follow, thus completing the proof of the theorem, once we show that

$$Tp = \lambda p \qquad \text{for every } p \in P. \tag{13.15}$$

Suppose, for some $p \in P$ and $\varepsilon > 0$, that $|Tp - \lambda p| > 2\varepsilon$ in some set $A \subset X$ of volume $|A| > 0$. We may further arrange that A is contained in the interior s° of a top-dimensional simplex s of X, and that λ is bounded: $\lambda < a$ in A. Let x_0 denote a point of density for A with respect to the Riemannian volume measure on s°, or equivalently with respect to Lebesgue measure on s°. Because p is continuous, there is an open neighbourhood U of x_0 in X such that

$$p(x_0) - \varepsilon/a < p < p(x_0) + \varepsilon/a \qquad \text{in } U. \tag{13.16}$$

Since T is linear and positive, cf. (13.13), it follows that

$$(p(x_0) - \varepsilon/a)T1 \leqslant Tp \leqslant (p(x_0) + \varepsilon/a)T1 \qquad \text{in } U,$$

i.e., by (13.14),

$$|Tp - \lambda p(x_0)| \leqslant \lambda\varepsilon/a < \varepsilon \qquad \text{a.e. in } A \cap U.$$

By (13.16), $|\lambda p - \lambda p(x_0)| < \varepsilon$ a.e. in $A \cap U$, and so altogether $|Tp - \lambda p| < 2\varepsilon$ a.e. in $A \cap U$. This leads to a contradiction because $|A \cap U| > 0$. We conclude that (13.15) and (13.10), and hence (c) hold true. □

Corollary 13.2. *If $\varphi : X \to N$ is a harmonic morphism and $\psi : N \to P$ is a harmonic map between smooth Riemannian manifolds without boundary, then the composition $\psi \circ \varphi : X \to P$ is a harmonic map.*

This follows by invoking Theorem 12.1. Indeed, ψ pulls germs of convex functions on P back to germs of subharmonic functions on N; and φ pulls such germs back to germs of subharmonic functions on X, according to Theorem 13.1 (i).

Example 13.1. In the cases $N = \mathbb{R}$ or S^1 it was noted in the proof of Theorem 13.2, (a) $\Longrightarrow$ (c), that a function $\varphi : X \to N$ is a harmonic morphism iff it is harmonic. These harmonic morphisms are Hölder continuous, but generally not Lipschitz continuous, cf. Remark 6.4.

Example 13.2. Every nonconstant holomorphic map $\pi : (M, g) \to (N, h)$ of a Kähler manifold to a Riemann surface is a harmonic morphism (thus satisfying any of the conditions of Theorem 13.2) [Fug 1978], [EL 1983, Corollary 8.17]. That is true for any metric h in its conformal equivalence class.

Example 13.3. Consider the nonconstant harmonic morphisms $\pi : M \to N$ in the case where $M = (M, g)$ is a Riemannian 4-manifold and N a Riemann surface. Mostly, these are fibrations with singularities [Wo 1992]. For instance, the singular fibrations of Kodaira [BPV 1984, V, §§6–14]. Here M is a connected complex surface, and π a proper holomorphic map, with all fibres connected. If M is Kählerian, then (via Example 13.2) π is a harmonic morphism.

Example 13.4. Return to the Riemannian branched cover $\pi : M \to (N, h)$ of Example 8.11, and to the polyhedral structure on M constructed there, relative to compatible triangulations rendering π simplicial: Thus π *is horizontally conformal* (because its restriction to every m-simplex is), and even horizontally isometric. Give M the harmonic structure $\mathcal{H}_M$ determined by Theorem 7.1. Then $\pi : (M, d_M, \mathcal{H}_M) \to (N, h)$ *is a harmonic map.* Indeed, to evaluate the restriction E_{U_x} of the energy functional on a small open set U_x, it suffices to evaluate the sum of integrals over $s^\circ \cap U_x$ for all m-simplexes s of M, because $\dim A \leqslant m - 2$. But π is harmonic there, being isometric. By Theorem 13.2, we conclude further that π *is a harmonic morphism*, i.e., $\mathcal{H}_M = \pi^* \mathcal{H}_{(N,h)}$.

As a variant of this construction, let X be a connected normal complex analytic space and $\pi : X \to (N, h)$ a finite proper holomorphic map onto a Kähler manifold, as in Subexample 8.11 (iv). Again, π is a harmonic morphism. (By Proposition 11.3, π is a local E-minimizer; and is horizontally conformal.)

Combining Theorem 12.1 and Theorem 13.1 (iii) we obtain:

Proposition 13.2. *Let $\pi : X \to Y$ be a surjective proper harmonic morphism between admissible Riemannian polyhedra, and $\varphi : Y \to (N, h)$ a map into a Riemannian manifold*:

$$\begin{array}{ccc} X & & \\ {\scriptstyle\pi}\downarrow & \searrow^{\psi} & \\ Y & \xrightarrow[\varphi]{} & N \end{array}$$

Then φ is harmonic iff $\psi = \varphi \circ \pi$ is harmonic.

Example 13.5. Take $X = (M, g)$, $Y = M/K$ and $\pi : M \to M/K$ as in Example 8.12 with Brelot sheaf $\pi_* \mathcal{H}_M$ on M/K. Then $\varphi : M/K \to N$ is harmonic iff $\psi : M \to M/K$ is.

Subexample 13.5. Maps of singular spaces S^m/K have been much studied ([HL 1971, Chapter II], [ER 1993, Chapter IV]) as orbit spaces of actions of compact subgroups K of low cohomogeneity (the dimension of its maximal orbits) in the isometry group $SO(m+1)$ of S^m.

Remark 13.2. The following simple example shows that Theorem 13.2 does not admit a natural extension to the situation where the Riemannian target manifold N is allowed to have a *boundary*.

Let $X = [-1, 1]$, $Y = [0, 1]$ (both with the Euclidean metric). The map $\varphi \in \mathrm{Lip}(X, Y)$ given by $\varphi(x) = |x|$ is a harmonic morphism (because a harmonic function on a subinterval of $[0, 1]$ containing 0 must be constant), but *not* a harmonic map (Definition 12.1). In fact, the map $\varphi_\varepsilon(x) = \max\{|x|, \varepsilon\}$ has a smaller energy than φ (and $\varphi_\varepsilon = \varphi$ off $[-\varepsilon, \varepsilon]$). We could modify this example by replacing $X = [-1, 1]$ with $\mathbb{R}$ and extending φ to $\mathbb{R}$ so as to become periodic with period 2. Higher dimensional examples to the same effect are derived from these examples when X, Y are replaced by $X \times Z$, $Y \times Z$, and φ by $\varphi \times \mathrm{id}$, Z being any Euclidean space.

Remark 13.3. Let (X, g) and (Y, h) be admissible Riemannian polyhedra, and suppose that $bY = \emptyset$ and that h is continuous and simplexwise smooth. In each of the cases where the harmonic maps are characterized by pulling germs of convex functions on Y back to germs of subharmonic functions on X (as in Theorem 12.1 and in Proposition 12.4), *every harmonic morphism* $\varphi : X \to Y$ *is a harmonic map*, because every convex function on an open subset of Y is subharmonic, by Proposition 7.9.

The converse is false, even for $(X, g) = \mathbb{R}$, except of course when (Y, h) is a 1-dimensional manifold (cf. Example 13.1). This is shown by the following example: Let $X = \mathbb{R}$, $Y =$ the union of the x-axis and the positive y-axis in the xy-plane (Cornea's tripod, cf. Remark 12.2). The injection $x \mapsto (x, 0)$ of X in Y is a geodesic of speed 1, and hence a harmonic map by Lemma 12.5; but φ is *not* a harmonic morphism. In fact, any map $\varphi \in W^{1,2}_{\mathrm{loc}}(X, Y)$ of an admissible Riemannian polyhedron (X, g) into a 1-dimensional Riemannian polyhedron (Y, h) is *weakly horizontally conformal* in the sense that (cf. (13.3))

$$|\nabla(v \circ \varphi)|^2 = \lambda(|\nabla v|^2 \circ \varphi) \qquad \text{a.e. in } X \tag{13.17}$$

for any Lip continuous function $v : Y \to \mathbb{R}$, the dilation λ being defined as $\lambda = 0$ in $\varphi^{-1}(q)$ for any vertex q of Y. The point is that although $|\nabla v|^2 \circ \varphi$ need not be defined in $\varphi^{-1}(q)$, $|\nabla(v \circ \varphi)|^2$ equals 0 a.e. in $\varphi^{-1}(q)$, by Remark 5.2, and so (13.17) must be regarded as valid in $\varphi^{-1}(q)$ (where $\lambda = 0$). And for any open 1-simplex t of N, (13.17) does hold a.e. in $\varphi^{-1}(t)$ for a suitable function $\lambda \geqslant 0$, $\lambda \in L^1_{\mathrm{loc}}(\varphi^{-1}(t))$ (as in Theorem 13.2, applied to the Riemannian manifold t without boundary).

Together with Remark 13.2, this shows that *Theorem 13.2 does not extend to targets other than manifolds without boundary.*

Remark 13.4. For a discrete analogue of harmonic morphisms between 1-dimensional Riemannian polyhedra, see Urakawa [U 2000]. In that context there is no counterpart to harmonic maps; and harmonic morphisms are the same as horizontally conformal maps.

14

Appendix: Energy according to Korevaar-Schoen

Here we describe the main steps leading to the definition and key properties [KS 1993, Chapter 1] of the energy functional of a map $\varphi : M \to Z$ of a Riemannian domain (M, g) (cf. Chapter 1 above, Maps between polyhedra) to a metric space (Y, d_Y), starting from the approximate energy density $e_\varepsilon(\varphi)$ of φ, as in the beginning of Chapter 9.

Let μ_g denote the Riemannian volume measure of (M, g), and d_M its Riemannian distance function. In the beginning of Chapter 9 we have defined the metric space $L^2(X, Y)$, to which the special case $X = M$ applies, and likewise the space $W^{1,2}(M, Y)$ of maps $\varphi \in L^2(M, Y)$ of *finite energy* $E(\varphi)$ according to Korevaar-Schoen (cf. Definition 9.1). Throughout this Appendix we maintain the numbering system of [KS 1993, Chapter 1].

Subpartitioning Lemma

For any map $\varphi \in L^2(M, Y)$ define

$$E_\varepsilon(\varphi) = \int_M e_\varepsilon(\varphi)\, d\mu_g,$$

and view it as a linear form on $C_c(M)$ by

$$E_\varepsilon(\varphi)(f) = \int f(x) e_\varepsilon(\varphi)(x)\, d\mu_g(x).$$

If $f \in C_c(M)$, let

$$\omega(f, \varepsilon)(x) = \max(|f(x) - f(y)| : d_M(x, y) < \varepsilon)$$

denote the oscillation of f in the ball $B(x, \varepsilon)$ in M. For any $C > 0$ define $f_\varepsilon^C : M \to \mathbb{R}$ by

$$f_\varepsilon^C(x) = (1 + C\varepsilon)\big(f(x) + \omega(f, 2\varepsilon)(x)\big).$$

Lemma 1.3.1. *Take any $\varphi \in L^2(M,Y)$. For any nonnegative function $f \in C_c(M)$ there is a constant $C > 0$ (depending only on the Ricci curvature of M) such that for sufficiently small $\varepsilon > 0$*

$$(E_\varepsilon(\varphi)(f))^{\frac{1}{2}} \leqslant \sum_i \lambda_i (E_{\lambda_i \varepsilon}(\varphi)(f_\varepsilon^C))^{\frac{1}{2}}.$$

Here (λ_i) is any finite sequence of positive numbers with $\sum \lambda_i = 1$.

This subpartitioning lemma is a key step in Korevaar-Schoen's study of energy of a map. When M is an interval, it is motivated through the definition of path length by piecewise smooth geodesic path approximation. (See Chapter 12, Geodesics as harmonic maps, for details of that construction.)

For the proof it is more convenient to use averages over spheres than over balls in the definition of $e_\varepsilon(\varphi)$. That permits reduction to proving the stated inequality in the case where (M,g) is a Euclidean domain, a relatively straightforward matter.

As applications:

Theorem 1.5.1. *Suppose $\varphi \in W^{1,2}(M,Y)$, i.e., φ has finite energy. The measure $e_\varepsilon(\varphi)(x)\, d\mu_g(x)$ then converges weakly as $\varepsilon \to 0$ to some energy density measure $de(\varphi)$, having the total mass $E(\varphi)$. For small $\varepsilon > 0$,*

$$E_\varepsilon(\varphi)(f) \leqslant E(\varphi)(f_\varepsilon^C), \qquad f \in C_c(M).$$

Theorem 1.6.1. (Lower semicontinuity of energy.) *Let $(\varphi_k) \subset W^{1,2}(M,Y)$ be a sequence which converges in $L^2(M,Y)$ to some map φ. Assume there is a constant $E < \infty$ such that*

$$\int_M e(\varphi_k)\, d\mu_g \leqslant E.$$

Then $\varphi \in W^{1,2}(M,Y)$, and its energy density measure $de(\varphi)$ satisfies

$$de(\varphi) \leqslant \liminf_{k \to \infty} de(\varphi_k),$$

i.e., $E_\varepsilon(\varphi)(f) \leqslant \liminf_k E(\varphi_k)(f)$ for every $f \in C_c^+(M)$.

Theorem 1.6.2. (Consistency when $Y = \mathbb{R}$.) $W^{1,2}(M,\mathbb{R}) = W^{1,2}(M)$ *in the sense of Chapter 5. Furthermore, the energy density measures are (dimensional) constant multiples of each other.*

Again, it is easy to reduce the proof to the case where (M, g) is a Euclidean domain. If u is a function of finite energy in the sense of Chapter 5, smoothing it (by convolution, for instance) gives a sequence (u_t) which converges to u in $W^{1,2}_{\text{loc}}(M)$. In particular, $|\nabla u_t|^2 d\mu_g$ converges to $|\nabla u|^2 d\mu_g$ weakly as measures. By Theorem 1.6.1 we know that

$$de(u) \leqslant \liminf_{t\to\infty} de(u_t).$$

Application of Jensen's inequality gives, after some estimations, the reverse inequality, so

$$de(u) = \lim_{t\to\infty} de(u_t),$$

from which Theorem 1.6.2 now follows.

Directional energies

Let Z be a Lipschitz vector field on the closure $\overline{M}$ of M, and $\zeta = \zeta(x,t)$ its flow:

$$\frac{d\zeta(x,t)}{dt} = Z(\zeta(x,t)) \qquad \text{with } \zeta(x,0) = x \text{ for all } x \in \overline{M}.$$

For $\varphi \in L^2(M,Y)$ define the integrable function

$${}^{Z}e_\varepsilon(\varphi)(x) = \frac{d_Y^2(\varphi(x), \varphi(\zeta(x,\varepsilon)))}{\varepsilon^2}$$

and the energy in the direction Z

$${}^{Z}E(\varphi) = \sup_{f\in C_c(M,[0,1])} \limsup_{\varepsilon\to 0} \int f\, {}^{Z}e_\varepsilon(\varphi)\, d\mu_g.$$

As in Theorem 1.5.1, there is

Theorem 1.7. *Let $\varphi \in L^2(M,Y)$ with ${}^{Z}E(\varphi) < \infty$. Then as $\varepsilon \to 0$, ${}^{Z}e_\varepsilon(\varphi)(x)\, d\mu_g(x)$ converges weakly to a measure $d\,{}^{Z}e(\varphi)$ with total mass ${}^{Z}E(\varphi)$.*

Theorem 1.8.1. *If $E(\varphi) < \infty$ then ${}^ZE(\varphi) < \infty$ and there is a constant C (depending only $\dim M = m$) so that*

$$d({}^Ze(\varphi)) \leqslant C|Z|^2_{L^\infty}\, de(\varphi).$$

Theorem 1.9.6. *If $E(\varphi) < \infty$ then the measures $d\,{}^Ze(\varphi)$ are absolutely continuous with respect to $\mu_g : d\,{}^Ze(\varphi) = {}^Ze(\varphi)(x)\, d\mu_g(x)$ for some L^1 function ${}^Ze(\varphi)$. Furthermore, ${}^Ze_\varepsilon(\varphi)$ converges a.e. to ${}^Ze(\varphi)$.*

Theorem 1.10. *If $E(\varphi) < \infty$ then $de(\varphi)$ is absolutely continuous with respect to μ_g.*

Trace maps

Say that the boundary ∂M of a Riemannian domain M is *Lipschitz* if every point of ∂M has a neighbourhood in $\overline{M}$ on which there is a smooth vector field Z (called *transverse*) with $\varrho > 0$, $t_0 > 0$ such that for all $x \in \partial M$,

$$\begin{aligned} \zeta(x,t) &\in M && \text{if } 0 < t < t_0,\\ \zeta(x,t) &\notin M && \text{if } -t_0 < t < 0,\\ d_Y(\zeta(x,t), \partial M) &> \varrho|t| && \text{if } |t| < t_0. \end{aligned}$$

If φ is of class $W^{1,2}(M,Y)$ and M has Lipschitz boundary, a *trace map* $\mathrm{tr}_{\partial M}(\varphi) = \mathrm{tr}(\varphi)$ in $L^2(\partial M, Y)$ is defined: With respect to $(m-1)$-dimensional measure on ∂M ($m = \dim M$), almost everywhere the map $\mathrm{tr}(\varphi)(x) = \lim_{t\to 0^+} \varphi(\zeta(x,t))$ exists, and satisfies:

Lemma 1.12.1. *That trace map is independent of the transverse field Z, and for compact subsets B of ∂M,*

$$\mathrm{tr}_B(\varphi) = \mathrm{tr}_{\partial M}(\varphi)|_B.$$

A Riemannian domain with Lipschitz boundary is called a *Riemannian Lipschitz domain.*

Theorem 1.12.2. *If (M,g) is a Riemannian Lipschitz domain and $(\varphi_i) \subset W^{1,2}(M,Y)$ a sequence with uniformly bounded energies $E(\varphi_i)$, and which converges in L^2 to a map φ, then the trace maps $\mathrm{tr}(\varphi_i)$ converge in $L^2(\partial M, Y)$ to $\mathrm{tr}(\varphi)$. Moreover, two maps $\varphi, \psi \in W^{1,2}(M,Y)$ have the same trace iff the function $d_Y(\varphi,\psi) \in W^{1,2}(M) : x \mapsto d_Y(\varphi(x), \psi(x))$ has trace 0.*

Theorem 1.12.3. *Let M be a Riemannian Lipschitz domain expressible as a disjoint union $M = M_1 \cup M_2$ of two such subdomains, with $\partial M_1 \cap \partial M_2$ also Lipschitz. If $\varphi_i \in W^{1,2}(M_i, Y)$, $i = 1, 2$, and $\operatorname{tr}_{\partial M_1}(\varphi_1) = \operatorname{tr}_{\partial M_2}(\varphi_2)$ on $\partial M_1 \cap \partial M_2$, then the map $\varphi : M \to Y$ defined by*

$$\varphi(x) = \begin{cases} \varphi_i(x), & \text{for } x \in M_i \\ \operatorname{tr}_{\partial M_i}(\varphi_i)(x), & \text{for } x \in \partial M_1 \cap \partial M_2 \end{cases}$$

has finite energy $E(\varphi) = E_{M_1}(\varphi_1) + E_{M_2}(\varphi_2)$.

Theorem 1.13. (Precompactness.) *Let (M, g) be a Riemannian domain and (Y, d_Y) a complete metric space for which closed balls are compact. If $(\varphi_i) \subset W^{1,2}(M, Y)$ is a sequence with*

$$\int_M d_Y^2(\varphi_i(x), q)\, d\mu_g(x) + E(\varphi_i) \leqslant C$$

for some point q in Y and constant C, then a subsequence of (φ_i) converges in $L^2(M, Y)$ to a map $\varphi \in W^{1,2}(M, Y)$.

15

Appendix: Minimizers with small energy decay

By Tomasz Serbinowski†

Introduction and results

The goal of this note is to prove a theorem analogous to the Schoen-Uhlenbeck "small energy implies regularity" theorem [SchU 1982]. The proof is somewhat similar to the one known in the classical setting. However, the original proof of Schoen and Uhlenbeck requires an isometric embedding of Y into a Euclidean space $\mathbb{R}^k$. That presents a problem, since to carry it to the case when Y is a metric space one needs a suitable ambient space. This is achieved by embedding Y "isometrically" into an NPC cone CY. (NPC stands for 'globally nonpositive curvature'.) This embedding allows one to use successfully the regularity theory for harmonic maps into NPC spaces in order to get useful information about minimizing maps into Y.

The main result is the following theorem (for the definition of the curvature bound, see the next section):

Theorem 15.1. *Let (Y, d) be a locally compact geodesic space whose curvature is bounded above by 1. Let $u : M \to Y$ be an energy minimizing map from a smooth Riemannian domain M into Y. There exists a positive constant ε_0 depending only on the domain M (and its metric) so that for any ball $B_\sigma(x) \subset M$, if*

$$\sigma^{2-n} \int_{B_\sigma(x)} |\nabla u|^2 d\mu_g \leqslant \varepsilon_0$$

then the map u is Lipschitz continuous on the ball $B_{\sigma/2}(x)$.

Remark 15.1. Throughout the text C denotes different constants (to avoid labeling them $C_1, C_2, \cdots$) that depend on the domain M and its metric g.

†Written under the supervision of, and in cooperation with, Nicholas J. Korevaar.

Embedding Y into an NPC cone

In this chapter (Y, d) will denote a geodesic space, that is a metric space in which every pair of points can be joined by a distance realizing path called a geodesic (it will always be assumed that geodesics are parametrized proportionally to the length). For such a space there is a global notion of curvature bound, based on triangle comparison.

Definition 15.1. The geodesic space (Y, d) is said to have curvature bounded above by 1, if given a point $Q \in Y$ and a geodesic $\gamma : [0,1] \to Y$ satisfying

$$d(Q, \gamma(0)) + d(\gamma(0), \gamma(1)) + d(\gamma(1), Q) < 2\pi$$

the following inequality holds for all $t \in [0,1]$

$$\begin{aligned}\cos d(\gamma(t), Q) \geqslant & \frac{\sin[td(\gamma(0),\gamma(1))]}{\sin d(\gamma(0),\gamma(1))} \cos d(\gamma(1), Q) + \\ & \frac{\sin[(1-t)d(\gamma(0),\gamma(1))]}{\sin d(\gamma(0),\gamma(1))} \cos d(\gamma(0), Q).\end{aligned} \tag{15.1}$$

One can give a more geometric meaning to the inequality (15.1). Let $P = \gamma(0)$ and $R = \gamma(1)$. Then the triangle $\triangle PQR$ has circumference less than 2π. Therefore there exists a (unique up to isometry) triangle $\triangle P'Q'R'$ on S^2 with corresponding sides equal to those of $\triangle PQR$. Inequality (15.1) states that secants of the triangle $\triangle PQR$ from Q are not longer than the corresponding secants of $\triangle P'Q'R'$. (Note that lengths of secants of a triangle on S^2 can be explicitly computed.)

It is not hard to see that a geodesic space Y whose curvature is bounded above by 1 has unique geodesics of length less then π.

Let $\mathcal{C}Y$ be the cone over Y. As a topological space

$$\mathcal{C}Y = Y \times [0, \infty)/Y \times \{0\}.$$

For a point $P \in Y$ and $t \in [0, \infty)$ one denotes by $[P, t]$ the equivalence class of the pair (P, t), that is

$$[P, t] = \{(Q, s) \in Y \times [0, \infty) : (P = Q \text{ and } t = s) \;\text{ or }\; t = s = 0\}.$$

It is easy to see that $\mathcal{C}Y$ can be turned into a metric space itself with the distance function D defined by the law of cosines

$$D^2([P, t], [Q, s]) = t^2 + s^2 - 2ts \cos \min\{d(P, Q), \pi\}. \tag{15.2}$$

Proposition 15.1. *Let (CY, D) be defined as above. Then*

(a) *if Y is a geodesic space then so is CY,*

(b) *if Y has unique geodesics of lengths less than π then CY has unique geodesics of any length,*

(c) *if Y has curvature bounded above by 1 then CY is an NPC space.*

Proof. Assuming that Y is a geodesic space, one can explicitly write down a geodesic joining a pair of points $\mathcal{P}_0 = [P_0, t_0]$ and $\mathcal{P}_1 = [P_1, t_1]$ in CY. If $d(P_0, P_1) \geqslant \pi$ then such a geodesic is unique and given by

$$\lambda \longmapsto \mathcal{P}_\lambda = \begin{cases} [P_0, (1-\lambda)t_0 - \lambda t_1] & \text{for } 0 \leqslant \lambda \leqslant \frac{t_0}{t_0+t_1} \\ [P_1, \lambda t_1 - (1-\lambda)t_0] & \text{for } \frac{t_0}{t_0+t_1} \leqslant \lambda \leqslant 1. \end{cases} \tag{15.3}$$

Assume therefore that $d = d(P_0, P_1) < \pi$. For any $\lambda \in [0,1]$, there exists a unique number $0 \leqslant d_\lambda \leqslant d$ satisfying

$$\frac{\sin d_\lambda}{\sin(d - d_\lambda)} = \frac{\lambda t_1}{(1-\lambda)t_0}.$$

Fix some geodesic joining P_0 and P_1 in Y (there is at least one), and let $P_\lambda \in Y$ be a point on that geodesic such that $d(P_\lambda, P_0) = d_\lambda$. Then

$$\lambda \longmapsto \mathcal{P}_\lambda = [P_\lambda, ((1-\lambda)t_0^2 + \lambda t_1^2 - \lambda(1-\lambda)D^2(\mathcal{P}_0, \mathcal{P}_1))^{1/2}] \tag{15.4}$$

is a geodesic in CY.

One verifies that in fact every geodesic in CY is of that form. That proves parts (a) and (b) of the Proposition.

The proof of part (c) is done along the lines of Berestovskii's proof of a similar statement about spherical cones [Be 1983, Theorem 2]. First, one verifies that for the standard sphere $S^2 \subset \mathbb{R}^3$ the above cone construction results in Euclidean space $\mathbb{R}^3$, that is $CS^2 \equiv \mathbb{R}^3$. Consider a geodesic triangle in CY with vertices $\mathcal{P}_0 = [P_0, t_0], \mathcal{P}_1 = [P_1, t_1]$ and $\mathcal{Q} = [Q, s]$. Assume (for simplicity) that the circumference of the triangle $\triangle P_0QP_1$ in Y is less than 2π, and let $\triangle P_0'Q'P_1'$ be a triangle on S^2 with corresponding sides equal to those of $\triangle P_0QP_1$.

Let $\mathcal{P}_\lambda = [P_\lambda, t_\lambda]$ be given by (15.4). Let $P_\lambda' \in S^2$ be the point on the geodesic joining P_0' and P_1' corresponding to P_λ (that is being d_λ away from P_0'). Then $\lambda \mapsto \mathcal{P}_\lambda' = [P_\lambda', t_\lambda]$ is a line segment (geodesic) in $\mathbb{R}^3$ and one shows that

$$D(\mathcal{P}_\lambda, \mathcal{Q}) \leqslant D_{\mathbb{R}^3}(\mathcal{P}_\lambda', \mathcal{Q}')$$

by combining the distance formula (15.2) with the inequality

$$d(P_\lambda, Q) \leqslant d_{S^2}(P'_\lambda, Q')$$

expressing the fact that Y has curvature bounded by 1. □

Remark 15.2. Metric spaces with curvature bounded above have been studied by a number of mathematicians beginning with A. D. Alexandrov around 1950. The more common definition of the curvature bound found in the literature uses relative deficit of a triangle rather than secants. Equivalence of the two follows from a version of Toponogov's theorem proven by Alexandrov [A 1951, Theorem 1]. Various cone constructions are due to Berestovskii (see for example [Be 1983]). The author does not know to whom to attribute the construction used here; he himself heard it in a lecture by G. Perelman.

Let $\mathcal{O}$ denote the vertex of the cone CY

$$\mathcal{O} = [P, 0] = Y \times \{0\}.$$

For any point $[P, t]$ in CY we then have

$$D([P, t], \mathcal{O}) = t.$$

In analogy with the case $Y = S^2$ and $CY = \mathbb{R}^3$ one defines the "norm" of $[P, t]$ to be the distance to $\mathcal{O}$

$$|[P, t]| = D([P, t], \mathcal{O}) = t.$$

The original space Y can be embedded "isometrically" into the cone CY, via the mapping $P \mapsto [P, 1]$. This embedding is isometric in the sense that the metrics infinitesimally coincide (lengths of paths computed with the use of either metric are the same). Similarly, if $u : M \to Y$ is a finite energy map, then $[u, 1] : M \to CY$ is also a finite energy map and their energies and energy densities (computed with respect to the metric d and D, respectively) are the same. Therefore one can identify Y with its image in CY

$$Y = \{[P, t] \in CY : t = 1\}.$$

Any map $u : M \to CY \setminus \{\mathcal{O}\}$ can be represented by a pair of maps

$$\hat{u} : M \to Y \quad \text{and} \quad |u| : M \to [0, \infty)$$

such that $u = [\hat{u}, |u|]$ (note that $|u| = D(u, \mathcal{O})$). This decomposition enables one to identify the Sobolev space $W^{1,2}(M, Y)$ with the subset of the space $W^{1,2}(M, \mathcal{C}Y)$ consisting of maps with images in Y (almost everywhere):

$$W^{1,2}(M, Y) = \{u \in W^{1,2}(M, \mathcal{C}Y) : |u| = 1 \text{ a.e.}\}.$$

In what follows we will need to know the exact expression for the "norm" along a geodesic in $\mathcal{C}Y$. For a pair of points $\mathcal{P}, \mathcal{Q} \in \mathcal{C}Y$, and a number $\lambda \in [0, 1]$, denote by $\mathcal{P}_\lambda$ the point on the unique geodesic joining $\mathcal{P}$ and $\mathcal{Q}$ which is a fraction λ away from $\mathcal{P}$

$$D(\mathcal{P}_\lambda, \mathcal{P}) = \lambda D(\mathcal{Q}, \mathcal{P}).$$

(Though $\mathcal{C}Y$ has no linear structure, one uses the notational convention that $\mathcal{P}_\lambda = (1 - \lambda)\mathcal{P} + \lambda\mathcal{Q}$.)

Proposition 15.2. *Let $\mathcal{P}$ and $\mathcal{Q}$ be a pair of points in $\mathcal{C}Y$. For a given number $\lambda \in [0, 1]$*

$$|\mathcal{P}_\lambda|^2 = (1 - \lambda)|\mathcal{P}|^2 + \lambda|\mathcal{Q}|^2 - \lambda(1 - \lambda)D^2(\mathcal{P}, \mathcal{Q}). \tag{15.5}$$

Proof. Assuming that $\mathcal{P} = [P, t]$ and $\mathcal{Q} = [Q, s]$, the point $\mathcal{P}_\lambda$ is given by either (15.3) or (15.4). In either case one easily verifies that (15.5) holds. □

Hölder continuity of the minimizer

The proof rests upon a replacement argument. Given a minimizing map u into the space Y one can construct a unique harmonic map v into the cone $\mathcal{C}Y$ having the same boundary values as u. Clearly, the total energy of v is then not greater than that of u, but it might be considerably smaller. One shows that it is not the case and concludes that the maps u and v are close in a suitable sense. Knowing the decay of the energy for the maps into NPC spaces allows one to deduce a similar decay for the energy of u. Showing that this decay implies Hölder continuity of the map is standard.

As in the preceding chapter it is enough to prove the theorem for a unit ball. Let $u : (M, g) \to Y$ be an energy minimizing map, and let $B_1 \subset M$ be some unit ball compactly contained in M. Without loss of generality one may assume that the ball B_1 is covered by a single coordinate chart. Though all results are stated and hold for an arbitrary smooth Riemannian metric g, the proofs, in general, are given for the Euclidean one. This is done only for the sake of simplicity.

The following lemma that allows one to compare u to a harmonic map into $\mathcal{C}Y$ is the key step in the proof.

Lemma 15.1. *Let u be as above. For $1/2 \leqslant R \leqslant 3/4$ let $v_R : B_R \to CY$ denote the energy minimizing map into the NPC cone CY satisfying*

$$v_R = u \quad \text{on } \partial B_R.$$

There exists a constant C depending only on the metric g and the dimension of the domain such that for some $R \in [1/2, 3/4]$

$$E_R(u) - E(v_R) \leqslant C(E(u))^{9/8}. \tag{15.6}$$

(Here $E_R(u) = E(u_{|B_R})$.)

Remark 15.3. Note that the lemma states the existence of an R so that (15.6) holds, and not that the inequality holds for all values of R. Note also that in the proof one may assume that $E(u)$ is small (otherwise the statement holds for C large enough).

Before proceeding with the proof of Lemma 15.1 it is necessary to establish some estimates for the energy and energy density of a smoothed version of u. One may smooth out the map by the process of averaging which works well in NPC spaces.

Let $h : [0,1] \to \mathbb{R}_+$ be a smooth nonincreasing function satisfying

$$\begin{aligned} h(r) &= \hbar \qquad \text{for } 0 \leqslant r \leqslant R \\ h(r) &= 0 \qquad \text{for } r \geqslant R + \tau \end{aligned}$$

where $R \in [1/2, 3/4]$. (It is assumed that $\hbar << \tau$ are both constants, to be determined later.)

For a finite energy map (not necessarily an energy minimizing one) $u : B_1 \to Y \subset CY$, one defines $u^h(x)$ to be the average of u over the ball $B_{h(|x|)}(x)$. More precisely, $u^h(x) = Q \in CY$ is the point in the cone CY that minimizes the functional

$$P \longmapsto \frac{1}{\omega_n} \int_{B_1} D^2\big(u(x + h(|x|)z), P\big) dz.$$

Note that the averaging process takes place in the cone CY which is an NPC space; that guarantees the existence of a unique minimizer of the above functional [KS 1993, Lemma 2.5.1].

Lemma 15.2. *Let $u^h : B_1 \to CY$ be as defined above. Then*

(i) u^h *is a finite energy map,*

(ii) $u^h = u$ *on* $B_1 \setminus B_{R+\tau}$,

(iii)

$$\int_{B_{R+\tau}\setminus B_R} |\nabla u^h|^2 d\mu_g \leqslant C \int_{B_{R+2\tau}\setminus B_{R-\tau}} |\nabla u|^2 d\mu_g,$$

(iv) u^h *is Lipschitz on the ball* B_R *with* $\mathrm{Lip}(u) \leqslant CE(u)(1-R)^{2-n}\hbar^{-2}$.

Proof. The second claim follows directly from the definition of the map u^h and the fact that $h(|x|) = 0$ outside the ball $B_{R+\tau}$. The three other claims can be deduced from the following estimate, found in [KS 1993, Proposition 2.5.2],

$$D^2(u^h(x), u^h(x+\varepsilon y)) \leqslant \frac{1}{\omega_n} \int_{B_1} D^2\big(u(x+h(|x|)z), u(x+\varepsilon y + h(|x+\varepsilon y|)z)\big)dz.$$

□

Proof of Lemma 15.1. The beginning of the proof follows closely the one given by Schoen–Uhlenbeck in [SchU 1982]. Suppose that $E(u) = \varepsilon$ is small (smaller than some ε_0 that will be determined later). The two constants τ and $\hbar$ can then be tied to ε as follows

$$\tau = \varepsilon^{1/8}, \quad \hbar = \varepsilon^{1/4}.$$

One may divide the interval $[1/2, 3/4]$ into subintervals I_j of width (approximately) 3τ:

$$[1/2, 3/4] = \bigcup I_j.$$

Let N be the number of subintervals. Then $N \approx 1/(12\tau)$. One chooses a particular subinterval so that the total energy of u over the annulus $\{|x| \in I_j\}$ is relatively small. Namely, as

$$\int_{\{|x|\in[1/2,3/4]\}} |\nabla u|^2 d\mu_g = \sum_j \int_{\{|x|\in I_j\}} |\nabla u|^2 d\mu_g \leqslant \varepsilon,$$

there exists I_j such that

$$\int_{\{|x|\in I_j\}} |\nabla u|^2 d\mu_g \leqslant \varepsilon/N = 12\tau\varepsilon = 12\varepsilon^{9/8}.$$

Then $I_j = [R-\tau, R+2\tau]$ for some $R \in [1/2, 3/4]$. It will be shown that the estimate (15.6) holds for $R+\tau$.

Denote by u^h as before a smoothed (by averaging) version of u. Using u^h one defines an auxiliary map $\hat{u} : B_{R+\tau} \to CY$ which is energy minimizing on B_R and coincides with u^h on the annulus $B_{R+\tau} \setminus B_R$. By projecting this map back onto Y one gets a competitor for u. In particular, if $\Pi : CY \setminus \{\mathcal{O}\} \to Y$ denotes a projection and $E_r(u)$ denotes the energy of a map over the ball B_r, then

$$E_{R+\tau}(u) \leqslant E_{R+\tau}(\Pi \circ \hat{u}). \tag{15.7}$$

The above inequality in itself is not interesting. However it gains some power when combined with the following estimate

$$D(\hat{u}, Y) \leqslant C\varepsilon^{1/4}. \tag{15.8}$$

The proof of (15.8) relies on the Poincaré inequality and the monotonicity of the scaled energy. Let $x \in B_{R+\tau}$. The Poincaré inequality (applied here to a map into the NPC space CY) gives the estimate for the oscillation of the map in terms of its energy. Writing h in place of $h(|x|)$ one gets

$$h^{-n} \int_{B_h(x)} D^2(u(z), u^h(x))dz \leqslant Ch^{2-n} \int_{B_h(x)} |\nabla u|^2 dz,$$

due to the fact that $u^h(x)$ is the average value of u on the ball B_h.

On the other hand, as the ball $B_{1/4}(x)$ is contained in B_1, the monotonicity of the energy (u is a minimizing map) implies that

$$h^{2-n} \int_{B_h(x)} |\nabla u|^2 dz \leqslant C4^{n-2} \int_{B_{1/4}(x)} |\nabla u|^2 dz \leqslant CE(u) = C\varepsilon.$$

In particular $D(u^h(x), Y) \leqslant C\varepsilon^{1/2}$. Lemma 15.2 gives an estimate for the oscillation of the map u^h over B_R. Namely

$$\sup_{B_R} D(u^h(x), u^h(0)) \leqslant C\varepsilon^{1/4}.$$

Therefore the same is true about the harmonic map $\hat{u}$ which agrees with u^h on ∂B_R. That implies (15.8).

Now it readily follows that

$$E_{R+\tau}(\Pi \circ \hat{u}) \leqslant \frac{1}{(1 - C\varepsilon^{1/4})^2} E_{R+\tau}(\hat{u}).$$

Combining that with the estimate (15.7) yields

$$E_R(\hat{u}) \geqslant (1 - C\varepsilon^{1/4})E_{R+\tau}(u) - C\varepsilon^{9/8}. \tag{15.9}$$

(One uses here Lemma 15.2 and the choice of subinterval I_j to claim that the energy of $\hat{u}$ over the annulus $B_{R+\tau} \setminus B_R$ is bounded by $C\varepsilon^{9/8}$.)

The inequality (15.9) can now be used to get the desired estimate for the energy of $v_{R+\tau}$. By gluing together $v_{R+\tau}$ and u^h one gets a competitor for $\hat{u}$. More precisely, consider another auxiliary map $w : B_R \to CY$ defined by

$$w(x) = \begin{cases} v_{R+\tau}\left(\frac{R+\tau}{R-\tau}x\right), & x \in B_{R-\tau} \\ u^h\left((2R - |x|)\frac{x}{|x|}\right) & x \in B_R \setminus B_{R-\tau}. \end{cases}$$

As all three maps w, u^h and $\hat{u}$ coincide on ∂B_R the minimizing property of $\hat{u}$ implies that

$$E_R(\hat{u}) \leqslant E_R(w) = \int_{B_{R-\tau}} |\nabla w|^2 dx + \int_{B_R \setminus B_{R-\tau}} |\nabla w|^2 dx. \tag{15.10}$$

One can estimate the integrals in (15.10) in terms of energies of $v_{R+\tau}$ and u^h:

$$\begin{aligned} \int_{B_{R-\tau}} |\nabla w|^2 dx &= \left(\frac{R+\tau}{R-\tau}\right)^{2-n} \int_{B_{R+\tau}} |\nabla v_{R+\tau}|^2 dx \\ &= \left(\frac{R+\tau}{R-\tau}\right)^{2-n} E_{R+\tau}(v_{R+\tau}), \end{aligned} \tag{15.11}$$

and

$$\begin{aligned} \int_{B_R \setminus B_{R-\tau}} |\nabla w|^2 dx &\leqslant \left(\frac{R+\tau}{R-\tau}\right)^{n+2} \int_{B_{R+\tau} \setminus B_R} |\nabla u^h|^2 dx \\ &\leqslant C\varepsilon^{9/8}. \end{aligned} \tag{15.12}$$

Putting together inequalities (15.10), (15.11) and (15.12), one gets

$$E_R(\hat{u}) \leqslant E_{R+\tau}(v_{R+\tau}) + C\varepsilon^{9/8}. \tag{15.13}$$

Finally, combining (15.9) and (15.13) results in

$$E_{R+\tau}(u) - E_{R+\tau}(v_{R+\tau}) \leqslant C\varepsilon^{9/8}$$

which is the claim of the lemma. □

Proof of Theorem 15.1

The way to prove ε-regularity is to show the energy decay. That is, one wants to prove that if the energy of a minimizing map u is small enough then there exists θ such that

$$\theta^{2-n} E_\theta(u) \leqslant \tfrac{1}{2} E_1(u). \tag{15.14}$$

Assuming that the inequality (15.14) holds for some θ (depending on the domain metric but not on the map u), by a standard line of reasoning one shows that the scaled energy decays at a certain rate

$$r^{2-n} E_r(u) \leqslant C r^{2\alpha}.$$

Hence by Morrey's lemma the map u is Hölder continuous with exponent α.

It remains to prove that (15.14) holds for some θ. Let $u : B_1 \to Y$ be a minimizing map into Y. In Lemma 15.1 it was shown that the total energy of u does not differ much from the energy of the harmonic map $v_R : B_R \to CY$ into the cone (for a suitable choice of R). If it can be shown that the energy densities are close as well, then the energy decay for u will be a consequence of the energy decay for v_R.

The following lemma (about maps into NPC spaces) shows that the energy densities of u and v_R are in fact close to each other.

Lemma 15.3. *Suppose that $u, v : M \to (Z, d)$ are two finite energy maps into an NPC space Z. Furthermore, suppose that v is a minimizing map, and in addition that u and v coincide on the boundary ∂M. Then*

$$\int_M (|\nabla u| - |\nabla v|)^2 d\mu_g \leqslant 2(E(u) - E(v)). \tag{15.15}$$

In particular if u is almost minimizing then the energy densities of u and v are L^2-close.

Proof. The lemma follows quite easily from the quadrilateral comparison for NPC spaces. Consider the midpoint map $w = (u+v)/2$. For a unit field ω, and a small positive ε, the distance $d(w(x), w(x+\varepsilon\omega))$ can be estimated as follows [KS 1993, Corollary 2.1.3]:

$$\begin{aligned} d^2(w(x), w(x+\varepsilon\omega)) \leqslant & \frac{1}{2} d^2(u(x), u(x+\varepsilon\omega)) + \frac{1}{2} d^2(v(x), v(x+\varepsilon\omega)) - \\ & \frac{1}{4} |d(u(x), u(x+\varepsilon\omega)) - d(v(x), v(x+\varepsilon\omega))|^2. \end{aligned}$$

By averaging over all directions $\omega \in S^{n-1}$ and letting $\varepsilon \to 0$ one gets in the limit

$$\int_M (|\nabla u| - |\nabla v|)^2 d\mu_g \leqslant 2E(u) + 2E(v) - 4E(w).$$

As the maps u and v coincide on ∂M so do the maps v and w. In view of minimality of v this implies that $E(v) \leqslant E(w)$, and the lemma follows. □

With the help of the last lemma, the existence of θ so that inequality (15.14) holds becomes straightforward. First using Lemma 15.1 one chooses $R \in [1/2, 3/4]$ so that

$$E_R(u) - E_R(v_R) \leqslant C(E_1(u))^{9/8}.$$

For a fixed $\theta \leqslant 1/4$ the energy of u over the ball B_θ is estimated by the energy of v_R over that ball:

$$\begin{aligned} E_\theta(u) &= \int_{B_\theta} |\nabla u|^2 d\mu_g \\ &\leqslant 2\int_{B_\theta} (|\nabla u| - |\nabla v_R|)^2 d\mu_g + 2\int_{B_\theta} |\nabla v_R|^2 d\mu_g. \end{aligned} \tag{15.16}$$

The first integral is handled with the help of Lemma 15.3 and (15.6)

$$\begin{aligned} \int_{B_\theta} (|\nabla u| - |\nabla v_R|)^2 d\mu_g &\leqslant \int_{B_R} (|\nabla u| - |\nabla v_R|)^2 d\mu_g \\ &\leqslant 2(E_R(u) - E_R(v_R)) \leqslant C(E_1(u))^{9/8}. \end{aligned} \tag{15.17}$$

Estimating the second integral uses interior regularity for harmonic maps into NPC spaces. Namely,

$$\sup_{B_{1/4}} |\nabla v_R|^2 \leqslant CE_{1/2}(v_R) \leqslant CE_R(v_R) \leqslant CE_1(u),$$

and hence

$$\int_{B_\theta} |\nabla v_R|^2 d\mu_g \leqslant C\theta^n E(u).$$

Putting it all together one gets

$$\theta^{2-n} E_\theta(u) \leqslant C[\theta^{2-n}(E_1(u))^{1/8} + \theta^2] E_1(u).$$

Choose θ depending on C so that $C\theta^2 < 1/4$ and then choose ε_0 depending on C and θ so that $C\theta^{2-n}\varepsilon_0^{1/8} < 1/4$. Then inequality (15.14) holds provided $E_1(u) \leqslant \varepsilon_0$. That completes the proof. □

Lipschitz continuity of the minimizer

The following heuristic argument outlines the main ideas involved in showing Lipschitz continuity of the minimizer. (Complete proof of all mentioned results can be found in [Se 1995].)

If $u : M \to N$ is a harmonic map from a Riemannian domain (M, g) into a smooth manifold whose sectional curvature is not greater than 1, then it is a consequence of the elliptic Bochner formula of Eells–Sampson [ES 1964] that the energy density $e = |\nabla u|^2$ satisfies

$$\frac{1}{2}\Delta e - |\nabla e^{\frac{1}{2}}|^2 + e^2 + Ce \geqslant 0. \tag{15.18}$$

Assuming that image$(u) \subset B_\varrho(Q)$ (with $\varrho < \frac{\pi}{2}$) one can estimate e in terms of $R(x) = d(u(x), Q)$

$$e \leqslant -\frac{\Delta \cos R}{\cos R}.$$

[JK 1979, Lemma 4, (29)]. Substituting this estimate into (15.18) yields the inequality

$$\frac{1}{2}\Delta e - |\nabla e^{\frac{1}{2}}|^2 - e\frac{\Delta \cos R}{\cos R} + Ce \geqslant 0$$

which can be formally rewritten as

$$\frac{e^{\frac{1}{2}}}{\cos R}\operatorname{div}\Big(\cos^2 \nabla\frac{e^{\frac{1}{2}}}{\cos R}\Big) + Ce \geqslant 0.$$

Note that e is assumed to be only integrable ($e \in L^1$), so that the above inequality holds only in a formal sense. One can, however, multiply it by η^2 ($\eta \in C_c^1(M)$) and integrate by parts to get

$$\int_M \cos^2 R\,\nabla\frac{e^{\frac{1}{2}}}{\cos R}\nabla\Big(\eta^2\frac{e^{\frac{1}{2}}}{\cos R}\Big)d\mu_g \leqslant C\int_M e\eta^2 d\mu_g. \tag{15.19}$$

After differentiating $\eta^2 e^{\frac{1}{2}}/\cos R$, one uses Young's inequality to get

$$\begin{aligned}(1-\varepsilon)\int_M \eta^2\cos^2 R\Big|\nabla\frac{e^{\frac{1}{2}}}{\cos R}\Big|^2 d\mu_g \\ \leqslant \frac{1}{\varepsilon}\int_M e|\nabla\eta|^2 d\mu_g + C\int_M e\eta^2 d\mu_g.\end{aligned} \tag{15.20}$$

This suggests that $e^{\frac{1}{2}} / \cos R \in W^{1,2}_{\mathrm{loc}}(M)$. If that is the case, then (15.19) implies that e is locally bounded (via the Moser iteration technique), and hence that u is locally Lipschitz.

Now let $u \in W^{1,2}(M,Y)$ be an energy minimizing map into a space with curvature bounded above by 1. As u is already known to be (Hölder) continuous, one may assume that image(u) is contained in a small geodesic ball in Y. To prove that the map u is Lipschitz continuous, one constructs a suitable variation and uses minimality of u to deduce the analogue of (15.19). To do so, one derives (roughly speaking) an estimate for the distance between u and its translated version. Let $\omega \in \mathbb{R}^n$ be a unit vector. Consider a map $u_{s\omega}(x) = u(xs\omega)$. If M is a Euclidean domain, then $u_{s\omega}$ is also minimizing. In that case the distance between the two maps $d(x) = d(u(x), u_{s\omega}(x))$ satisfies (weakly) a differential inequality (established for maps into smooth manifolds by Jäger and Kaul [JK 1979])

$$\operatorname{div}\left(\cos R \cos R_{s\omega} \nabla \sqrt{\frac{2(1-\cos d)}{\cos R \cos R_{s\omega}}}\right) \geqslant 0 \tag{15.21}$$

where $R = d(u(x), Q)$ and $R_{s\omega} = d(u_{s\omega}(x), Q)$, [Se 1995, Corollary 2.3].

Note that in view of the already established Hölder continuity of u, the two maps u and $u_{s\omega}$ are close to each other, and hence $2(1 - \cos d) \approx d^2$. Now it is easy to believe that dividing (15.21) by s and averaging over all directions $\omega \in S^{n-1}$ may yield (15.19).

In the non-Euclidean case, the map $u_{s\omega}$ is energy minimizing but with respect to a pull-back metric. The difference between the original and pull-back metric results in an error term. Handling that error term makes the non-Euclidean case considerably harder.

Bibliography

[A 1951] Alexandrov, A.D., *A theorem on triangles in a metric space and some of its applications*, (In Russian), Trudy Math. Inst. Steklov **38** (1951), 5–23.

[A 1957] Alexandrov, A.D., *Über eine Verallgemeinerung der Riemannschen Geometrie*, Schr. Forschungsinst. Math. Berlin **1** (1957), 33–84.

[AB 1990] Alexander, S.B., and R.L. Bishop, *The Hadamard-Cartan theorem in locally convex metric spaces*, L'Enseignement Math. **36** (1990), 309–320.

[AB 1996] Alexander, S.B., and R.L. Bishop, *Comparison theorems for curves of bounded geodesic curvature in metric spaces of curvature bounded above*, Diff. Geom. Appl. **6** (1996), 67–86.

[ABB 1987] Alexander, S.B., I.D. Berg, and R.L. Bishop, *The Riemannian obstacle problem*, Illinois J. Math. **31** (1987), 167–184.

[ABB 1993] Alexander, S.B., I.D. Berg, and R.L. Bishop, *Geometric curvature bounds in Riemannian manifolds with boundary*, Trans. Amer. Math. Soc. **339** (1993), 703–716.

[Ad 1975] Adams, R.A., *Sobolev Spaces*, Pure Appl. Math., Academic Press, 1975.

[AG 1991] Alinhac, S., and P. Gérard, *Opérateurs pseudo-différentiels et théorème de Nash-Moser*, InterEditions/CNRS, 91.

[AH 1996] Adams, D.R., and L.I. Hedberg, *Function Spaces and Potential Theory*, Grundlehren, Band 314, Springer, Berlin, 1996.

[AKS 1962] Aronszajn, N., A. Krzywicki, and J. Szarski, *A unique continuation theorem for exterior differential forms on Riemannian manifolds*, Arkiv f. Mat. **4** (1962), 417–453.

[B 1984] Baldes, A., *Die Äquatorabbildung ins Ellipsoid und Harmonische Abbildungen singulärer Riemannscher Mannigfaltigkeiten*, Dissertation, Bonn, 1984.

[B 1986] Baldes, A., *Degenerate elliptic operators, diagonal systems and variational integrals*, Manuscripta Math. **55** (1986), 467–486.

[Bal 1995] Ballmann, W., *Lectures on Spaces of Nonpositive Curvature*, DMV Seminar, Band 25, Birkhäuser, Boston, 1995.

[Bar 1972] Barth, T.J., *The Kobayashi distance induces the standard topology*, Proc. Amer. Math. Soc. **35** (1972), 439–441.

[Bau 1966] Bauer, H., *Harmonische Räume und ihre Potentialtheorie*, Lecture Notes in Math. No. 22, Springer, Berlin, 1966.

[Bau 1984] Bauer, H., *Harmonic Spaces, a Survey*, Conferenze del Seminario di Matematico Bari No. 197, Laterza, Bari, 1984.

[BC 1993] Bao, D., and S.S. Chern, *On a notable connection in Finsler geometry*, Houston J. Math. **19** (1993), 135–180.

[BD 1959] Beurling, A., and J. Deny, *Dirichlet spaces*, Proc. Nat. Acad. Sci. U.S.A. **45** (1959), 208–215.

[BE 1981] Baird, P., and J. Eells, *A conservation law for harmonic maps*, Geometry Symposium, Utrecht 1980. Lecture Notes in Math. No. 894, Springer, Berlin, 1981, pp. 1–25.

[Be 1976] Berestovskii, V.N., *Introduction of a Riemannian structure in certain metric spaces*, Siberian Math. J. **16** (1976), 499–507.

[Be 1983] Berestovskii, V.N., *Borsuk's problem on the metrization of a polyhedron*, Soviet Math. Doklady **27** (1983), 56–59.

[Ben 1974] Bendikov, A., *Spatially homogeneous Markov processes on abelian groups and harmonic structures*, Russian Math. Surveys **29** (1974), no. 5, 215–216.

[Ben 1995] Bendikov, A., *Potential Theory on Infinite-Dimensional Abelian Groups*, W. de Gruyter, Berlin, 1995.

[Berg 1975] Berg, C., *Fonctions harmoniques sur* $\mathbf{T}^{\infty}$, C.R. Acad. Sci. Paris **280** (1975), 1519–1521.

[Berg 1976] Berg, C., *Potential theory on the infinite dimensional torus*, Inventiones Math. **32** (1976), 49–100.

[Berg 1977] Berg, C., *On Brownian and Poissonian convolution semigroups on the infinite dimensional torus*, Inventiones Math. **38** (1977), 227–235.

[Bers 1960] Bers, L., *Quasiconformal mappings and Teichmüller's theorem*, Analytic Functions, Princeton Univ. Press, 1960, pp. 89–119.

[BeS–C 1997] Bendikov, A., and L. Saloff–Coste, *Elliptic diffusions on infinite products*, J. reine angew. Math. **493** (1997), 171–220.

[BeS–C 1999] Bendikov, A., and L. Saloff–Coste, *Potential theory on infinite products and locally compact groups*, Potential Anal. **11** (1999), 325–358.

[BGS 1985] Ballmann, W., M. Gromov, and V. Schroeder, *Manifolds of nonpositive curvature*, Progress in Math. Vol. 61, Birkhäuser, Boston, 1985.

[BHH 1987] Barthel, G., F. Hirzebruch, and T. Höfer, *Geradenkonfigurationen und Algebraische Flächen*, Aspects of Mathematics, Band 4, Vieweg, Braunschweig/Wiesbaden, 1987.

[BJS 1964] Bers, L., F. John, and M. Schechter, *Partial Differential Equations*, Academic Press, New York, 1964.

[BL 1996] Bao, D., and B. Lackey, *A Hodge decomposition theorem for Finsler spaces*, C.R. Acad. Sci. Paris **323** (1996), 51–56.

[BM 1991] Biroli, M., and U. Mosco, *Formes de Dirichlet et estimations structurelles dans les milieux discontinus*, C.R. Acad. Sci. Paris A **313** (1991), 593–598.

[BM 1995] Biroli, M., and U. Mosco, *A Saint-Venant type principle for Dirichlet forms on discontinuous media*, Annali Mat. P.A. (4) **169** (1995), 125–181.

[BNik 1993] Berestovskij, V.N., and I.G. Nikolaev, *Multidimensional generalized Riemannian spaces*, Geometry IV, Encycl. Math. Sci. **70**, Springer, Berlin, 1993, pp. 165–243.

[BNir 1955] Bers, L., and L. Nirenberg, *On a representation theorem for linear elliptic systems with discontinuous coefficients and its applications*, Convegno internazionale sulle equazione lineari alle derivate parziali, Trieste, Ed. Cremonese, Roma, 1955, pp. 111–140.

[Bo 1967] Bony, J.–M., *Détermination des axiomatiques de théorie du potentiel dont les fonctions harmoniques sont différentiables*, Ann. Inst. Fourier (Grenoble) **17** (1967), 353–382.

[Bo 1970] Bony, J.–M., *Opérateurs elliptiques dégénérés associés aux axiomatiques de la théorie du potentiel*, Potential Theory, C.I.M.E, Stresa 1969, Ed. Cremonese, Roma, 1970, pp. 69–119.

[Bor 1954] Borel, A., *Kählerian coset spaces of semi-simple Lie groups*, Proc. Nat. Acad. Sci **40** (1954), 1147–1151.

[Bou 1965] Bourbaki, N., *Intégration*, 2e éd., Hermann, Paris, 1965.

[Bou 1974] Bourbaki, N., *Topologie générale*, chap. IX: Utilisation des nombres réels en topologie générale, Hermann, Paris, 1974.

[Bou 1981] Bourbaki, N., *Groupes et algèbres de Lie*, 2e éd., Hermann, Paris, 1981.

[BPV 1984] Barth, W., C. Peters, and A. Van de Ven, *Compact Complex Surfaces*, Springer, Berlin, 1984.

[Br 1957] Brelot, M., *Extension axiomatique des fonctions sous-harmoniques*, C.R. Acad. Sci. Paris **245** (1957), 1688–1690.

[Br 1958a] Brelot, M., *Une axiomatique générale du problème de Dirichlet dans les espaces localement compacts*, Sém. Théorie du Potentiel **1** (1958), 6.01–6.16.

[Br 1958b] Brelot, M., *Extension axiomatique des fonctions sous-harmoniques*, C.R. Acad. Sci. Paris **246** (1958), 2334–2337.

[Br 1959] Brelot, M., *Axiomatique des fonctions harmoniques et surharmoniques dans un espace localement compact*, Sém. Théorie du Potentiel **2** (1959), 1.01–1.40.

[Br 1960] Brelot, M., *Lectures on Potential Theory*, Part IV, Tata Institute of Fundamental Research, Bombay, 1960. Reissued 1967.

[Br 1962a] Brelot, M., *Introduction axiomatique à l'effilement*, Ann. Mat. Pura Appl. **57** (1962), 77–95.

[Br 1962b] Brelot, M., *Quelques propriétés et applications nouvelles de l'effilement*, Sém. Théorie du Potentiel Paris **6** (1962), 1.27–1.40.

[Br 1969] Brelot, M., *Axiomatique des fonctions harmoniques*, Sér. Math. Sup. Montréal Eté 1965. 2e éd. no. 14, 1969.

[Br 1970] Brelot, M., *Historical Introduction*, Potential Theory, 1–21, CIME, Edizione Cremonese, Roma 1970.

[Bra 1979] Brand, N., *Necessary conditions for the existence of branched coverings*, Inventiones Math. **54** (1979), 1–10.

[Bri 1991] Bridson, M., *Geodesics and curvature in metric simplicial complexes*, Group theory from a geometrical viewpoint, World Sci., 1991, pp. 373–463.

[Bro 1989] Brown, K.S., *Buildings*, Springer, Berlin, 1989.

[BruTi 1972] Bruhat, F., and J. Tits, *Groupes reductifs sur un corps local. I. Données radicielles valuées*, Publ IHES **41** (1972), 5–251.

[BS 1994] Bao, D., and Z. Shen, *On the volume of unit tangent spheres in a Finsler manifold*, Results in Math. **26** (1994), 1–17.

[Bus 1947] Busemann, H., *Intrinsic area*, Ann. Math. **48** (1947), 234–267.

[BW 1992] Baird, P., and J.C. Wood, *Harmonic morphisms, Seifert fibre spaces and conformal foliations*, Proc. London Math. Soc. **64** (1992), 170–196.

[BZ 1974] Bagby, T., and W.P. Ziemer, *Pointwise differentiability and absolute continuity*, Trans. Amer. Math. Soc. **191** (1974), 129–148.

[C 1909] Carathéodory, C., *Untersuchungen über die Grundlagen der Thermodynamik*, Math. Ann. **67** (1909), 355–386.

[Ca 1946] Cartan, H., *Théorie générale du balayage en potentiel newtonien*, Ann. Univ. Grenoble, Sect. Sci. Math. Phys. **22** (1946), 221–280.

[Ca 1957] Cartan, H., *Quotient d'un espace analytique par un groupe d'automorphismes*, Symp. S. Lefschetz, Princeton, 1957, pp. 90–102.

[Caf 1982] Cafarelli, L., *Regularity theorems for weak solutions of some nonlinear systems*, Commun. Pure Appl. Math. **35** (1982), 833–838.

[CC 1965] Constantinescu, C., and A. Cornea, *Compactifications of harmonic spaces*, Nagoya Math. J. **25** (1965), 1–57.

[CC 1972] Constantinescu, C., and A. Cornea, *Potential Theory on Harmonic Spaces*, Grundlehren, Band 158, Springer, Berlin, 1972.

[CD 1993] Charney, R., and M. Davis, *Singular metrics of nonpositive curvature on branched covers of Riemannian manifolds*, Amer. J. Math. **115** (1993), 929–1009.

[Ce 2000] Centore, P., *Finsler Laplacians and minimal energy maps*, Internat. J. Math. **11** (2000), 1–13.

[Ch 1948] Chern, S.S., *Local equivalence and Euclidean connections in Finsler spaces*, Sci. Rep. Nat. Tsiang Hua Univ. Ser. A5 (1948), 95–121. Selected Papers Vol. II, 194–212.

[Cha 1988] Chang, S. X.–D., *Two dimensional minimizing integral currents are classical minimal surfaces*, J. Amer. Math. Soc. **1** (1988), 699–778.

[Chee 1980] Cheeger, J., *On the Hodge theory of Riemannian pseudomanifolds*, Proc. Symp. Pure Math. **36** (1980), 91–146.

[Chee 1983] Cheeger, J., *Spectral geometry of singular Riemannian spaces*, J. Diff. Geom. **18** (1983), 575–657.

[Chee 1999] Cheeger, J., *Differentiability of Lipschitz functions on metric measure spaces*, Geom. Funct. Anal. **9** (1999), 428 - 517.

[Chen 1995] Chen, J.–Y., *On energy minimizing mappings between and into singular spaces*, Duke Math. J. **79** (1995), 77–99.

[Chi 1990] Chiang, Y.–J., *Harmonic maps of V-manifolds*, Ann. Glob. Anal. Geom. **8** (1990), 315–344.

[Chow 1939] Chow, W.L., *Über Systeme von linearen partiellen Differentialgleichungen erster Ordnung*, Math. Ann. **117** (1939), 98–105.

[Co 1992] Corlette, K., *Archimedean superrigidity and hyperbolic geometry*, Ann. Math. **135** (1992), 165–182.

[CP 1988] De Cecco, G., and G. Palmieri, *Distanza intrinseca su una varietà Riemanniana di Lipschitz*, Rend. Sem. Mat. Univ.Torino **46** (1988), 157–170.

[CP 1990] De Cecco, G., and G. Palmieri, *Length of curves on Lip manifolds*, Rend. Mat. Acc. Lincei (9) **1** (1990), 215–221.

[CP 1991] De Cecco, G., and G. Palmieri, *Integral distance on a Lipschitz Riemannian manifold*, Math. Z. **207** (1991), 223–243.

[CP 1993] De Cecco, G., and G. Palmieri, *Distanza intrinseca su una varietà finsleriana di Lipschitz*, Rend. Accad. Naz. Sci. **17** (1993), 129–151.

[CP 1995] De Cecco, G., and G. Palmieri, *LIP manifolds: from metric to Finslerian structure*, Math. Z. **218** (1995), 223–237.

[CR 1992] Chiang, Y.–J., and A. Ratto, *Harmonic maps on spaces with conical singularities*, Bull. Soc. Math. France **120** (1992), 251–262.

[CT 1968] Church, P.T., and J.G. Timourian, *Fibre bundles with singularities*, J. Math. Mech. **18** (1968), 71–90.

[CW 1971] Coifman, R.R., and G. Weiss, *Analyse harmonique non commutative sur certains espaces homogènes*, Lecture Notes in Math. No. 242, Springer, Berlin, 1971.

[De 1970] Deny, J., *Méthodes hilbertiennes en théorie du potentiel*, Potential Theory, C.I.M.E, Stresa 1969, Ed. Cremonese, Roma, 1970, pp. 121–201.

[Di 1970] Dieudonné, J., *Éléments d'Analyse*, tome III, Gauthier–Villars, Paris, 1970.

[DJ 1991] Davis, M.W., and T. Januszkiewicz, *Hyperbolization of polyhedra*, J. Diff. Geom. **34** (1991), 347–388.

[DL 1954] Deny, J., and J.L. Lions, *Les espaces du type de Beppo Levi*, Ann. Inst. Fourier (Grenoble) **5** (1954), 305-370.

[Dol 1982] Dolgachev, I., *Weighted projective varieties*, Group Actions and Vector Fields, Lecture Notes in Math. No. 956, Springer, Berlin, 1982, pp. 34–71.

[Doob 1956] Doob, J.L., *Probability methods applied to the first boundary value problem*, Proceedings 3rd Berkeley Symposium on Mathematics, Statistics, and Probability, Univ. of California Press, Berkeley, 1956, pp. 49–80.

[Doob 1984] Doob, J.L., *Classical Potential Theory and Its Probabilistic Counterpart*, Grundlehren, Band 262, Springer, Berlin, 1984.

[DS 1989] Donaldson, S.K., and D.P. Sullivan, *Quasiconformal 4-manifolds*, Acta Math. **163** (1989), 181–252.

[DT 1958] Dold, A., and R. Thom, *Quasifaserungen und unendliche symmetrische Produkte*, Ann. Math. **67** (1958), 239–281.

[DuL 1971] Dupont, J.L., and G.Lusztig, *On manifolds with* $w_1^2 = 0$, Topology **10** (1971), 81–92.

[E 1997] Eells, J., *Minimal representations of homology classes*, unpublished.

[Ed 1976] Edmonds, A.L., *Branched coverings and orbit maps*, Mich. Math. J. **23** (1976), 289–301.

[EdFi 1976] Edmonds, A.L., and R. Fintushel, *Singular circle fiberings*, Math. Z. **151** (1976), 89–99.

[EF 1991] Eells, J., and M.J. Ferreira, *On representing homotopy classes by harmonic maps*, Bull. London Math. Soc. **23** (1991), 160–162.

[EiSt 1952] Eilenberg, S., and N. Steenrod, *Foundations of Algebraic Topology*, Princeton, 1952.

[EL 1978] Eells, J., and L. Lemaire, *A report on harmonic maps*, Bull. London Math. Soc. **10** (1978), 1–68.

[EL 1980] Eells, J., and L. Lemaire., *On the construction of harmonic and holomorphic maps between surfaces*, Math. Ann. **252** (1980), 27–52.

[EL 1983] Eells, J., and L. Lemaire, *Selected topics in harmonic maps*, CBMS Regional Conf. Ser. Vol 50, Amer. Math. Soc., Providence, 1983.

[EL 1988] Eells, J., and L. Lemaire, *Another report on harmonic maps*, Bull. London Math. Soc. **20** (1988), 385–524.

[EP 1984] Eells, J., and J.C. Polking, *Removable singularities of harmonic maps*, Indiana Univ. Math. J. **33** (1984), 859–871.

[ER 1993] Eells, J., and A. Ratto, *Harmonic Maps and Minimal Immersions with Symmetries*, Annals of Math. Studies No. 130, Princeton Univ. Press, 1993.

[ES 1964] Eells, J., and J.H. Sampson, *Harmonic mappings of Riemannian manifolds*, Amer. J. Math. **86** (1964), 109–160.

[EV 1998] Eells, J., and A. Verjovsky, *Harmonic and Riemannian foliations*, Bol. Soc. Mat. Mexicana (3) **4** (1998), 1–12.

[Far 1990] Faran, J.J., *Hermitian Finsler metrics and the Kobayashi metric*, J. Diff. Geom. **31** (1990), 601–625.

[Fe 1969] Federer, H., *Geometric Measure Theory*, Grundlehren Band 153, Springer, Berlin, 1969.

[FHK 1998] Franchi, B., P. Hajłasz, and P. Koskela, *Definitions of Sobolev classes on metric spaces*.

[FLW 1996] Franchi, B., G. Lu, and R. Wheeden, *A relationship between Poincaré type inequalities and representation formulas in spaces of homogeneous type*, Internat. Math. Res. Notes **1** (1996), 1–14.

[FOT 1994] Fukushima, M., Y. Oshima, and M. Takeda, *Dirichlet Forms and Symmetric Markov Processes*, W. de Gruyter, Berlin, 1994.

[FP 1978] Feyel, D., and A. de La Pradelle, *Construction d'un espace harmonique de Brelot associé à un espace de Dirichlet de type local vérifiant une hypothèse d'hypoellipticité*, Inventiones Math. **44** (1978), 109–128.

[Fr 1982] Freedman, M.H., *The topology of four-dimensional manifolds*, J. Diff. Geom. **17** (1982), 357–453.

[Fug 1971] Fuglede, B., *The quasi topology associated with a countably subadditive set function*, Ann. Inst. Fourier (Grenoble) **21** (1971), no. 1, 123–169.

[Fug 1972] Fuglede, B., *Finely Harmonic Functions*, Lecture Notes in Math., No. 289, Springer, Berlin, 1972.

[Fug 1975] Fuglede, B., *Sur la fonction de Green pour un domaine fin*, Ann. Inst. Fourier (Grenoble) **25** (1975), no. 3–4, 201–206.

[Fug 1978] Fuglede, B., *Harmonic morphisms between Riemannian manifolds*, Ann. Inst. Fourier (Grenoble) **28** (1978), no. 2, 107–144.

[Fug 1979a] Fuglede, B., *Harmonic morphisms*, Complex analysis, Joensuu 1978. Lecture Notes in Math. No. 747, Springer, Berlin, 1979, pp. 123–131.

[Fug 1979b] Fuglede, B., *Harnack sets and openness of harmonic morphisms*, Math. Ann. **241** (1979), 181–186.

[Fug 1982a] Fuglede, B., *Fonctions BLD et fonctions finement surharmoniques*, Lecture Notes in Math. No. 906, Springer, Berlin, 1982, pp. 126–157.

[Fug 1982b] Fuglede, B., *Localization in fine potential theory and uniform approximation by subharmonic funtions*, J. Funct. Anal. **49** (1982), 57–72.

[Fug 2000] Fuglede, B., *The beginnings of harmonic morphisms*, Harmonic Morphisms, Harmonic Maps and Related Topics, Brest Proc. 1997. Eds. C.K. Anand et al., Chapman & Hall/CRC Research Notes in Math. 413, London, 2000, pp. 3–12.

[Fuk 1980] Fukushima, M., *Dirichlet Forms and Markov Processes*, North-Holland, Amsterdam, 1980.

[GG 1982] Giaquinta, M., and E. Giusti, *On the regularity of the minima of variational integrals*, Acta Math. **148** (1982), 31–46.

[GH 1982] Giaquinta, M., and S. Hildebrandt, *A priori estimates for harmonic mappings*, J. reine angew. Math. **336** (1982), 124–164.

[Gi 1957] De Giorgi, E., *Sulla differenziabilitá e l'analiticitá delle estremali degli integrali multipli regolari*, Mem. Accad. Sci. Torino, Cl. Sci. Fis. Mat. Nat. (3) **3** (1957), 25–43.

[Gie 1964] Giesecke, B., *Simpliziale Zerlegung abzählbarer analytischer Räume*, Math. Z. **83** (1964), 177–213.

[GKM 1968] Gromoll, D., W. Klingenberg, and W. Meyer, *Riemannsche Geometrie im Grossen*, Lecture Notes in Math. No. 55, Springer, Berlin, 1968.

[GLP 1981] Gromov, M., J. Lafontaine, and P. Pansu, *Structures métriques pour les variétés riemanniennes*, Cedic/Fernand Nathan, Paris, 1981.

[GMacP 1980] Goresky, M., and R. MacPherson, *Intersection homology theory*, Topology **19** (1980), 135–162.

[Gre 1998] Gregori, G., *Sobolev spaces and harmonic maps between singular spaces*, Calc. Var. P.D.E. **7** (1998), 1–18.

[GReck 1965] Grauert, H., and H. Reckziegel, *Hermitische Metriken und normale Familien holomorpher Abbildungen*, Math. Z. **89** (1965), 108–125.

[GRem 1958] Grauert, H., and R. Remmert, *Komplexe Räume*, Math. Ann. **136** (1958), 245–318.

[GRem 1984] Grauert, H., and R. Remmert, *Coherent Analytic Sheaves*, Springer, Berlin, 1984.

[GrHa 1978] Griffith, P., and J. Harris, *Principles of Algebraic Geometry*, Wiley–Interscience, New York, 1978.

[Gro 1986] Gromov, M., *Partial Differential Relations*, Ergeb. Math. 3. Folge, Band 19, Springer, Berlin, 1986.

[Gro 1987] Gromov, M., *Hyperbolic groups*, Essays in Group Theory, Vol. 8, MSRI series, Ed. S.M. Gersten, Springer, 1987, pp. 75–264.

[Gro 1988] Gromov, M., *Dimension, Non-Linear Spectra, and Width*, Lecture Notes in Math., No. 1317, Springer, Berlin, 1988, pp. 132–184.

[Gro 1991] Gromov, M., *Foliated Plateau problem, Part II: Harmonic maps of foliations*, Geometry and Functional Analysis **1** (1991), 253–320.

[Gro 1996] Gromov, M., *Carnot-Carathédory Spaces Seen from Within*, Sub-Riemannian Geometry, Progr. in Math. Vol. 144, Ed. A. Belläiche and J.–J. Risler, Birkhäuser, Boston, 1996, pp. 85–323.

[Gro 1999] Gromov, M., *Metric Structures for Riemannian and Non-Riemannian Spaces*, with Appendices by M. Katz, P. Pansu, and S. Semmes, Progr. in Math. vol. 152, Birkhäuser, Basel, 1999. (This is a much expanded version of [GLP 1981]).

[GRok 1970] Gromov, M., and V. Rokhlin, *Embeddings and immersions in Riemannian geometry*, Russian Math. Surveys **25**:4 (1970), 1–57. Transl. from Uspekhi Mat. Nauk.

[GRos 1965] Gunning, R.C., and H. Rossi, *Analytic functions of several complex variables*, Prentice–Hall, Englewood Cliffs, N.J., 1965.

[GS 1992] Gromov, M., and R. Schoen, *Harmonic maps into singular spaces and p-adic superrigidity for lattices in groups of rank one*, Publ. IHES N° 76 (1992), 165–246.

[GT 1998] Gilbarg, D., and N.S. Trudinger, *Elliptic Partial Differential Equations of Second Order*, Grundlehren Band 224, (2nd edition), Springer, Berlin, 1998.

[GW 1973] Greene, R.E., and H. Wu, *On the subharmonicity and plurisubharmonicity of geodesically convex functions*, Indiana Univ. Math. J. **22** (1973), 641–653.

[GW 1975] Greene, R.E., and H. Wu, *Embedding of open Riemannian manifolds by harmonic functions*, Ann. Inst. Fourier (Grenoble) **25** (1975), no. 1, 215–235.

[GW 1976] Greene, R.E., and H. Wu, *C^∞ convex functions and manifolds of positive curvature*, Acta Math. **137** (1976), 209–245.

[Hae 1984] Haefliger, A., *Introduction to piecewise linear intersection homology*, A. Borel et al., Intersection Cohomology, Progress in Math., vol. 50, Birkhäuser, Boston, 1984, pp. 1–21.

[Haj 1996] Hajłasz, P., *Sobolev spaces on an arbitrary metric space*, Potential Anal. **5** (1996), 403–415.

[Ham 1990] Hamenstädt, U., *Some regularity theorems for Carnot-Carathéodory metrics*, J. Diff. Geom. **32** (1990), 819–850.

[Har 1975] Hardt, R.M., *Stratification of real analytic mappings and images*, Inventiones Math. **28** (1975), 193–208.

[HeKo 1998] Heinonen, J., and P. Koskela, *Quasiconformal maps in metric spaces with controlled geometry*, Acta Math. **181** (1998), 1–67.

[Hel 1969] Helms, L.L., *Introduction to Potential Theory*, Wiley–Interscience, New York, 1969.

[Her 1962] Hervé, R.–M., *Recherches axiomatiques sur la théorie des fonctions surharmoniques et du potentiel*, Ann. Inst. Fourier (Grenoble) **12** (1962), 415–571.

[Her 1964] Hervé, R.–M., *Un principe du maximum pour les sous-solutions locales d'une équation uniformément elliptique de la forme $Lu = -\sum_i \partial/\partial x_i \left(\sum_j a_{ij}\partial u/\partial x_j\right) = 0$*, Ann. Inst. Fourier (Grenoble) **14** (1964), no. 2, 493–508.

[Her 1965] Hervé, R.–M., *Quelques propriétés des fonctions surharmoniques associées à une équation uniformément elliptique de la forme $Lu = -\sum_i \partial/\partial x_i \left(\sum_j a_{ij}\partial u/\partial x_j\right) = 0$*, Ann. Inst. Fourier (Grenoble) **15** (1965), no. 2, 215–224.

[Herm 1977] Hermann, R., *Differential Geometry and the Calculus of Variations*, 2nd edition, Math. Sci. Press, Brookline, Mass., 1977.

[HH 1969] Hervé, R.–M., and M. Hervé, *Les fonctions surharmoniques associées à un opérateur elliptique du second ordre à coefficients discontinus*, Ann. Inst. Fourier (Grenoble) **19** (1969), 305–359.

[HH 1972] Hervé, R.–M., and M. Hervé, *Les fonctions surharmoniques dans l'axiomatique de M. Brelot associées à un opérateur elliptique dégénéré*, Ann. Inst. Fourier (Grenoble) **22** (1972), no. 2, 131–145.

[Hir 1964] Hironaka, H., *Resolution of singularities of an algebraic variety over a field of characteristic zero, I, II*, Ann. Math. **74** (1964), 109–326.

[Hir 1975] Hironaka, H., *Triangulation of algebraic sets*, Proc. Symp. Pure Math., Vol. 29, 1975, pp. 165–185.

[HK 1998] Hajłasz, P., and J. Kinnunen, *Hölder quasicontinuity of Sobolev functions on metric spaces*, Rev. Mat. Iberoamer (to appear).

[HKM 1993] Heinonen, J., T. Kilpeläinen, and O. Martio, *Nonlinear Potential Theory of Degenerate Elliptic Equations*, Oxford Math. Monographs, 1993.

[HKW 1975] Hildebrandt, S., H. Kaul, and K.O. Widman, *Harmonic mappings into Riemannian manifolds with non-positive sectional curvature*, Math. Scand. **37** (1975), 257–263.

[HKW 1976] Hildebrandt, S., H. Kaul, and K.O. Widman, *Dirichlet's boundary value problem for harmonic mappings of Riemannian manifolds*, Math. Z. **147** (1976), 225–236.

[HKW 1977] Hildebrandt, S., H. Kaul, and K.O. Widman, *An existence theorem for harmonic mappings of Riemannian manifolds*, Acta Math. **138** (1977), 1–16.

[HL 1971] Hsiang, W.–Y., and H.B. Lawson, *Minimal submanifolds of low cohomogeneity*, J. Diff. Geom. **5** (1971), 1–38. Correction in F. Uchida, J. Diff. Geom. **15** (1980), 569–574.

[Hö 1967] Hörmander, L., *Hypoelliptic second order differential equations*, Acta Math. **119** (1967), 147–171.

[HW 1975] Hildebrandt, S., and K.O. Widman, *Some regularity results for quasilinear elliptic systems of second order*, Math. Z. **142** (1975), 67–86.

[HW 1977] Hildebrandt, S., and K.O. Widman, *On the Hölder continuity of weak solutions of quasilinear elliptic systems of second order*, Ann. Scu. Norm. Sup. Pisa, Ser. IV **4** (1977), 145–178.

[Ish 1979] Ishihara, T., *A mapping of Riemannian manifolds which preserves harmonic functions*, J. Math. Kyoto Univ. **19** (1979), 215–229.

[J 1994] Jost, J., *Equilibrium maps between metric spaces*, Calc. Var. **2** (1994), 173–204.

[J 1995] Jost, J., *Convex functionals and generalized harmonic maps of non-positive curvature*, Comm. Math. Helv. **70** (1995), 659–673.

[J 1996] Jost, J., *Generalized harmonic maps between metric spaces*, Geometric Analysis and Calculus of Variations, Internat. Press, Cambridge, Mass., 1996, pp. 143–174.

[J 1997a] Jost, J., *Generalized Dirichlet forms and harmonic maps*, Calc. Var. P.D.E. **5** (1997), 1–19.

[J 1997b] Jost, J., *Nonpositive Curvature: Geometric and Analytic Aspects*, Birkhäuser, Boston, 1997.

[J 1997c] Jost, J., *Nonlinear Dirichlet Forms*, Lecture Notes N° 1, Max Planck Institut, Leipzig, 1997.

[Jä 1966] Jänich, K., *Differenzierbare Mannigfaltigkeiten mit Rand als Orbiträume differenzierbarer G-Mannigfaltigkeiten ohne Rand*, Topology **5** (1966), 301–320.

[JaNg 1975] Janssen, K., and Nguyen–Xuan–Loc, *On the balayage of Green functions on finely open sets*, Z. Wahrscheinlichkeitstheorie verw. Geb. **31** (1975), 147–155.

[Je 1986] Jerison, D., *The Poincaré inequality for vector fields satisfying Hörmander's condition*, Duke Math. J. **53** (1986), 503–523.

[JeS–C 1987] Jerison, D., and A. Sánchez–Calle, *Subelliptic second order differential equations*, Complex Analysis III. Proceedings, Lecture Notes in Math. No. 1277, Springer, Berlin, 1987, pp. 46-77.

[JK 1979] Jäger, W., and H. Kaul, *Uniqueness and stability of harmonic maps and their Jacobi fields*, Manuscripta Math. **28** (1979), 269–291.

[JoNi 1961] John F., and L. Nirenberg, *On functions of bounded mean oscillation*, Comm. Pure Appl. Math. **14** (1961), 415–426.

[JX 1998] Jost, J., and C.–J. Xu, *Subelliptic harmonic maps*, Trans. Amer. Math. Soc. **350** (1998), 4633–4649.

[K 1988] Kazdan, J., *Unique continuation in geometry*, Comm. Pure Appl. Math. **41** (1988), 667-681.

[Ka 2000] Kasue, A., *Convergence of Riemannian manifolds and harmonic maps*, Preprint, Osaka City University.

[Kar 1970] Karcher, H., *Anwendungen der Alexandrowschen Winkelvergleichssätze*, Manuscripta Math. **2** (1970), 77–102.

[KilM 1992] Kilpeläinen, T., and J. Malý, *Supersolutions to degenerate elliptic equations on quasi open sets*, Comm. P.D.E. **17** (1992), 371–405.

[KiSi 1969] Kirby, R.C., and L.C. Siebenmann, *On the triangulation of manifolds and the Hauptvermutung*, Bull. Amer. Math. Soc. **75** (1969), 742–749.

[KK 1994] Kasue, A., and H. Kamura, *Spectral convergence of Riemannian manifolds*, Tôhoku Math. J. **46** (1994), 147–179.

[KK 1996] Kasue, A., and H. Kamura, *Spectral convergence of Riemannian manifolds*, Tôhoku Math. J. **48** (1996), 71–120.

[KM 1996] Kinnunen, J., and O. Martio, *The Sobolev capacity on metric spaces*, Ann. Acad. Sci. Fenn. (A.I., Math.) **21** (1996), 367–382.

[KN 1969] Kobayashi, S., and K. Nomizu, *Foundations of Differential Geometry*, Vol. II, Interscience, New York, 1969.

[Kob 1970] Kobayashi, S., *Hyperbolic Manifolds and Holomorphic Mappings*, Pure and Appl. Math. No. 2, Dekker, New York, 1970.

[Kob 1976] Kobayashi, S., *Intrinsic distances, measures and geometric function theory*, Bull. Amer. Math. Soc. **82** (1976), 357–416.

[Kob 1998] Kobayashi, S., *Hyperbolic Complex Spaces*, Grundlehren Band 318, Springer, Berlin, 1998.

[KoSu 1998] Kotani, M., and T. Sunada, *Standard realizations of crystal lattices via harmonic maps*, Preprint, to appear in Trans. Amer. Math. Soc.

[KoSu 2000] Kotani, M., and T. Sunada, *Spectral geometry of crystal lattices*, Preprint.

[KS 1993] Korevaar, N.J., and R. M. Schoen, *Sobolev spaces and harmonic maps for metric space targets*, Comm. Anal. Geom. **1** (1993), 561–659. Reprinted as Chapter X in *Lectures on harmonic maps*, by R. Schoen and S.T. Yau, Conf. Proc. and Lecture Notes in Geometry and Topology, Vol. II (1997), 204–310.

[KS 1997] Korevaar, N.J., and R. M. Schoen, *Global existence theorems for harmonic maps to non-locally compact spaces*, Comm. Anal. Geom. **5** (1997), 333–387.

[Ku 1996] Kuwert, E., *Harmonic maps between flat surfaces with conical singularities*, Math. Z. **221** (1996), 421–436.

[La 1980] Lassoued, L., *Espaces de Dirichlet et axiomatique Brelot*, Sém. Théorie du Potentiel Paris No. 5. Lecture Notes in Math. No. 814, Springer, Berlin, 1980, pp. 117–152.

[Leb 1996] Lebeau, E., *Applications harmoniques entre graphes finis et un théorème de superrigidité*, Ann. Inst. Fourier (Grenoble) **46** (1996), 1183–1203.

[Leh 1986] Lehtola, P., *An axiomatic approach to non-linear potential theory*, Ann. Acad. Sci. Fenn. (A.I., Math.) **62** (1986), 1–40.

[Lei 1988] Leite, M.L., *Harmonic mappings of surfaces with respect to degenerate metrics*, Amer. J. Math. **110** (1988), 399–412.

[Lem 1978] Lemaire, L., *Applications harmoniques de surfaces riemanniennes*, J. Diff. Geom. **13** (1978), 51–78.

[Lem 1982] Lemaire, L., *Boundary value problems for harmonic and minimal maps of surfaces into manifolds*, Ann. Scuola Norm. Sup. Pisa (4) **9** (1982), 91–103.

[LiT 1995] Li, P., and G. Tian, *On the heat kernel of the Bergmann metric on algebraic varieties*, J. Amer. Math. Soc. **8** (1995), 857–877.

[Lin 1997] Lin, F.H., *Analysis on singular spaces*, Geometry, Analysis, and Mathematical Physics, edited by Li Ta–Tsien, World Sci. Publications, 1997, pp. 114–126.

[Ło 1964] Łojasiewicz, S., *Triangulation of semi-analytic sets*, Ann. Scuola Norm. Sup. Pisa **18** (1964), 449–474.

[Ło 1991] Łojasiewicz, S., *Introduction to Complex Analytic Geometry*, Birkhäuser, Boston, 1991.

[LS 1998] Liu, Z.–D., and Z. Shen,, *Riemannian geometry of conical singular sets*, Ann. Glob. Anal. Geom. **16** (1998), 29–62.

[LSW 1963] Littman, W., G. Stampacchia, and H.F. Weinberger, *Regular points for elliptic equations with discontinuous coefficients*, Ann. Scuola Norm. Sup. Pisa (3) **17** (1963), 43–77.

[LV 1977] Luukkainen, J., and J. Väisälä, *Elements of Lipschitz topology*, Ann. Acad. Sci. Fenn. (A.I., Math.) **3** (1977), 85–122.

[MacD 1962] MacDonald, I.G., *Symmetric products of an algebraic curve*, Topology **1** (1962), 319–343.

[Mae 1980] Maeda, F.–Y., *Dirichlet Integrals on Harmonic Spaces*, Lecture Notes in Math. No. 803, Springer, Berlin, 1980.

[Mas 1975] Masur, H., *On a class of geodesics in Teichmüller space*, Ann. Math. **102** (1975), 205–221.

[May 1998] Mayer, U.F., *Gradient flows on nonpositively curved metric spaces and harmonic maps*, Commun. Anal. Geom. **6** (1998), 199–253.

[McC 1977] McCrory, C., *A characterization of homology manifolds*, J. London Math. Soc. **16** (1977), 149–159.

[ME 1956] Morrey, C.B., and J. Eells, *A variational method in the theory of harmonic integrals I*, Ann. Math. **63** (1956), 91–128.

[Mei 1984] Meier, M., *On quasilinear elliptic systems with quadratic growth*, Preprint.

[Mi 1974] Miller, K., *Nonunique continuation for uniformly parabolic and elliptic equations in self-adjoint divergence form with Hölder continuous coefficients*, Archive Rational Mechanics and Analysis **54** (1974), 105–117.

[Mil 1959] Milnor, J., *Sommes de variétés différentiables et structures différentiables des sphères*, Bull. Soc. Math. France **87** (1959), 439–444.

[Miy 1968] Miyahara, Y., *On some properties of a Teichmüller mapping*, TRU Math. **4** (1968), 36–43.

[Mol 1988] Molino, P., *Riemannian Foliations*, Progr. in Math., vol. 73, Birkhäuser, Boston, 1988.

[Mor 1938] Morrey, C.B., *On the solutions of quasi-linear elliptic partial differential equations*, Trans. Amer. Math. Soc. **43** (1938), 126–166.

[Mor 1959] Morrey, C.B., *Second order elliptic equations in several variables and Hölder continuity*, Math. Z. **72** (1959), 146–164.

[Mor 1966] Morrey, C.B., *Multiple Integrals in the Calculus of Variations*, Grundlehren, Band 130, Springer, Berlin, 1966.

[Mos 1960] Moser, J., *A new proof of de Giorgi's theorem concerning the regularity problem for elliptic differential equations*, Comm. Pure Appl. Math. **13** (1960), 457–468.

[Mos 1961] Moser, J., *On Harnack's theorem for elliptic differential equations*, Comm. Pure Appl. Math. **14** (1961), 577–591.

[MS 1979] Macías, R.A., and C. Segovia, *Lipschitz functions on spaces of homogeneous type*, Adv. in Math. **33** (1979), 257–270.

[N 1979] Nikolaev, I.G., *Solution of Plateau's problem in spaces with curvature at most K*, Siberian Math. J. **20** (1979), 246-252.

[N 1980] Nikolaev, I.G., *Parallel translation and smoothness of the metric spaces of bounded curvature*, Soviet Math. Doklady **21** (1980), 263–265.

[Nag 1988] Nag, S., *The Complex Analytic Theory of Teichmüller Spaces*, Canad. Math. Soc. Series, Wiley, New York, 1988.

[Nash 1956] Nash, J., *The imbedding problem for Riemannian manifolds*, Ann. Math. **63** (1956), 20–63.

[Nash 1958] Nash, J., *Continuity of solutions of parabolic and elliptic equations*, Amer. J. Math. **80** (1958), 931-954.

[NSW 1985] Nagel, A., E.M. Stein, and S. Wainger, *Balls and metrics defined by vector fields I: Basic properties*, Acta Math. **155** (1985), 103–147.

[O'B 1971] O'Byrne, B., *On Finsler geometry and applications to Teichmüller spaces*, Annals of Math. Studies No. 66, Princeton Univ. Press, 1971, pp. 317–328.

[OS 1994] Otsu, Y., and T. Shioya, *The Riemannian structure of Alexandrov spaces*, J. Diff. Geom. **39** (1994), 629–658.

[P 1991] Plaut, C., *Almost Riemannian spaces*, J. Diff. Geom. **34** (1991), 515–537.

[P 1998] Plaut, C., *Metric spaces of curvature $\geqslant k$*, Handbook of Geometric Topology, Elsevier, 2000.

[Pf 2000] Pflaum, M.J., *Contributions to the Geometry and Analysis of Stratified Spaces*, Habilitationsschrift, Humboldt Universität, Berlin, 2000.

[Ra 1987] Ratto, A., *Harmonic Maps of Spheres and Equivariant Theory*, Thesis, Univ. of Warwick, 1987.

[Rei 1961] Reinhart, B.L., *Closed metric foliations*, Mich. Mat. J. **8** (1961), 7–9.

[Rei 1983] Reinhart, B.L., *Differential Geometry of Foliations*, Ergebnisse der Math. No. 99, Springer, Berlin, 1983.

[Resh 1968] Reshetnyak, Y.G., *Non-expanding maps in a space of curvature no greater than K*, Siberian Math. J. **9** (1968), 918–927.

[Roy 1971a] Royden, H.L., *Remarks on the Kobayashi metric*, Lecture Notes in Math. No. 185, Springer, Berlin, 1971, pp. 125–137.

[Roy 1971b] Royden, H.L., *Automorphisms and isometries of Teichmüller space*, Annals of Math. Studies No. 66, Princeton Univ. Press, 1971, pp. 369–383.

[RS 1960] Remmert, R., and K. Stein, *Eigenliche holomorphe Abbildungen*, Math. Z. **73** (1960), 159–189.

[RV 1961] Remmert, R., and T. van de Ven, *Über holomorphe Abbildungen projektiv-algebraischer Mannigfaltigkeiten auf komplexe Räume*, Math. Ann. **142** (1961), 453-486.

[SaU 1981] Sacks, J., and K. Uhlenbeck, *The existence of minimal immersions of 2-spheres*, Ann. Math. **113** (1981), 1-24.

[S–C 1984] Sánchez-Calle, A., *Fundamental solutions and geometry of the sum of squares of vector fields*, Inventiones Math. **78** (1984), 143–160.

[Sc 1983] Scott, P., *The geometries of 3-manifolds*, Bull. London Math. Soc. **15** (1983), 401-487.

[Sch 1997] Schoen, R.M., *Analytic aspects of the harmonic map problem*, Conf. Proc. and Lecture Notes in Geometry and Topology Vol. II, Internat. Press, 1997, 155-203. Earlier version in MSRI Publ. Vol 2, Cambridge, Mass., 1984, 321–358.

[SchU 1982] Schoen, R.M., and K. Uhlenbeck, *A regularity theory for harmonic maps*, J. Diff. Geom. **17** (1982), 307–335 and **18** (1983), 329.

[Se 1994] Serbinowski, T., *Boundary regularity of harmonic maps to nonpositively curved metric spaces*, Comm. Anal. Geom. **2** (1994), 1-15.

[Se 1995] Serbinowski, T., *Harmonic Maps into Metric Spaces with Curvature Bounded Above*, Thesis, Univ. Utah, 1995.

[Serre 1980] Serre, J.P., *Trees*, Springer, Berlin, 1980.

[Sh 1974] Shafarevich, I.R., *Basic Algebraic Geometry*, Grundlehren Band 213, Springer, Berlin, 1974.

[Si 1971] Siebenmann, L.C., *Topological manifolds*, Internat. Congr. Math., 1970, vol 2, 1971, pp. 133-163.

[Sp 1966] Spanier, E.H., *Algebraic Topology*, McGraw-Hill, New York, 1966.

[St 1995a] Sturm, K.–T., *Analysis on local Dirichlet spaces. II: Upper Gaussian estimates for the fundamental solutions of parabolic equations*, Osaka J. Math. **32** (1995), 275–312.

[St 1995b] Sturm, K.–T., *On the geometry defined by Dirichlet forms*, Progr. in Probability, Vol. 36, Birkhäuser, Basel, 1995, 231–242.

[St 1996] Sturm, K.–T., *Analysis on local Dirichlet spaces. III: The parabolic Harnack inequality*, J. Math. Pures Appl. **75** (1996), 273–297.

[St 1997] Sturm, K.–T., *Monotone approximation of energy functionals for mappings into metric spaces. I*, J. reine angew. Math. **486** (1997), 129–151.

[St 1998] Sturm, K.–T., *Diffusion processes and heat kernels on metric spaces*, Ann. Prob. **26** (1998), 1–55.

[St 1999a] Sturm, K.–T., *Monotone approximation of energy functionals for mappings into metric spaces. II*, Potential Anal. **11** (1999), 359–386.

[St 1999b] Sturm, K-T., *Nonlinear Markov operators, discrete heat flow, and harmonic maps between singular spaces*, Preprint.

[St 1999c] Sturm, K-T., *Nonlinear heat flow associated with symmetric Markov kernels and energy minimizing maps between singular spaces.*, Preprint.

[Str 1986] Strichartz, R., *Sub-Riemannian geometry*, J. Diff. Geom. **24** (1986), 221–263.

[Su 1979] Sullivan, D.P., *Hyperbolic geometry and homeomorphisms*, Proc. Georgia Conv. Geometry Topology 1977, Acad. Press, New York, 1979, pp. 543-555.

[Te 1983] Teleman, N., *The index of signature operators on Lipschitz manifolds*, Publ. IHES No. 58 (1983), 39-78.

[Thom 1966] Thom, R., *Propriétés différentielles locales des ensembles analytiques*, Sém. Bourbaki **1964/65** (1966), no. 3, 281.01–281.12.

[Thom 1969] Thom, R., *Ensembles et morphismes stratifiés*, Bull. Amer. Math. Soc. **75** (1969), 240–284.

[Ti 1974] Tits, J., *On buildings and their applications*, Proc. Internat. Congr. Math. Vancouver, vol. 1, Amer. Math. Soc., Providence, R.I., 1974, pp. 209–220.

[U 2000] Urakawa, H., *A discrete analogue of the harmonic morphism and Green kernel comparison theorems*, Harmonic Morphisms, Harmonic Maps, and Related Topics. Brest Proc. 1997. Eds. C.K. Anand et al., Chapman & Hall/CRC Research Notes in Math. 413 (2000), 97–108, London.

[V 1986] Varouchas, J., *Sur l'image d'une variété Kählérienne compacte*, Lecture Notes in Math., vol. 1188, Springer, 1986, pp. 245–259.

[Va 1935] Vasilesco, F., *Sur la continuité du potentiel à travers des masses et la démonstration d'un lemme de Kellogg*, C. R. Acad. Sci. Paris **200** (1935), 1173–1174.

[Vä 1966] Väisälä, J., *Discrete open mappings on manifolds*, Ann. Acad. Sci. Fenn. (A.I) (Math.) (1966), no. 392, 1–10.

[Ver 1984] Verona, A., *Stratified Mappings – Structure and Triangulability*, Lecture Notes in Math. No. 1102, Springer, Berlin, 1984.

[Vi 1999] Ville, M., *The singular fibres of harmonic morphisms from 4-manifolds to 2-surfaces are also minimal* (to appear).

[Vi 2000] Ville, M., *Harmonic morphisms from Einstein 4-manifolds to Riemann surfaces* (to appear).

[W 1965a] Whitney, H., *Local properties of analytic varieties*, Symposium M. Morse, Princeton, 1965, pp. 205–244.

[W 1965b] Whitney, H., *Tangents to an analytic variety*, Ann. Math. **81** (1965), 496–549.

[Wh 1986] White, B., *Infima of energy functionals in homotopy classes of mappings*, J. Diff. Geom. **23** (1986), 127-142.

[Wo 1992] Wood, J.C., *Harmonic morphisms and Hermitian structures on Einstein 4-manifolds*, Internat. J. Math. **3** (1992), 415–439.

[XZ 1997] Xu, C.–Y., and C. Zuily, *Higher interior regularity for quasilinear subelliptic systems*, Calc. Var. PDE **5** (1997), 323–343.

[Z 1999] Zhou, Z.–R., *Uniqueness of subelliptic harmonic maps.*, Ann. Glob. Anal. Geom. **17** (1999), 581–594.

Special symbols

Chapter 1

Chapter 2

Chapter 3

Chapter 4

Chapter 10

Chapter 11

Chapter 12

Chapter 13

Chapter 14

Chapter 15

Index